# Advances in Data Analytics, AI, and Smart Systems

**Series Editor**

Robin Qiu, Division of Engineering & Info Sci, Pennsylvania State University, Malvern, PA, USA

This book series investigates the transformative influence of data analytics and artificial intelligence (AI) on modern smart systems. Driven by the convergence of big data, AI, the Internet of Things (IoT), cloud computing, and Web 3.0, these technologies are revolutionizing industries and reshaping societal structures. The series critically examines how these innovations optimize processes, enhance decision-making, and drive cross-sector innovation.Beyond core themes like AI/IoT integration and interdisciplinary applications (e.g., healthcare, finance, smart cities), the series prioritizes emerging trends, including Generative AI, Artificial General Intelligence (AGI), blockchain, quantum computing, and the decentralized potential of Web 3.0. Crucially, it also highlights ethical and governance considerations, fostering research on responsible AI development, data security and privacy, and bias mitigation in algorithmic decision-making.

This series publishes monographs, contributed volumes, and professional books that provide firsthand insights into the latest research, pioneering innovations, and new perspectives on classical fields. Its target audience includes students, researchers, and industry professionals.

Ken Huang · Chunxiao Xing · Yuyan Duan
Editors

# Humanoid Robots and Physical AI

## Reshaping Work, Society, and Human Purpose

Springer

*Editors*
Ken Huang
DistributedApps.ai
Herndon, VA, USA

Chunxiao Xing
Tsinghua University
Beijing, China

Yuyan Duan
AI+
San Francisco, CA, USA

ISSN 3091-2733 ISSN 3091-2741 (electronic)
Advances in Data Analytics, AI, and Smart Systems
ISBN 978-3-032-33668-2 ISBN 978-3-032-33669-9 (eBook)
https://doi.org/10.1007/978-3-032-33669-9



This Springer imprint is published by the registered company Springer Nature Switzerland AG
The registered company address is: Gewerbestrasse 11, 6330 Cham, Switzerland

# Foreword by Vijay Bolina, CSO at Stealth, Former CISO at Google DeepMind, San Francisco, CA, USA

There are moments in technology when a field stops being an abstraction and becomes a force in the world. Artificial intelligence has already crossed one such threshold: it now writes, reasons, codes, plans, explains, and collaborates in ways that would have seemed speculative only a few years ago. Humanoid robotics marks the next threshold. It is the moment when intelligence leaves the screen, enters the room, and begins to act among us.

That transition changes everything.

For most of my career, I have worked on systems where the consequences of failure were measured in trust, continuity, national security, and human safety. I have investigated breaches carried out by nation-states, protected frontier AI systems, and helped build programs for securing the most advanced models of their time. At Google DeepMind, where I served as the first CISO of a frontier AI lab, I saw firsthand how quickly AI capabilities could move from research frontier to infrastructure dependency. I also saw something else: when powerful models become embedded in real systems, security is no longer a perimeter function. It becomes a property of the whole organism.

This book arrives at precisely the right time because humanoid robots are not simply another class of devices. They are full-stack cyber-physical agents. Their risk surface includes actuators, sensors, batteries, cameras, microphones, operating systems, cloud services, foundation models, policy layers, remote operators, supply chains, data pipelines, and the messy social environments in which they will operate. A compromised laptop can leak information. A compromised humanoid robot can move through a hospital, warehouse, factory, school, home, or critical facility with force, perception, and delegated authority. When models meet matter, cybersecurity becomes physical safety.

The chapters that follow treat this transformation with the seriousness it deserves. They begin with the foundations: why humanoid form factors matter, why human environments reward humanlike morphology, and why agentic AI changes robots from programmed tools into goal-directed collaborators. The discussion then moves into the hard physics of embodiment. Motors, actuators, structural materials, dexterous hands, batteries, thermal limits, and manufacturing tolerances are not side details. They are the ground truth that determines what intelligence can actually do. The most capable model in the world cannot reason its way around insufficient torque, exhausted batteries, overheating compute, or a manipulator that fails after sustained real-world use.

The book then turns to perception and interaction, where robots build their understanding of the world through cameras, depth sensors, tactile arrays, audio, proprioception, sensor fusion, and foundation-model interpretation. This is one of the great unsolved security and reliability problems of physical AI. A humanoid robot must decide what it sees, what it can touch, whom it might harm, when it is uncertain, and when it must stop. Learned perception gives these systems extraordinary flexibility, but it also introduces failure modes that traditional robotics and classical safety engineering were not designed to handle: adversarial examples, sensor spoofing, distribution shift, model hallucination, poisoned data, and overconfident action in ambiguous environments.

The sections on agentic architecture and control are especially important. Modern humanoids are neither pure neural networks nor old-fashioned deterministic machines. They are hybrids: high-level reasoning systems, vision-language-action models, skill libraries, planners, world models, memory systems, real-time controllers, simulation pipelines, safety monitors, and fallback mechanisms layered together. This layered architecture is the only plausible path toward useful general-purpose robots. It is also where many future failures will hide. A system can be locally safe at the motor controller, locally plausible at the language model, locally optimized in simulation, and still globally unsafe when those layers interact under pressure.

That is why the chapter on safety, security, and reliability is not a supporting chapter. It is the hinge of the entire book. Physical AI demands a fusion of disciplines that have too often lived apart: functional safety, cyber defense, AI assurance, robotics reliability, privacy engineering, vulnerability disclosure, incident response, human factors, and governance. We will need safety cases that account not only for mechanical hazards but also for prompt injection, malicious tool use, corrupted updates, sensor manipulation, insecure teleoperation, fleet compromise, and emergent behavior across multi-agent systems. We will need red teams that understand both cyber intrusion and physical consequence. We will need incident responders who can reason about a robot as a machine, an endpoint, an AI agent, and a social actor all at once.

The economic chapters widen the aperture. Humanoid robots may reshape manufacturing, logistics, healthcare, eldercare, hazardous work, and service industries. They may ease labor shortages and improve productivity, but they may also intensify inequality, displace vulnerable workers, and concentrate wealth among those who own the capital and data flywheels. The strategic competition between the United States and China is not merely about robot shipments or component costs. It is about who defines the platforms, standards, supply chains, data ecosystems, and norms for embodied intelligence.

The later themes of governance, ethics, psychology, coexistence, and human purpose are not philosophical ornaments placed after the technical work is done. They are deployment requirements. Robots that share space with people will change how we experience trust, privacy, work, care, dependence, companionship, aging, childhood, disability, dignity, and agency. A society can adopt powerful technology before it knows what that technology is doing to its institutions. We have done that before. We should be wiser this time.

My own view is that humanoid robotics will test whether the AI field has learned the central lesson of cybersecurity: capability without resilience creates fragility at scale. The question is not whether these systems will become useful. They will. The question is whether we can build and govern them with enough discipline that usefulness does not outrun accountability.

This book is valuable because it does not treat humanoid robots as a miracle, a threat, or a market category alone. It treats them as systems: technical, economic, social, geopolitical, and human. That is the right frame. The future of physical AI will not be decided by a single model release, a viral demo, or a cost curve. It will be decided by integration quality, safety culture, security architecture, manufacturing reality, public trust, and the choices we make before deployment becomes irreversible.

We are entering an era in which intelligence will increasingly have hands. Those hands will build, carry, clean, care, inspect, repair, and sometimes fail. They may extend human capability in extraordinary ways. They may also expose our systems, institutions, and assumptions to forms of risk we are only beginning to understand.

The work ahead is not to slow imagination. It is to match imagination with responsibility.

That is why this book matters.

Vijay Bolina
Former Chief Information Security Officer, at Google DeepMind,
Former Interim Chief Information Security Officer, at Isomorphic Labs
Stealth
Boston, MA, USA

Google DeepMind
San Francisco, CA, USA

# Foreword by Matthew R. Versaggi, MS. MBA, White House Presidential Innovation Fellow, Washington DC, USA

AI has always advanced through thresholds. First, it was a field of mathematical ideas and theoretical possibilities. Then it became a set of practical tools that could classify, predict, recommend, generate, and optimize. More recently, with the rise of agentic AI, it became something more dynamic: systems capable of planning, coordinating, using tools, and pursuing goals across complex digital environments.

In the previous volume by Ken Huang, *Agentic AI: Theories and Practices*, readers were introduced to this important transition. That book offered a timely foundation for understanding AI agents, their architectures, their applications, and the governance challenges that follow when software systems become increasingly autonomous. It helped clarify a fast-moving field at a moment when organizations, researchers, entrepreneurs, and policymakers were all trying to understand what agentic AI would mean for industry and society.

This new book extends that conversation into an even more consequential domain: the physical world.

As someone who has worked deeply across artificial intelligence, healthcare, emerging technologies, and national innovation efforts, including my service as a White House Presidential Innovation Fellow, I have seen how quickly powerful technologies can move from research environments into real operational settings. In healthcare, the promise of AI is not abstract. It affects diagnosis, care delivery, workforce burden, patient safety, and trust. In quantum computing and advanced computational systems, we see a similar pattern: technical breakthroughs create new capabilities, but they also demand new forms of literacy, governance, and responsibility.

Humanoid robotics brings these questions into sharper focus. These systems are not simply AI agents with a different interface. They are embodied agents. They perceive, move, manipulate, respond, and interact in environments designed for human beings. They must reason under uncertainty, operate under physical constraints, and coexist with people whose safety and trust cannot be treated as secondary considerations.

That is what makes this book both timely and necessary.

The central contribution of this volume is that it treats humanoid robotics not as a spectacle, but as a system. It begins with the foundations and vision of physical

AI, explaining why the convergence of agentic intelligence and humanlike embodiment represents a new stage in technological evolution. The early chapters examine hardware, actuation, batteries, materials, locomotion, dexterous manipulation, perception, sensing, and human-robot interaction. These are not merely engineering details. They determine whether intelligent machines can function reliably outside controlled demonstrations and inside factories, hospitals, warehouses, homes, and public spaces.

The book then moves into the architectures that make these systems intelligent. Vision-language-action models, foundation models for robotics, world models, planning systems, multi-agent coordination, memory, simulation, real-time control, and whole-body integration all become part of a larger story. The story is not about replacing robotics engineering with AI. Rather, it is about the layered integration of learned models, classical control, safety systems, and software infrastructure. This is precisely the kind of practical synthesis that complex technologies require before they can be responsibly deployed.

One of the strongest aspects of this book is its recognition that capability alone is not enough. Safety, security, and reliability sit at the center of the humanoid robotics challenge. This is especially important as robots become more autonomous and operate in closer proximity to human beings. A digital agent that fails may create operational, financial, or informational harm. An embodied agent that fails may also create physical harm. That difference matters.

The safety and security chapters therefore, deserve careful attention from every reader. Humanoid robots introduce a uniquely complex risk surface: sensors can be spoofed, perception models can be fooled, software supply chains can be compromised, remote control systems can be attacked, and autonomous planning systems can produce actions that are plausible in language but unsafe in the world. These risks are not reasons to reject the technology. There are reasons to design it with rigor from the beginning.

The economic and social chapters are equally important. Humanoid robots may help address labor shortages, improve productivity, reduce dangerous work, support aging populations, and enable new forms of service delivery. At the same time, they may disrupt labor markets, alter wage structures, create new dependencies, and raise difficult questions about human dignity, purpose, and social cohesion. As with agentic AI, the future will not be determined by technology alone. It will be shaped by policy, incentives, education, institutions, and public trust.

This is where the connection to the previous Agentic AI volume becomes especially meaningful. The earlier work helped readers understand autonomous digital agents: how they reason, how they coordinate, how they act, and how they reshape workflows. This book asks what happens when those agents acquire bodies? It asks what changes when the "tool use" of an AI system becomes the physical use of tools, when planning becomes movement, when perception becomes surveillance or safety monitoring, and when autonomy becomes presence in shared human environments.

In my view, this is one of the defining questions of the next era of AI.

The value of this book lies in its breadth and its practicality. Engineers will find a detailed technical map of the humanoid robotics stack. AI practitioners will see how agentic architectures evolve when real-time control, embodiment, and physical safety enter the equation. Healthcare leaders will recognize both the promise and the caution required for care settings. Business leaders will find an analysis of deployment economics and industry transformation. Policymakers will find a framework for thinking about regulation, labor, security, and governance before these systems become pervasive.

Most importantly, readers will find a disciplined way to think about a future that is arriving quickly. Humanoid robots will not be defined solely by viral demonstrations or bold market forecasts. They will be defined by whether they can operate safely, reliably, securely, and usefully in the real world. They will be judged not only by what they can do, but by whether people can trust them.

That trust must be earned.

It is my honor to contribute this foreword to a book that builds naturally on the foundations laid by Agentic AI: Theories and Practices while moving the discussion into one of the most important frontiers of our time. This volume invites researchers, practitioners, entrepreneurs, policymakers, and students to engage seriously with the next phase of AI: not merely intelligent systems that assist us from behind a screen, but embodied systems that may work beside us in the world.

The promise is extraordinary. The responsibility is equally great.

Matthew R. Versaggi, MS. MBA
AI and Quantum SME
Distinguished Engineer
White House Presidential Innovation Fellow
Special AI Cohort of '24
Washington, DC, USA

# Preface

When I finished editing *Agentic AI: Theories and Practices*, I believed I had written about the most consequential technology of our time. I still believe that. What I underestimated was how quickly that technology would escape the screen.

The agents documented in that volume—reasoning, planning, tool-using, autonomous systems—existed entirely in the digital realm. They browsed the web, wrote code, managed workflows, coordinated with one another across APIs and message queues. They were powerful, and they were, in an important sense, weightless. They could not pick up a glass of water. They could not steady an elderly person on a staircase. They could not reach across the physical world and rearrange it.

That constraint is dissolving.

## From Digital Agents to Physical Ones

The question that animated this volume was simple to state and difficult to answer: what happens when agentic AI acquires a body?

The answer is not merely "a robot." Robots have existed for decades—precise, powerful, and confined. What is new is the union of embodied hardware with the reasoning and adaptive capability that large-scale AI has made possible. A humanoid robot animated by an agentic intelligence is something qualitatively different from any machine that preceded it. It can receive an instruction in plain language, interpret context, plan a sequence of physical actions, recover when those actions fail, and improve with experience. It can operate in spaces and situations its designers did not explicitly anticipate. That combination—morphological flexibility meeting cognitive agency—is what makes the present moment feel discontinuous rather than incremental.

I began this book because no existing volume captured that discontinuity whole. The robotics literature, rich as it is, tends to treat cognition as a subsystem. The AI literature, equally rich, tends to treat embodiment as an afterthought. The subject demands a treatment that holds both in view simultaneously—and then looks beyond them to the world those combined systems will inhabit and reshape.

## The Ecosystem We Are Talking About

The field this book addresses is not a single technology but a convergence of many. On the hardware side: actuator design, materials science, energy storage, and precision manufacturing have reached a threshold where agile, dexterous, bipedal machines can be produced at costs that make commercial deployment credible. On the software side: foundation models, world models, imitation learning, and multi-agent coordination have reached a threshold where machines can handle the unstructured complexity of real environments. On the infrastructure side: cloud compute, edge inference, 5G connectivity, and simulation environments have reached a threshold where humanoid fleets can be trained, updated, and managed at scale.

These thresholds did not arrive simultaneously by accident. They arrived together because they share common enabling technologies—particularly the transformer architecture and the scaling laws that govern it—and because investment in each reinforces the others. The result is an ecosystem with genuine momentum: every deployment generates training data that improves the models; every model improvement expands the range of deployable applications; every application justifies further investment in hardware and infrastructure. We are not watching a technology approach a cliff edge from which it will eventually fall back. We are watching a technology climb a curve that is, for the foreseeable future, still ascending.

The actors in this ecosystem span geographies and sectors in ways that make simple narratives impossible. Boston Dynamics, Figure AI, Agility Robotics, and Tesla Optimus represent the American commercial frontier. Unitree, UBTECH, Fourier Intelligence, and dozens of newer entrants represent a Chinese industrial ecosystem that moved from follower to peer with remarkable speed. Academic laboratories, national research programs, and defense institutions are active participants. The capital flowing into this space—measured in the tens of billions of dollars annually as of 2026—reflects a collective judgment that the competition for leadership in physical AI is among the most consequential industrial races of the decade.

Understanding that ecosystem—its structure, its incentives, its fault lines, and its shared dependencies—is prerequisite to understanding what the technology will actually do in the world, as opposed to what its most enthusiastic or most alarmed commentators claim it will do.

## The Future We Can Envision

I want to be honest about the uncertainty that surrounds any claim about where this goes.

The optimistic case is not difficult to articulate. Humanoid robots working in logistics, manufacturing, elder care, construction, and hazardous environments could free hundreds of millions of people from physically demanding, repetitive, and dangerous work. They could extend the productive lives of aging societies confronting severe demographic pressure. They could make goods and services available at costs that genuinely expand access. In this vision, the productivity gains from physical AI are broadly shared, the transition is managed with care, and humanity arrives at a future with more capacity for the activities—creative, relational, intellectual, civic—that people report finding most meaningful.

The cautionary case is equally easy to articulate. Displacement concentrated among workers who lack the capital, credentials, or geographic mobility to absorb it. Productivity gains captured at the top of ownership structures. Safety incidents that harm the people these machines were meant to serve. Security vulnerabilities exploited at scale. Governance frameworks that arrive years after the damage they were meant to prevent. A global race to deploy in which competitive pressure consistently overrules precaution.

Neither case is inevitable. Both are possible. The trajectory of this technology is not written in its physics or its mathematics; it is written in the decisions that engineers, executives, regulators, workers, and citizens make in the next five to ten years. This book is addressed to all of them.

What I am certain of is this: the emergence of capable, autonomous, physically embodied AI is not a problem to be solved or a threat to be contained. It is a condition to be navigated—with clear eyes, serious analysis, and a genuine commitment to the proposition that technological progress and human flourishing are not competing projects.

That conviction is why I wrote this book. I hope it serves you well.

This volume would not exist without the generosity of many. Colleagues wrote forewords, co-authored chapters, and gave their time as reviewers—some named, some anonymous—each bringing expertise and care that shaped every page. I am deeply grateful to all of them, and I have dedicated a full acknowledgements section to recognizing their contributions individually.

One thing each of them made clear to me, and that I want to honor by stating it plainly here: the views expressed in this book are their own. They do not represent the positions of their employers, institutions, or any affiliated organizations. I reiterate that not as a legal formality but because they asked me to—and because it reflects the intellectual independence that made their contributions worth having.

Spring 2026

Ken Huang
DistributedApps.ai
Fairfax, VA, USA

Herndon, USA

# Acknowledgments

This book began with a simple conviction and an impossible premise. The conviction was that humanoid robots and physical AI represent a genuine threshold in the history of technology, the moment intelligence steps off the screen and begins to act among us, and that this moment deserves a serious, honest, end-to-end treatment rather than a collection of demos and forecasts. The impossible premise was that any single person could write such a treatment alone. They cannot. A book that moves from actuators and battery chemistry to vision-language-action models, from sim-to-real transfer to labor economics, from cybersecurity to the philosophy of human purpose, can only be the product of a community of people willing to lend their expertise, their time, and their candor. This book is that community's work, and these pages are my attempt to thank them properly.

## To the Authors of the Forewords

I am deeply grateful to **Vijay Bolina** for opening this volume. As the world's first Chief Information Security Officer of a frontier AI lab at Google DeepMind, and through his work on the security of the most advanced AI systems of their time, Vijay has spent his career at exactly the place where this book lives, the seam where models meet matter. His foreword does something I could not have done as well myself: it states, with the authority of someone who has investigated nation-state breaches and protected frontier systems, why a compromised humanoid robot is a category of risk unlike anything that came before, and why security must be treated as a property of the whole organism rather than a perimeter. His framing, that capability without resilience creates fragility at scale, became a quiet north star for the entire manuscript.

I am equally indebted to **Matthew R. Versaggi**, whose foreword draws a through-line from the earlier volume *Agentic AI: Theories and Practices* into the embodied world this book explores. As an AI and quantum subject-matter expert, a Distinguished Engineer, a senior leader in AI healthcare, and a White House Presidential Innovation Fellow, Matthew brought a rare combination of technical depth and institutional wisdom. His insistence that we treat humanoid robotics as a system rather than a spectacle, and his reminder that trust must be earned rather than assumed, shaped how every chapter approaches its subject. I am thankful not only for his words at the front of this book but for his careful reading of it as a reviewer.

## To My Fellow Editors

This volume had three editors, and I was fortunate in both of my partners. **Professor Chunxiao Xing** of Tsinghua University, Foreign Academician of the Russian Academy of Engineering and Dean of the Tsinghua-Kelan Advanced Intelligent Database Research Institute, brought decades of research leadership and an unmatched perspective on the global landscape of intelligent systems. His scholarly rigor and his view across hundreds of publications and dozens of national research programs raised the standard of the entire project and kept our treatment honest about the state of the art.

**Lynn Duan** brought the vantage point of someone who builds and funds this future in practice. As a product leader and early-stage investor in robotics and AI ventures, and as the founder of a thriving community of founders and investors, Lynn kept the book tethered to the realities of deployment, capital, and adoption. Her instinct for what matters to the people actually shipping these systems sharpened our chapters on economics, business models, and the path to scale. Working alongside both of you has been one of the genuine privileges of my career.

## To the Chapter Co-Authors

A book of this scope is only as strong as the people who wrote its chapters, and I was lucky to assemble contributors who are practitioners and researchers at the frontier of their fields. Their willingness to write with precision, to revise patiently through many drafts, and to disagree productively is the reason this book has substance rather than just ambition.

My thanks to **Grace Huang**, whose finance and product-strategy lens runs through the foundations, hardware, perception, agentic-architecture, and integration chapters, and who repeatedly grounded our technical claims in the operational and risk realities of enterprise adoption. To **Bo Tian** of Tesla, whose firsthand work on the Optimus AI computer gave the hardware and perception chapters a level of board-level and inferencing detail that no secondary source could provide.

To **Manish Bhatt**, whose leadership in AI safety, from the CyberSecEval series to the OWASP AI Security Verification Standard and the emerging AI Vulnerability Scoring System, anchored the perception, agentic-architecture, and integration chapters in a serious treatment of risk. To **Tongzhou Mu** of Rhoda AI, whose research on world models and robot simulation, and whose work behind the widely used ManiSkill framework, brought authority to our treatment of agentic architectures and sim-to-real. To **Chuer Pan** of Stanford University, whose research on contact-rich manipulation learned from human demonstration deepened the integration and control chapters.

To **Jerry Huang** of Kleiner Perkins, whose breadth across machine learning, security, scalable systems, and applied mathematics shaped the safety, economics, governance, psychology, and convergence chapters, and whose investor's eye kept our forecasts disciplined. To **Bhavya Gupta** of Stanford University, whose work in security governance and on the OWASP Agentic AI risk efforts strengthened the safety and governance chapters. To **Víctor Mayoral-Vilches** of Alias Robotics, whose specialized research in robot cybersecurity and robot teardown gave the safety chapter a rigor that the field badly needs. And to **Modar Alaoui** of ALM Ventures, founder of the Humanoids Summit, whose view from the center of the humanoid ecosystem enriched the chapter on human-robot coexistence and the broader narrative of industrial transformation.

## To the Reviewers

Good reviewers are a gift, because their work is largely invisible and their effect is everywhere. This book was strengthened immeasurably by people who read it not to praise it but to find where it was wrong, unclear, or overconfident, and who told us so plainly.

I owe a particular and unusual debt to the anonymous reviewers drawn from leading humanoid and robotics companies, who cannot be named here but whose firsthand, frontline knowledge corrected errors and tempered claims that no outside reader could have caught. Their generosity in reviewing this work, often on their own time and without public credit, reflects the best of a field that takes its responsibilities seriously. To each of you: thank you. You know who you are, and we are in your debt.

Among the named reviewers, I again thank **Vijay Bolina** and **Matthew R. Versaggi**, who not only wrote forewords but read the manuscript with the same seriousness they brought to their own words. My thanks to **Jasmine Lombardi**, whose more than twenty-five years across customer success, professional services, and global systems implementation, and whose leadership of large customer-success organizations at a major robotics company, gave us an indispensable real-world check on how these systems are actually delivered, supported, and trusted by the people who use them. And to **Nikhil Abraham**, cofounder and CEO of CloudChef and a previous founder whose generative-AI company was acquired by Adobe, whose experience deploying autonomous robots inside real commercial kitchens, including a Michelin-starred operation, kept our chapters honest about the distance between a demonstration and a working system.

## To Springer and the Production Team

This book owes a great deal to the team at Springer, whose belief in the project and steady professionalism turned an ambitious manuscript into a finished volume. I am especially grateful to Jialin Yan, Sneha Arunagiri, Brian Halm, and Ruth Milewski, whose editorial guidance and production oversight shaped this book from proposal to publication and carried it through every stage with meticulous care. Their confidence in the scope of this project, their patience with a subject that kept evolving as we wrote about it, and their attention to the countless details that readers never see but always benefit from made all the difference. A book that tries to capture a moving target requires a publisher willing to move with it, and the dedication of this team made the difference between a timely book and a merely intended one.

## To Our Families

Every contributor to this volume borrowed the time to write it from someone, an evening, a weekend, a vacation that became a writing retreat. On behalf of all of us, I want to thank the families, partners, and friends who absorbed that cost without complaint and often with encouragement. You did not write these chapters, but they could not have been written without you, and the patience you extended to us is woven into every page.

## To the Reader

Finally, I want to thank you, the reader, for taking this subject as seriously as we have tried to. Humanoid robots will not be defined by a single model release, a viral demonstration, or a cost curve. They will be defined by integration quality, safety culture, security architecture, manufacturing reality, public trust, and the choices we make before deployment becomes irreversible. Many of those choices are yours. If this book helps you make them with more clarity, more rigor, and more responsibility, then everyone named on these pages will consider their contribution well spent.

*With gratitude,*
**Ken Huang**
*Editor-in-Chief*
*On behalf of all the editors, authors, and contributors*

# Reviewers

1 Various anonymous reviewers from top established humanoid company and robotic companies.

2 Vijar Bolina

Vijay Bolina, with over 25 years of experience in cybersecurity and national security, is currently the Chief Information Security Officer at a stealth frontier AI lab focused on the risks of the next frontier in physical AI—where models meet matter. Prior to this role, Vijay was the world's first frontier AI lab CISO at Google DeepMind and also served as the Head of Cybersecurity Research, where he was the first to pioneer security for the world's most advanced AI systems, including Gemini. In this role he also led the research program that gave birth to ground breaking technologies like BIGSLEEP, CODEMENDOR, and SECGEMINI. His career includes leading teams for high-stakes investigations at Mandiant into some of the world's most notorious data breaches, such as the North Korean hack of Sony Pictures, the theft of next-generation chip designs from NVIDIA, and the Uber breach. Previously, he led engineering teams at Booz Allen Hamilton to build large-scale systems for detecting foreign hackers and also served as a technical lead for hacking mission-critical systems for the US Government. He began his career in the 90s hacking scene, writing software at an early age and auditing networks for the US Government and their allies while attending high school.

Vijay holds a Bachelor of Science in Computer Science from UC Davis and a Master of Science in Computer Science from The George Washington University. A veteran of the early security research community, his focus areas include frontier AI offense and defense, national security, critical infrastructure, incident response, and the security of systems where digital intelligence intersects with physical matter.

3 Matthew R. Versaggi, MS, MBA

Matthew R. Versaggi, MS, MBA
AI SME, Distinguished Engineer,
Sr Leader in AI Healthcare,
White House Presidential Innovation Fellow, Special AI Cohort '24

Matthew is a Senior Leader in the Artificial Intelligence space with Federal Government and Fortune-5 Healthcare experience; a successful Entrepreneur of 16 years with 60+ clients nationwide; an established Strategist in the AI business/technology space; a noted Public Speaker of 20+ years globally, a decorated Educator of decades in the academic and professional domains, and a successful athletic coach of 15 years specializing in the Pole Vault.

4 Jasmine Lombardi

Jasmine Lombardi brings more than 25 years of experience in customer success, professional services, software development, project management, global systems implementation, and business operations. She leads a team of over 110 customer success professionals across North America, Europe, the UK, and Australia. Prior to joining a major robotics company, Jasmine held senior leadership roles at Barracuda Networks, Intronis, Aspen Technology, Mzinga, Prospero Technologies, and PAREXEL International, where she built a strong foundation in delivering customer value and operational excellence. She has earned numerous awards, including Mass High Tech's 2014 Women to Watch, CRN Women of the Channel and Power 100 (2016), and a Gold Stevie Award for Customer Service Leader of the Year (2016). Most recently, she was honored as one of the Top 50 Women Leaders in Boston for 2024. Jasmine holds a Bachelor of Science in Computer Science from the National University of Singapore, a Master of Science in Information Systems from Boston University, and is a certified Project Management Professional (PMP).

5 Nikhil Abraham

Nikhil Abraham is the cofounder CEO of CloudChef—democratizing gourmet meals through autonomous humanoid robots. CloudChef robots are already working inside several leading commercial kitchens across the country including a Michelin-starred operation. CloudChef is backed by top Silicon Valley investors and food executives. Nikhil previously cofounded the generative AI platform Rephrase AI (acquired by Adobe) and is a graduate from IIT Bombay.

# Contents

# Editors and Contributors

## About the Editors

**Ken Huang** is a globally recognized expert in AI and Web3 security, a prolific author, and a leading figure in shaping industry standards. He is the CEO and Chief AI Officer (CAIO) of DistributedApps.ai, specializing in Generative AI training and consulting. Ken is deeply involved in driving the development of secure AI systems. He serves as a Research Fellow and Co-Chair of the AI Safety Working Groups at the Cloud Security Alliance (CSA), leading Agentic AI initiatives. He is also a Co-Chair of the AI STR Working Group at the World Digital Technology Academy under the UN Framework, and a core contributor to the OWASP GenAI project, focusing on Agentic AI security. His expertise extends to his contributions to OWASP's Top 10 Risks for LLM Applications and his active participation in the NIST Generative AI Public Working Group. He is also a member of Open AI forum.

Ken has authored and edited numerous influential books, including:

*Agentic AI: Theories and Practices* (Springer, June 2025)
*Securing AI Agents: Foundations, Frameworks, and Real-World Deployment* (Springer, October 2025)
*Beyond AI: ChatGPT, Web3, and the Business Landscape of Tomorrow* (Springer, 2023)
*Generative AI Security: Theories and Practices* (Springer, 2024)
*Practical Guide for AI Engineers* (Volumes 1 and 2, DistributedApps.ai, 2024)
*The Handbook for Chief AI Officers: Leading the AI Revolution in Business* (DistributedApps.ai, 2024)
*Web3: Blockchain, the New Economy, and the Self-Sovereign Internet* (Cambridge University Press, 2024)
*Web3 Applications Security and New Security Landscape: Theories and Practices* (Springer, 2024)

His co-authored book, "Blockchain and Web3: Building the Cryptocurrency, Privacy, and Security Foundations of the Metaverse" (Wiley, 2023), was recognized as a must-read by TechTarget.

A sought-after speaker, Ken has presented at prestigious global forums, including Davos WEF, ACM, IEEE, CSA AI Summit, the Depository Trust & Clearing Corporation (DTCC), and World Bank conferences.

Explore Ken's books on Amazon: https://www.amazon.com/author/kenhuang

**Professor Chunxiao Xing** is a Foreign Academician of the Russian Academy of Engineering. He currently serves as the Dean of the Tsinghua-Kelan Advanced Intelligent Database Research Institute, and Vice Dean of the Institute for **Medical-Robotic Engineering** Interdisciplinary Research at Tsinghua University.

His research focuses on artificial intelligence, blockchain technology, big data, smart cities, and e-government. He has published over 380 academic papers, holds 43 invention patents and 33 software copyrights, and has led more than 40 major national research projects including National 973 Program, 863 Program, and National Key R&D projects.

Professor Xing has received numerous honors including the National Science and Technology Progress Award (Second Class) and recognition as one of the "Metaverse 30" and "Blockchain 60" innovation leaders in China. He holds leadership positions in several professional organizations including the China Computer Federation, the Zhongguancun Blockchain Industry Alliance, and the Chinese Institute of Electronics Blockchain Branch.

**Lynn Duan** is a product leader and angel investor at the forefront of AI and robotics innovation. As an early-stage investor, she has backed a robotics startup supported by Salesforce CEO Marc Benioff, Skild's founder, and Nat Friedman, as well as a healthcare LLM venture backed by Anthropic's co-founder. Known for writing first checks, Lynn has already seen one portfolio company advance to Series B funding.

Since 2019, Lynn has served as Product Leader at Treasure Data, a pre-IPO customer data platform company, where she has architected roughly one-third of the company's product suite. She pioneered the company's AI transformation by creating its first AI Agent Framework, which has generated 30+ enterprise-grade agents and driven 30+ new customer engagements. Her work spans enterprise big data, data security, and AI-driven marketing solutions.

In 2023, Lynn founded AI+, a thriving AI startup community with 50+ contributing members that has hosted 150+ events over two years, connecting 2,000+ founders annually with 200+ investors and corporate partners. She holds degrees from the University of Michigan (Psychology & Statistics, International Relations) and Georgia Tech (M.S. in Human-Computer Interaction with a Certificate in Management of Technology).

## List of Contributors

**Modar Alaoui** ALM Ventures, Mountain View, CA, USA

**Manish Bhatt** Amazon Kuiper Security/OWASP, Seattle, WA, USA

**Yuyan Duan** AI+ Founders Club, Silicon Valley, CA, USA

**Bhavya Gupta** Stanford University, Stanford, CA, USA

**Grace Huang** PIMCO, Austin, TX, USA

**Jerry Huang** Kleiner Perkins, San Francisco, CA, USA

**Ken Huang** DistributedApps.Ai, Fairfax, VA, USA

**Víctor Mayoral-Vilches** Alias Robotics, Vitoria-Gasteiz, Spain

**Tongzhou Mu** Rhoda AI, Mountain View, CA, USA

**Chuer Pan** Stanford University, Stanford, CA, USA

**Bo Tian** Tesla, Palo Alto, CA, USA

**Chunxiao Xing** Tsinghua University, Beijing, China

# Foundations and Vision

1

Ken Huang, Grace Huang, Yuyan Duan, and Chunxiao Xing

## 1.1 Introduction

The year 2026 represents a watershed moment in the evolution of artificial intelligence and robotics. For decades, AI existed solely in digital realms—processing text, generating images, beating humans at chess and Go—while physical robots remained largely confined to structured factory environments executing preprogrammed sequences. The convergence of these domains into embodied AI or physical AI—intelligent systems that perceive, reason, and act in the three-dimensional world—marks a fundamental discontinuity in technological capability with profound implications for civilization.

At the forefront of this transformation stand humanoid robots: bipedal, anthropomorphic machines designed to navigate human spaces, manipulate human tools, and collaborate with human workers. The humanoid form is not merely aesthetic; it is a pragmatic design response to human-built environments. Our built environment—from doorways to stairways, steering wheels to keyboards—reflects human morphology. A robot designed for broad operation in this world may benefit from approximating selected human physical characteristics, although the degree of humanoid morphology required depends on the task and environment.

K. Huang (✉)
DistributedApps.ai, Fairfax, VA, USA
e-mail: ken@distributedapps.ai

G. Huang
PIMCO, Austin, TX, USA

Y. Duan
AI+ Founders Club, Silicon Valley, CA, USA

C. Xing
Tsinghua University, Beijing, China

 
K. Huang et al. (eds.), *Humanoid Robots and Physical AI*, Advances in Data Analytics, AI, and Smart Systems, https://doi.org/10.1007/978-3-032-33669-9_1

Yet hardware alone does not create transformation. The revolution lies in combining humanoid embodiment with agentic AI—systems capable of autonomous goal-directed behaviour, real-time adaptation, and continuous learning from experience. Where traditional robots followed rigid programs or required constant human supervision, agentic humanoid systems understand natural language instructions, plan multi-step actions, recover from failures autonomously, and improve through deployment. This synthesis of physical capability and cognitive agency creates systems that can support more interactive forms of human–robot collaboration than traditional command-execution robots.

The pace of transformation has accelerated dramatically. At CES 2026 in January, Boston Dynamics unveiled its production-ready electric Atlas, with the company announcing "the final product version is being built now" and "all Atlas deployments are already fully committed for 2026, with fleets scheduled to ship to Hyundai's Robotics Metaplant Application Center and Google DeepMind" (Boston Dynamics, 2026a). This transition from research platform to commercial product signals the field's maturation.

Simultaneously, market dynamics reveal explosive growth. The global humanoid robot market is "projected to grow from USD 2.92 billion in 2025 to USD 15.26 billion by 2030, registering a compound annual growth rate (CAGR) of 39.2%" (MarketsandMarkets, 2025). China's robotics sector alone saw "610 investment deals totaling 50 billion yuan ($7 billion) in the first nine months of 2025—representing a 250% increase year-over-year" (Automate.org, 2025). These figures indicate strong investor expectations for commercialisation, but they remain forward-looking and should be interpreted alongside technical, regulatory, and adoption uncertainties.

The implications extend far beyond technology and economics. As humanoid robots transition from factories to hospitals, warehouses to homes, they will reshape fundamental aspects of human society: the nature and meaning of work, economic structures and wealth distribution, caregiving for ageing populations, education of children, and ultimately how humans define purpose in an age of increasingly capable machines. Understanding these technologies—their capabilities, limitations, and trajectories—becomes essential for anyone seeking to navigate the transformations ahead.

This chapter provides the foundational knowledge for that understanding. We examine the convergence of embodiment and intelligence (Sect. 1.2), establish precise definitions and taxonomies (Sect. 1.3), survey the current state of the art as of 2026 (Sect. 1.4), present detailed case studies of leading platforms (Sect. 1.5), examine adoption constraints and thresholds (Sect. 1.6), and provide a roadmap to the chapters ahead (Sect. 1.7).

## 1.2 The Convergence of Embodiment and Intelligence

The transformation of robotics from fixed industrial machinery to intelligent, adaptable humanoid systems represents the culmination of multiple technological streams maturing simultaneously. Understanding this convergence requires examining both its historical roots and the fundamental shift it represents in what artificial intelligence can accomplish. This section traces the evolution from traditional automation through the emergence of embodied AI, explores what happens when intelligence takes physical form, examines the rise of agentic systems capable of autonomous goal-directed behaviour, and explains why the humanoid form factor has emerged as a prominent design candidate for robotics in human-built environments. Together, these four dimensions illuminate how we arrived at the current watershed moment where robots are transitioning from tools that execute commands to collaborators that understand objectives.

### 1.2.1 From Industrial Automation to Humanoid Systems

The story of modern robotics begins not with humanoid forms but with the Uni mate—a programmable robotic arm installed in a General Motors plant in 1961 that performed spot welding and materials handling. For over five decades, industrial robotics followed this paradigm: specialised, stationary machines performing narrowly defined, repetitive tasks in structured environments with millimetre precision.

This model achieved remarkable success within its domain. As of 2026, millions of industrial robots operate in factories worldwide, with automotive and electronics manufacturing leading adoption. These systems demonstrate extraordinary precision, repeatability, and endurance. However, they remain fundamentally limited: they require extensive setup and programming by skilled engineers, operate in carefully controlled environments separated from human workers by safety cages, lack flexibility to handle variations in parts or processes, and cannot adapt to environmental changes without human intervention.

The recognition that truly general-purpose robotics would require human-like morphology emerged gradually from multiple domains. The Defense Advanced Research Projects Agency (DARPA) Robotics Challenge (2012–2015) tested robots on tasks requiring human-form capabilities: driving vehicles, opening doors, climbing stairs, manipulating tools designed for human hands. The event starkly revealed the gap between human effortless performance and robotic struggle—many advanced robots fell while attempting to walk across rubble or failed to turn simple valves. Yet it demonstrated that humanoid platforms offered the greatest versatility across diverse challenges compared to wheeled or quadrupedal alternatives.

By the early 2020s, multiple technological streams had matured sufficiently to enable practical humanoid robots. Advances in electric motors and actuators provided the strength and speed needed for dynamic movement. Machine learning

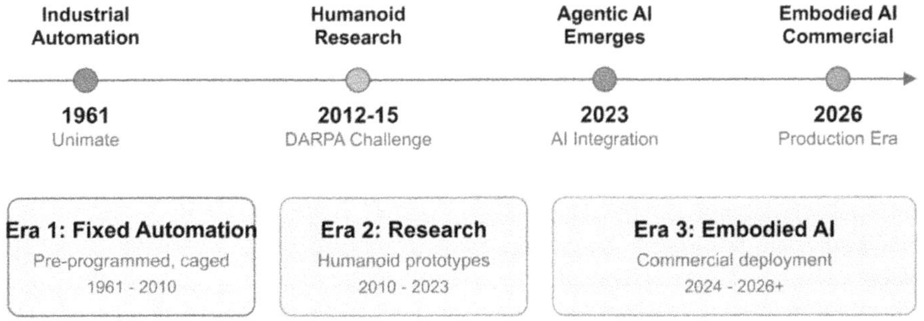


**Fig. 1.1** Evolution of robotics (1961–2026)

breakthroughs enabled robots to learn complex behaviours from demonstration or trial-and-error rather than requiring explicit programming. Computer vision using deep neural networks approached human-level object recognition. Most critically, large language models emerged enabling robots to understand natural language instructions and reason about tasks at abstract levels.

The transition from research curiosity to commercial viability accelerated dramatically between 2023 and 2026. This period witnessed Boston Dynamics transitioning Atlas from hydraulic to electric actuation and announcing production plans, Tesla advancing Optimus through multiple generations with aggressive cost targets, Chinese manufacturers achieving breakthrough pricing, including Unitree's $5,900 R1 and lower-cost compact humanoid platforms, and Figure AI completing an 11-month Figure 02 deployment at BMW Spartanburg (the Figure 02 programme has since wound down in favour of Figure 03). These developments mark a visible commercialisation inflection point, although deployment remains bounded and uneven.

Figure 1.1 shows the progression from early industrial automation to contemporary embodied AI systems, highlighting key milestones including the Unimate (1961), DARPA Robotics Challenge (2012–2015), and the commercial deployment era (2024–2026).

### 1.2.2 Embodied AI: When Intelligence Gets Physical

The term "embodied AI" or "physical AI" captures a fundamental shift in artificial intelligence from purely digital to physically instantiated systems. As Deloitte's 2026 Tech Trends report notes, we are "navigating the convergence of AI and robotics" where "AI goes physical" (Deloitte, 2026). This represents more than incremental progress—it is a qualitative transformation in what AI systems can accomplish.

One analysis describes 2026 as "the Great Physical Breakout" marking "the year of Embodied AI" when "AI began escaping digital confines and entering the physical world" (Rai, 2025). The characterisation captures the momentousness: AI that previously existed only in servers and smartphones now operates robot bodies, manipulates physical objects, and navigates three-dimensional environments.

The convergence creates compound effects. AI makes robots intelligent enough for autonomous operation in unstructured environments. Robots provide AI with physical embodiment enabling real-world learning and deployment. Data from physical robot interactions trains better AI models. Improved AI enables more sophisticated robot behaviours. This feedback loop may accelerate progress, although its pace depends on data quality, deployment volume, safety constraints, and the transferability of learned behaviours.

Technical evidence indicates a shift away from deterministic automation, in which robots execute fixed, pre-programmed paths, towards learning-based embodied systems that integrate perception, planning, and control. These systems increasingly use neural models to interpret environments, adapt to object variation, and select actions under uncertainty, although most deployments still require substantial site-specific integration, safety validation, and human supervision. This represents an important architectural shift in how robots are designed and deployed.

Emerging applications demonstrate practical value across bounded industrial settings. Humanoid and mobile robotic systems are being tested or deployed in manufacturing and logistics workflows such as parts handling, tote movement, and materials transport (Agility Robotics, 2025; BMW Group, 2025; Figure AI, 2025). These examples show movement beyond laboratory demonstrations, but most remain pilots or limited deployments whose scalability depends on reliability, safety certification, integration cost, and task economics.

### 1.2.3 Agentic AI: From Reactive to Goal-Directed Systems

Parallel to hardware embodiment, AI itself has transformed from reactive pattern-matching systems to agentic systems capable of autonomous goal-directed behaviour. The distinction is fundamental: reactive AI responds to inputs with predetermined outputs, while agentic AI pursues objectives through planned sequences of actions, adapting strategies based on environmental feedback.

Several industry leaders and analysts identify 2026 as an important year for agentic AI commercialisation. Senior voices from Amazon Web Services, Oracle, Google Cloud, and Cisco have converged on a shared outlook: 2026 is the year agentic AI graduates from experimental pilots to production-ready deployment at scale (Nextgov, 2025). The market reflects this: Gartner forecasts that by the end of 2026, 40% of enterprise applications will feature task-specific AI agents, up from less than 5% in 2025 (Gartner, 2025).

Agentic AI's defining characteristics include: goal-directed autonomy where systems pursue explicit objectives over extended time horizons, environmental modelling where systems build and update representations of their surroundings, adaptive planning where systems adjust strategies based on success or failure, proactive information gathering where systems actively seek data to reduce uncertainty, and natural communication where systems explain their reasoning and ask clarifying questions when instructions are ambiguous.

When agentic AI combines with humanoid embodiment, the results can be operationally meaningful in bounded settings, but they should not be mistaken for broad autonomy. At BMW's Spartanburg facility, Figure AI deployed its Figure 02 humanoid robots over an eleven-month period (2024–2025), processing more than 90,000 sheet-metal parts and contributing to the production of more than 30,000 BMW X3 vehicles across approximately 1,250 operational hours (Figure AI, 2025); the Figure 02 programme has since wound down in favour of Figure 03. The deployment demonstrated task-specific parts-handling capability in a structured manufacturing workflow, closer to Level 2 partial autonomy than general Level 3 autonomy, because the workcell, task scope, and exception handling remained tightly bounded.

However, "agentic" remains a spectrum rather than a binary property. Current systems achieve agency within bounded contexts (warehouses, manufacturing lines) but lack the open-ended flexibility of human intelligence. Understanding these boundaries is beneficial for realistic capability assessment and appropriate application selection.

### 1.2.4 Why the Humanoid Form Factor Matters

The choice of humanoid morphology for general-purpose robots reflects both practical necessity and theoretical considerations. At the pragmatic level, our entire civilisation constitutes a vast infrastructure optimised for human proportions and capabilities. Doorways average 80 cm wide and 2 m tall because humans fit those dimensions. Stairs have rise-and-run ratios matching human gait mechanics. Vehicles feature controls positioned for human reach and strength. Tools from screwdrivers to keyboards assume human hand anatomy and grip strength.

Redesigning this trillion-dollar infrastructure to accommodate non-humanoid robots would be economically prohibitive and socially disruptive. A wheeled robot cannot climb stairs limiting it to single floors or specially modified buildings. A quadruped robot cannot easily manipulate objects positioned for human standing height or grasp tools designed for opposed thumbs. Specialized form factors optimised for specific tasks fail at others, requiring either multiple specialised robots or extensive environmental modification.

The humanoid form offers compelling long-term value for environments built around human bodies: a platform that may eventually operate across many human spaces, use a subset of human-oriented tools, and perform multiple task classes with less environmental modification than many specialised alternatives. Near-term commercial deployments, however, are far more specialised. Companies are not yet buying humanoids for open-ended versatility; they are constraining them to tasks such as tote movement, sheet-metal loading, or parts transfer where the workcell can be engineered around the robot. The versatility argument is therefore best understood as an option value: the same morphology may support future task expansion after reliability, safety, and autonomy improve.

Beyond environmental compatibility, humanoid morphology facilitates human–robot interaction and collaboration. Humans possess sophisticated innate abilities to predict human motion trajectories, interpret body language and gesture, and coordinate actions with partners. These same cognitive systems are leveraged for human–robot collaboration when robots exhibit human-like form and motion. A humanoid robot's gestures convey intent in ways immediately interpretable to human collaborators, enabling more natural and effective teamwork than with mechanically alien forms.

The humanoid form also supports learning from human demonstrations. When robot and human share similar embodiment—similar ranges of motion and sensorimotor constraints—the transfer of skills from human demonstration to robot execution can become more tractable. Combined with modern machine learning techniques, this can reduce the burden of hand programming, although the speed and reliability of transfer depend on task complexity, sensor quality, and embodiment similarity.

However, the humanoid form imposes significant engineering challenges. Bipedal locomotion requires sophisticated control to maintain dynamic stability against gravity and ground reaction forces. The human body's complexity translates to robotic systems requiring up to 56 degrees of freedom (as in Boston Dynamics Atlas), with the production all-electric Atlas reaching 56 DoF, each demanding actuators, sensors, power distribution, and computational resources. The humanoid form offers no specialisation advantages; quadrupeds are more stable, wheeled platforms more energy-efficient, specialised manipulators more precise at specific tasks. The humanoid design accepts these costs in exchange for versatility.

## 1.3 Defining the Landscape

Before examining specific platforms and market dynamics, establishing precise terminology and frameworks is essential for navigating the diverse and rapidly evolving humanoid robotics field. The term "humanoid robot" encompasses significant variation across design philosophies, capability levels, and application domains. Similarly, "autonomy" and "agentic AI" require careful definition to distinguish genuine advances from marketing claims. This section provides that conceptual foundation through four key frameworks: a comprehensive taxonomy classifying humanoid robots by their physical characteristics and target applications, a clear distinction between agentic and traditional AI systems, a five-level autonomy framework enabling objective capability assessment, and a set of standardised performance metrics for comparing platforms. These definitional tools enable the rigorous analysis that follows in subsequent sections and chapters.

### 1.3.1 Humanoid Robot Taxonomy and Classifications

The term "humanoid robot" encompasses significant diversity. Precise taxonomy helps navigate this variation and understand different systems' capabilities, target applications, and design trade-offs.

**By Scale and Morphology**:

Full-scale anthropomorphic humanoids approximate adult human dimensions (150–180 cm height, 50–90 kg weight). Examples include Boston Dynamics Atlas and Unitree H1, alongside systems such as Tesla Optimus that target similar human-scale operation (Boston Dynamics, 2026b; Unitree Robotics, 2026). These systems target general-purpose applications requiring operation in standard human environments.

Compact humanoids employ smaller form factors (typically 80–135 cm height). Unitree's G1 at 132 cm standing height (folding to 69 cm for transport) exemplifies this category, targeting applications where full human scale is unnecessary but humanoid morphology remains beneficial. Lower mass enables faster movement and reduces energy consumption, though limiting payload capacity and reach.

Upper-body humanoids feature humanoid torso, arms, and head but employ wheeled or fixed lower bodies, trading mobility for manipulation capability and stability. These designs avoid the complexity of bipedal locomotion while retaining manipulation advantages of anthropomorphic arms and hands.

**By Degrees of Freedom (DoF)**:

DoF measures mechanical complexity and movement capability. Current humanoid robots span:

Limited DoF (12–20): Basic locomotion and manipulation, often with simplified grippers. These designs minimise cost and control complexity at the expense of dexterity and movement fluidity.

Moderate DoF (20–35): Capable locomotion plus reasonably dexterous manipulation. Many commercial platforms occupy this range, balancing capability against complexity.

High DoF (35–56+): Approaching or exceeding human-like fluidity. Boston Dynamics Atlas features 56 DoF, enabling "super-human range of motion" (Campbell, 2026). However, each additional DoF increases cost, control difficulty, and potential failure modes.

**By Actuation Technology**:

Electric motor systems dominate contemporary humanoid designs, offering precise control, relatively high efficiency (70–85% with gearing), compact packaging, and established manufacturing ecosystems. Boston Dynamics' shift

from hydraulic to electric Atlas signals industry consensus on electric actuation for commercial applications.
Hydraulic systems, while offering exceptional power density for explosive movements, impose operational complexity through pumps, fluid reservoirs, high-pressure hoses, and maintenance requirements that challenge commercial viability.
Hybrid approaches employing different actuation types for different body regions based on performance requirements represent emerging designs attempting to capture advantages of multiple technologies.

**By Application Domain**:

Industrial/Manufacturing: Robots like Figure AI's 2024–2025 Figure 02 deployment at BMW Spartanburg performing assembly tasks, parts sequencing, and quality inspection.
Logistics/Warehousing: Agility Robotics' Digit handling package sorting and transport, having "moved over 100,000 totes at GXO's Flowery Branch facility" (Agility Robotics, 2025).
Service/Healthcare: Robots assisting with patient care, eldercare, hospitality, and retail applications, particularly prominent in Japan and China addressing ageing demographics.
Research and Development: Academic and corporate platforms for studying locomotion, manipulation, learning algorithms, and human–robot interaction.

Figure 1.2 is a classification framework showing different categories of humanoid robots by scale (full-scale, compact, upper-body), degrees of freedom (12–20, 20–35, 35–56+), actuation technology (electric, hydraulic, hybrid), and application domain (industrial, logistics, service, research).

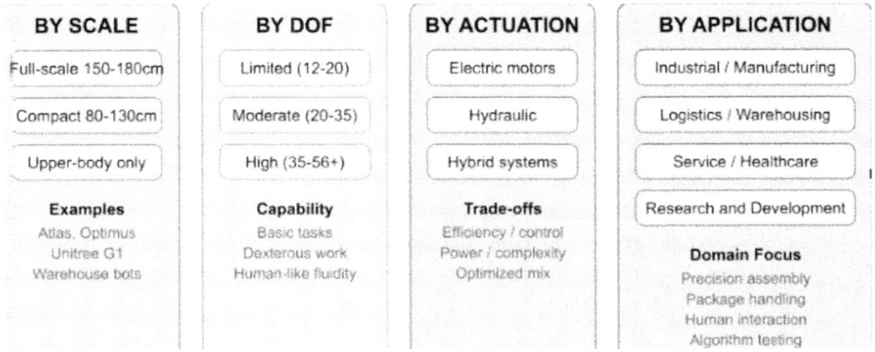


**Fig. 1.2** Humanoid robot taxonomy

### 1.3.2 Agentic AI Versus Traditional AI Systems

Understanding what makes contemporary humanoid robots qualitatively different from earlier generations requires grasping the distinction between agentic and traditional AI.

Traditional AI characteristics include:

- Reactive behaviour: Responding to inputs without persistent goals or long-term planning
- Narrow specialisation: Excelling at specific tasks but failing catastrophically outside trained domains
- Explicit programming: Requiring extensive human engineering of rules or labelled training data for each new task
- Passive operation: Waiting for human queries rather than proactively pursuing objectives
- Brittle performance: Failing unpredictably when encountering out-of-distribution situations

Agentic AI characteristics demonstrate qualitative differences:

- Goal-directed autonomy: Pursuing explicit objectives over extended time horizons through planned action sequences
- Environmental awareness: Building and maintaining models of surroundings, tracking changes, predicting future states
- Adaptive learning: Improving performance through experience, both offline training and online adaptation during deployment
- Proactive behaviour: Taking initiative to gather information, resolve ambiguities, or request assistance rather than simply failing
- Robust generalisation: Transferring learned capabilities to novel situations not seen during training
- Natural communication: Understanding and generating language for receiving instructions, asking questions, explaining actions

The market data confirms this shift: "The agentic AI market will surge from $7.8 billion today to over $52 billion by 2030" with "40% of enterprise applications embedding AI agents by the end of 2026" (Nextgov, 2025). This growth reflects genuine capability advances enabling practical applications.

However, current agentic systems still operate within bounded domains. A robot that autonomously performs warehouse tasks may fail completely in a hospital environment without retraining. True open-ended agency—the ability to tackle arbitrary novel tasks with human-like flexibility—remains a longer-term aspiration. Recognising this distinction prevents both over-optimism about current capabilities and under-appreciation of genuine progress.

### 1.3.3 Levels of Autonomy: A Framework for Assessment

Analogous to the Society of Automotive Engineers (SAE) levels of vehicle autonomy, robotics requires standardised frameworks for discussing autonomy. We adopt and extend existing frameworks for humanoid robots:

Level 0: Direct Control—No autonomous capability. Every motion controlled by human operator through teleoperation. The robot functions as mechanical extension of human control.
Level 1: Assisted Control—Robot performs low-level control (balance maintenance, grasp stabilisation) while humans provide high-level commands for each action. The system handles immediate reactive responses but lacks planning capability.
Level 2: Partial Autonomy—Robot autonomously performs specific, well-defined tasks in structured environments but requires human intervention for exceptions, transitions, or unusual situations. Planning and execution occur within narrow task boundaries. Most current commercial deployments operate at this level.
Level 3: Conditional Autonomy—Robot handles most situations within a defined operational domain autonomously, including common exceptions and variations. Humans monitor and intervene only for rare edge cases or when robot explicitly requests assistance. This represents the current frontier for commercial systems.
Level 4: High Autonomy—Robot operates independently within a well-defined domain, handling essentially all situations without human intervention. Performance may fall short of human capability in edge cases, but system recognises its limitations and operates safely. This remains primarily research-level as of 2026.
Level 5: Full Autonomy—Robot matches or exceeds human capability across all tasks and environments humans can handle, adapting to completely novel situations without prior training. This represents human-level general intelligence and remains distant.

As of 2026, most commercial humanoid robots operate between Levels 2 and limited Level 3 within bounded domains. Figure AI's BMW deployment, which handled parts within a 5-mm tolerance in a structured manufacturing workflow (BMW Group, 2025; Figure AI, 2025), demonstrates task-specific autonomy rather than general Level 3 autonomy. The same systems cannot autonomously handle arbitrary new tasks without additional training, integration, or human support.

Figure 1.3 shows a Five-level autonomy framework for humanoid robots, from Level 0 (Direct Control) through Level 5 (Full Autonomy), with current commercial systems operating primarily at Levels 2–3. Each level shows characteristic capabilities and limitations.

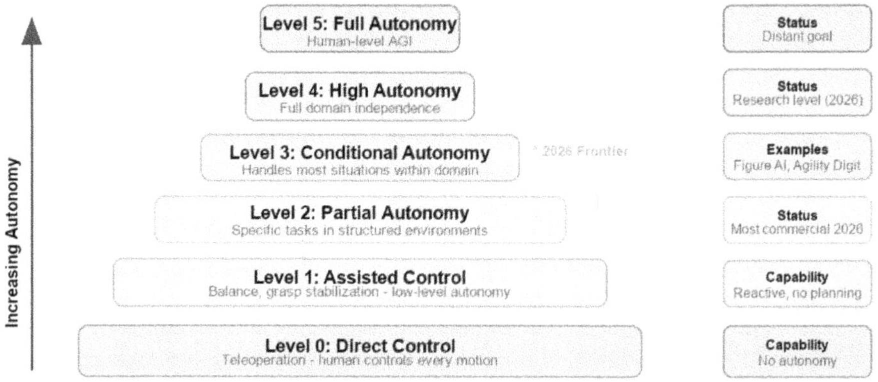


**Fig. 1.3** Levels of autonomy framework

### 1.3.4 Key Performance Metrics and Benchmarks

Evaluating and comparing humanoid robots requires standardised metrics spanning multiple performance dimensions:

Locomotion Metrics:

- Walking speed: Current platforms vary widely; Unitree's H1 demonstrates dynamic bipedal locomotion, including running, while other platforms emphasise controlled indoor walking over speed (Unitree Robotics, 2026)
- Energy efficiency: Cost of transport (joules per kilogram per metre), comparing energy consumption to weight and distance
- Terrain adaptability: Success rates on stairs, uneven surfaces, slopes, obstacles
- Robustness: Recovery from disturbances, pushes, or slips without falling

Manipulation Metrics:

- Grasp success rate: Percentage of successful grasps for standardised object sets
- Dexterity: Complexity of manipulation can be approximated through degrees of freedom, grasp success, force control, and task-level benchmarks. Vendor-reported hand designs suggest improving fine manipulation capability, but independent benchmarks are needed before comparing current systems to human dexterity.
- Force control: Precision in applying specified forces, critical for assembly and caregiving
- Task completion time: How quickly robots perform standardised tasks compared to human workers

Autonomy Metrics:

- Task success rate: Percentage of successfully completed task assignments without human intervention
- Mean time between failures: Hours of operation before requiring human assistance
- Recovery rate: Ability to recover from failures without human intervention
- Adaptation speed: Time or attempts required to learn new task variations

Economic Metrics:

- Acquisition cost: Purchase price per unit. Reported pricing varies from Unitree's R1 entry-level humanoid at $5,900 to premium platforms exceeding $100,000, with many prices reflecting announced or list-price figures rather than fully delivered fleet costs (Automate.org, 2025; China Daily, 2025)
- Operating cost: Hourly cost including energy, maintenance, depreciation
- Productivity: Tasks completed per hour relative to human workers
- Payback period: Time to recover initial investment through labour displacement or productivity gains

These metrics enable objective comparison and tracking of progress. However, real-world performance often diverges from controlled benchmark results. Figure AI's BMW deployment revealed that "the robot's forearm was the top hardware failure point" requiring redesign (Sinha, 2025), illustrating that field testing remains essential for understanding practical reliability.

## 1.4 State of the Art in Humanoid Robotics and Physical AI

The years 2024–2026 mark a decisive inflection point where humanoid robotics transitioned from research curiosity to serious commercial sector. Multiple indicators confirm this shift: production facilities coming online, extended real-world deployments completing, billions in investment flowing to the sector, and costs declining faster than projected. However, beneath the excitement and capital flows lie both genuine capability advances and persistent technical challenges that constrain near-term applications. This section provides an evidence-based assessment of where the field stands as of 2026 across four critical dimensions: market dynamics including investment flows and growth projections, cost trajectories and their implications for economic viability, honest evaluation of what contemporary robots can and cannot reliably accomplish, and the specific technical and operational challenges preventing full autonomy. Understanding both achievements and limitations grounds realistic expectations for deployment timelines and appropriate applications.

### 1.4.1 Market Landscape: Explosive Growth and Investment

The humanoid robotics market has experienced explosive growth during 2024–2026, transitioning from research niche to serious commercial sector attracting tens of billions in investment. Multiple data points document this transformation.

Market size projections vary across research firms but all indicate rapid growth. Industry analysts project the global humanoid robot market will expand from approximately $2.92 billion in 2025 to over $15 billion by 2030, reflecting a compound annual growth rate (CAGR) of 39.2% (MarketsandMarkets, 2025). Looking further ahead, Morgan Stanley anticipates the sector could reach $5 trillion in value by 2050, by which point over a billion humanoid robots may be in active use worldwide (Morgan Stanley, 2025).

Investment activity in 2025 reached unprecedented levels, particularly in China, where the first nine months of the year alone saw 610 funding deals totalling roughly 50 billion yuan—a 250% surge compared to the same period the prior year (Automate.org, 2025). Notable funding milestones included Figure AI and Agility Robotics reaching valuations of $2.6 billion and $1.2 billion respectively, Skild closing a $300 million Series A round, and UBTECH securing $1 billion in strategic financing (Automate.org, 2025).

This capital influx signals investor confidence in near-term commercialisation prospects. Strategic investments from major tech companies underscore the sector's legitimacy—Google DeepMind partnering with Boston Dynamics to embed advanced foundation models into the Atlas platform (Campbell, 2026), and Hyundai committing to an annual humanoid production target of up to 30,000 units by 2028 (The Register, 2026)—representing substantial corporate commitments beyond venture speculation.

Geographic distribution reveals important patterns. China remains the world's largest market for industrial robot installations and has become a major centre for humanoid robotics manufacturing, component supply chains, and cost competition (Automate.org, 2025). Chinese manufacturers have also pushed entry-level pricing downward, including Unitree's R1 humanoid at $5,900 and G1 models positioned at higher price tiers. These developments indicate manufacturing and supply-chain advantages, but they should not be treated as evidence that humanoid robot installations have reached industrial-robot scale.

The United States maintains leadership in AI-first approaches through companies like Figure AI (partnered with OpenAI and BMW), advanced dynamic capabilities through Boston Dynamics, and aggressive manufacturing scale ambitions through Tesla. This geographic distribution reflects broader patterns: China leveraging manufacturing prowess and government support, the US leveraging AI research leadership and venture capital ecosystems.

Production scaling demonstrates serious commercial intent, but confirmed capacity differs from aspirational targets. Agility Robotics has built RoboFab in Oregon for Digit production, and the company reports commercial deployments such as Digit moving more than 100,000 totes at GXO's Flowery Branch facility (Agility Robotics, 2025). Tesla, BYD, and other manufacturers have announced or

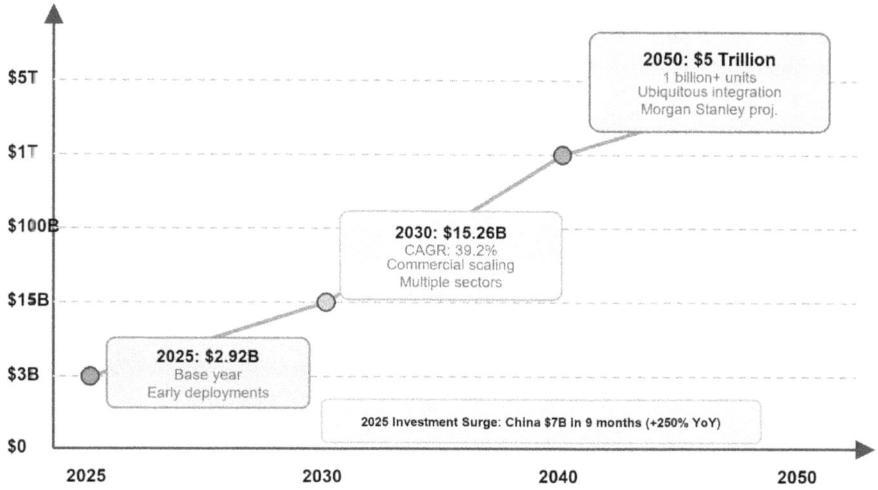


**Fig. 1.4** Market growth projections (2025–2050)

reportedly pursued ambitious humanoid production plans, but these targets should be distinguished from verified fleet deliveries and sustained customer deployments.

Figure 1.4 shows the humanoid robot market size projections with explosive growth from $2.92 billion (2025) to $15.26 billion (2030) at 39.2% CAGR, with long-term projections reaching $5 trillion by 2050.

### 1.4.2 Cost Curves: The Path to Economic Viability

Perhaps the most significant development enabling commercial deployment is the decline in estimated manufacturing costs. Industry analyses report rapid cost reductions as component supply chains mature, electric actuation improves, and Chinese manufacturers introduce lower-priced humanoid platforms, although exact percentage declines should be treated cautiously unless tied to primary cost studies.

Looking forward, the Bank of America Institute anticipates that per-unit material costs, currently around $35,000 in 2025, could fall to somewhere between $13,000 and $17,000 over the next decade (Deloitte, 2026). If this trajectory continues, costs could approach automobile price points by the mid-2030s, enabling consumer market entry beyond industrial applications.

Chinese manufacturers have achieved even more aggressive pricing. Unitree's R1 has been announced at $5,900, while G1 models have been reported around 99,000 yuan (Automate.org, 2025; China Daily, 2025). These price points indicate Chinese manufacturers' cost pressure on the market, although capability, reliability, software maturity, and fleet-support quality still require independent comparison with higher-cost Western platforms.

Tesla's stated consumer-market ambition is to push Optimus towards pricing that would eventually be accessible beyond large corporations. If achieved, this would materially expand the humanoid robotics market, but current price targets should be treated as aspirational until supported by delivered units, reliability data, and total-cost-of-ownership evidence.

### 1.4.3 Capabilities: What Robots Can (and Cannot) Do in 2026

Contemporary humanoid robots demonstrate impressive capabilities in specific domains while facing significant limitations in others. Realistic assessment requires acknowledging both.

**Mature Capabilities Commercially Deployed in 2026 Include**:

Bipedal locomotion on flat terrain: Multiple platforms walk reliably on even surfaces in controlled indoor environments. Boston Dynamics' Atlas can navigate "minus 20 to 40 degrees Celsius" temperatures (CGTN, 2026), demonstrating operational robustness.

Basic object manipulation: Pick-and-place operations for standard objects work reliably when objects have known properties and positions. Figure AI's parts handling "within a 5-millimeter tolerance in just 2 seconds" (BMW Group, 2025) demonstrates precision sufficient for manufacturing applications.

Natural language understanding: Robots can interpret some structured verbal instructions, especially when the task scope and operating environment are constrained. However, ambiguous, multi-step, or safety-critical instructions still require careful grounding, user confirmation, or supervisory logic.

Autonomous navigation in structured environments: Robots navigate warehouses, factories, and hospitals with known layouts reliably, avoiding static and dynamic obstacles through vision-based perception.

Advancing capabilities demonstrated in research or limited deployment but requiring refinement include:

Unstructured terrain navigation: Walking on outdoor terrain, stairs, irregular surfaces succeeds in controlled testing but reliability in fully unpredictable environments remains inconsistent.

Dexterous manipulation: Hand designs, force control, and learning-based manipulation continue to advance, but complex tasks such as tying knots, handling deformable objects, or manipulating delicate items remain research-level for most humanoid platforms.

Long-duration autonomy: Extended operation for 8+ hour shifts without human supervision remains elusive. Battery limitations constrain runtime to 2–5 h for most platforms. Beyond energy, accumulating errors and edge case failures require periodic human intervention.

Visual scene understanding in clutter: While robots recognise common objects, understanding complex scenes with occlusion, variable lighting, transparent or reflective surfaces, and novel object categories remains problematic.
Whole-body dynamic coordination in structured performance: Public demonstrations such as the 2026 Chinese Spring Festival Gala in robotic dance showcase illustrate that humanoids can execute synchronised choreography, martial-arts-style movements, and rapid formation changes in tightly controlled environments. However, these performances rely on pre-planned routines, known stage conditions, and extensive calibration; comparable fluid, adaptive whole-body control across unpredictable real-world tasks remains beyond current commercial capability.

**What Robots Cannot Do Yet**:

Common sense reasoning: Robots lack human intuitive understanding of physics (that cups hold liquids, glass breaks when dropped), psychology (inferring human intentions from behaviour), and everyday causality (that pressing a button might open a door).
Transfer learning: Robots trained on specific tasks often fail on minor variations, lacking human-like flexible adaptation. Training on warehouse boxes doesn't automatically transfer to hospital linens.
Whole-body dynamic control: While specialised systems like Atlas demonstrate athletic capabilities, general commercial platforms cannot match human fluid, adaptive whole-body movement across diverse activities.
Safety assurance: Guaranteed safe operation in proximity to humans across all possible scenarios remains unsolved. Current approaches err conservatively, limiting performance.

The International Federation of Robotics' "Top 5 Global Robotics Trends 2026" emphasises that while "robots still face technical limits such as short battery life, weak performance on rough terrain, and unreliable sensors in complex settings" (International Federation of Robotics, 2026), progress on each constraint continues steadily. The question is pace: whether critical capabilities mature in 3 years, 10 years, or longer profoundly affects economic and social planning.

### 1.4.4 The Intelligence Infrastructure: The "NVIDIA Moment"

While individual manufacturers refine hardware, one important contributor to 2026 momentum is the maturation of shared physical-AI infrastructure. NVIDIA's 2026 announcements around Cosmos, Isaac GR00T (Generalist Robot 00 Technology), Jetson, and simulation tooling illustrate how large vendors are trying to standardise robot training, simulation, and deployment pipelines (NVIDIA, 2026).

The release of NVIDIA Cosmos and Isaac GR00T N1.6 reflects a move towards reusable robot-learning infrastructure rather than bespoke, single-purpose programming. Cosmos provides world-model tooling intended to support physical reasoning and simulation, while integrations with open-source robotics frameworks such as LeRobot can broaden developer access to robot-learning workflows (NVIDIA, 2026).

On the hardware front, the introduction of the Jetson T4000 module and Jetson Thor computers provided the necessary onboard compute for these heavy AI workloads. Delivering 4× the energy efficiency of previous generations and 1,200 FP4 (4-bit floating point) TFLOPS of compute, these modules allow humanoids to run complex reasoning models locally without relying on high-latency cloud connections. This standardisation is evident in the ecosystem's adoption: from LG Electronics' smart home robots to Caterpillar's autonomous construction fleets and LEM Surgical's medical arms, the industry is coalescing around a shared compute architecture, reducing the time-to-market for new entrants.

## 1.5 Leading Platform Case Studies

Examining specific platforms and companies illuminates diverse technical approaches, strategic visions, and paths towards commercial viability pursued across the industry.

### 1.5.1 Boston Dynamics Atlas: From Research Icon to Commercial Product

Boston Dynamics' Atlas has long represented the pinnacle of dynamic humanoid robotics, famous for videos demonstrating backflips, parkour, and impressive balance recovery. However, Atlas remained primarily a research platform—expensive, hydraulically complex, and impractical for widespread deployment—until the company's strategic pivot.

In April 2024, Boston Dynamics announced the retirement of its hydraulic Atlas and introduction of a fully electric version designed explicitly for commercial applications. By CES 2026, this transition reached fruition: "Boston Dynamics is entering production, with the robotics company announcing that the final product version is being built now" with initial deployments to "Hyundai and Google DeepMind" (The Register, 2026).

The production Atlas features impressive specifications: the platform offers up to 56 degrees of freedom, giving it a range of motion that surpasses human capability, along with human-scale hands equipped with tactile sensing, a reach of roughly 2.3 m, a payload capacity of up to 50 kg, and the ability to operate across a wide temperature range from −20 to 40 °C (CGTN, 2026). These specs demonstrate that commercial focus has not sacrificed capability.

The strategic partnerships illuminate Boston Dynamics' commercialisation strategy. Hyundai has outlined plans to deploy Atlas in its automotive plants beginning in 2028 for parts sequencing tasks, with ambitions to expand its role by 2030 to include component assembly and operations involving repetitive motion and heavy loads (The Register, 2026). This provides a concrete deployment timeline and defined application scope.

The Google DeepMind partnership addresses the AI dimension: Boston Dynamics announced a collaboration with Google DeepMind focused on embedding state-of-the-art foundation models into Atlas, with the goal of substantially enhancing the robot's cognitive and decision-making abilities (Campbell, 2026). This combines Boston Dynamics' hardware and control expertise with DeepMind's AI leadership, potentially creating the most capable commercial humanoid platform.

However, challenges remain. Boston Dynamics has not disclosed pricing, though it likely substantially exceeds Chinese competitors given hardware sophistication. All Atlas units slated for 2026 have already been allocated to committed partners (Boston Dynamics, 2026a), indicating limited production capacity initially. Scaling from tens to thousands to tens of thousands of units will test whether Boston Dynamics can transition from boutique manufacturer to mass production.

The Atlas trajectory—from DARPA Challenge participant (2013), to YouTube sensation with athletic demonstrations (2016–2023), to announced commercial product (2024), to production commencement (2026)—illustrates the long development path for cutting-edge robotics. The 13+ year timeline from DARPA Robotics Challenge participation (2013) to commercial product (2026) underscores the difficulty of robotics commercialisation even for best-funded companies.

### 1.5.2 Tesla Optimus: Mass Manufacturing Ambitions

Tesla's entry into humanoid robotics, announced by Elon Musk in August 2021, brought unprecedented attention and resources to the field. Optimus aims to leverage Tesla's automotive manufacturing expertise, vertical integration, and AI capabilities developed for self-driving to achieve breakthrough price points enabling mass deployment.

Development has progressed through multiple generations. Gen 1 (2022) demonstrated basic walking, while Gen 2 (2023) showed improvements in speed, dexterity, and sensor capability. Later Optimus claims emphasise improved hands, locomotion, payload, and manufacturability, but many detailed Gen 3 specifications remain company-reported or media-reported rather than independently benchmarked.

The production timeline has faced delays typical of complex engineering projects, and public statements about demonstrations, limited production, and long-term scale should be treated as targets rather than verified delivery milestones. Tesla's importance lies in its manufacturing ambition and vertical integration strategy, but the chapter should distinguish announced goals from demonstrated fleet deployment.

Tesla's cost targets are perhaps most significant for industry trajectory. If humanoid robots eventually reach consumer or small-business price points, the addressable market would broaden substantially beyond large corporations. However, the feasibility of those targets remains uncertain because humanoid robots impose different actuation, dexterity, safety, and reliability requirements than automobiles.

Tesla's vertical integration strategy designs and manufactures most components in-house, from actuators to circuit boards to batteries to AI chips. This approach maximises cost optimisation and performance but requires enormous capital investment. The bet is that capturing entire value chain enables cost structures competitors cannot match.

Internal deployment can provide crucial testing data because factory environments expose robots to repeated manipulation tasks, part variation, safety constraints, and uptime requirements. However, whether Tesla's automotive AI and manufacturing stack translates effectively to humanoid manipulation must be demonstrated through sustained deployment rather than assumed.

### 1.5.3 Figure AI: AI-First Commercial Focus

Figure AI represents a new generation of robotics companies built from inception around modern AI capabilities rather than traditional robotics approaches. Founded in 2022 by Brett Adcock with backing from OpenAI, Microsoft, and other tech giants, Figure has pursued aggressive commercial focus.

The company's Figure 02 platform completed a landmark deployment at BMW: "The Figure 02 robots ran 10-hour shifts Monday through Friday, accumulating 1,250 hours of run-time, loaded more than 90,000 parts, and contributed to the production of more than 30,000 BMW X3 vehicles" (Figure AI, 2025) during an 11-month period ending in late 2025.

This deployment was an expensive but necessary learning experience rather than a smooth iteration story. The intensive use revealed failure modes: "The robot's forearm was the top hardware failure point at BMW, which informed the redesign of Figure 03" (Sinha, 2025). Figure AI retired the Figure 02 fleet after the deployment and rolled the lessons into Figure 03, that outcome is still valuable, but it underscores that commercial reliability is earned through real-world wear, interventions, repair data, and customer-site learning that simulation cannot fully capture.

However, the deployment also ended: "Currently, there are no Figure AI robots at BMW Group Plant Spartanburg, and there is no definite timetable established for bringing Figure robots to the plant" (Sinha, 2025). The retirement of the Figure 02 units indicates that the BMW deployment functioned as a high-value pilot that exposed reliability issues and informed Figure 03, but it does not yet demonstrate readiness for broad scaling.

Figure AI's approach emphasises several differentiators: deep integration with large language models (through OpenAI partnership) enabling natural language task specification, learning from demonstration where robots acquire skills by observing humans rather than through explicit programming, focus on specific commercial applications (manufacturing, logistics) rather than general capabilities, and partnership strategy leveraging industry leaders' domain expertise (BMW for automotive, OpenAI for AI).

The commercialisation challenges are evident: despite successful BMW deployment, the company has not announced immediate scaling to other customers or facilities. The gap between successful pilot and widespread deployment remains substantial. Whether Figure AI crosses this gap faster than competitors will significantly influence industry trajectory.

### 1.5.4 Agility Robotics: Logistics Specialisation

Agility Robotics, founded in 2015 as an Oregon State University spinout, has pursued perhaps the most focused strategy: designing humanoid robots specifically optimised for logistics applications rather than general-purpose use.

The Digit platform's commercial traction accelerated dramatically in 2025. The company's major milestone: "Digit has moved over 100,000 totes at GXO's Flowery Branch facility" (Agility Robotics, 2025), demonstrating sustained real-world productivity. Unlike short-term pilots, this represents extended deployment generating actual business value.

Recent partnerships expanded Digit's footprint: "Mercado Libre is introducing humanoid robots into its warehouse network, deploying Digit robots inside Mercado Libre's fulfilment centre in San Antonio, Texas" (Digital Commerce 360, 2025) for "moving totes and transporting materials," with plans to "evaluate broader use cases with an eye toward future expansion into facilities across Latin America."

Agility's manufacturing capability distinguishes it from many competitors: "The company now operates a dedicated RoboFab factory in Oregon designed to build hundreds of robots per year, with a roadmap to more than 10,000 units annually" (Agility Robotics, 2025). This represents the first humanoid-robot-specific mass manufacturing facility, designed from the ground up for efficient robot production rather than adapted from other purposes.

The specifications reflect logistics optimisation rather than general capability: "Digit stands at a height of 5 feet 9 inches, weighs 140 pounds, and can handle payloads up to 35 pounds" (Agility Robotics, 2025). The relatively lower payload compared to human capability (humans routinely handle 50+ lbs) reflects trade-offs for energy efficiency and cost. For many logistics tasks, 35 lbs suffices.

However, challenges persist: "As of 2025, Agility has sold around 100 Digit units" (Agility Robotics, 2025), indicating limited market penetration despite multiple years of commercial availability. The company must accelerate adoption significantly to justify manufacturing investments and valuations. Competing robots with broader capabilities or lower costs could limit Digit to niche applications.

The specialised approach—optimising for specific applications rather than general capabilities—represents a viable commercialisation strategy: achieve commercial revenue in focused domains rather than waiting for general AI capabilities to mature. Whether this strategy succeeds faster than general-purpose approaches pursued by Tesla or Boston Dynamics will significantly inform future robotics business models.

### 1.5.5 Unitree: Chinese Cost and Speed Leadership

Unitree Robotics exemplifies Chinese manufacturers' distinctive approach: aggressive cost optimisation through supply-chain advantages while achieving impressive dynamic performance. The company originally gained prominence through quadruped robots competing with Boston Dynamics Spot at a fraction of the cost before entering humanoid markets.

Unitree's H1 platform emphasises dynamic performance, with Unitree describing the system as capable of bipedal locomotion including running (Unitree Robotics, 2026). Its relatively lightweight design supports high-speed motion, although cross-platform speed and payload comparisons should be based on standardised benchmarks rather than vendor demonstrations alone.

However, Unitree's most significant impact may be its aggressive pricing strategy across multiple humanoid platforms. The R1 introduced an entry-level $5,900 price point, while the G1 has been reported at 99,000 yuan, or approximately $13,800, with China Daily reporting about 5,000 G1 units sold in the first half of 2025 (China Daily, 2025).

These sales figures, while modest compared to automobile markets, represent the largest commercial humanoid robot deployment as of mid-2025. Applications range from "concert performances to factory material handling" (China Daily, 2025), demonstrating versatility. The company "completed initial public offering (IPO) tutoring in November 2025, with stated plans to expand research capacity, scale production, and further integrate its industrial ecosystem following a successful listing" (China Daily, 2025).

Unitree's strategy leverages several advantages: China's world-leading supply chain for electromechanical components enabling component cost reduction, vertical integration of actuators and control systems, rapid iteration cycles with multiple product releases annually, and government support through Made in China 2025 industrial policy initiatives.

However, questions remain about software and AI sophistication relative to Western competitors with stronger ties to frontier AI research (OpenAI, DeepMind, Anthropic). Unitree's strength lies in electromechanical engineering and manufacturing rather than cutting-edge AI, potentially limiting autonomous capability. Whether cost advantages compensate for AI gaps will determine competitiveness in AI-intensive applications.

The broader significance is strategic: Unitree forces all competitors to accelerate cost reduction. Western manufacturers claiming $30,000–$50,000 price points face pressure when Chinese competitors deliver at $5,900 (R1) to roughly $13,800 (G1). This competition may accelerate the timeline towards mass affordability.

### 1.5.6 The Broader Ecosystem: Emerging Competitors and Specialised Players

While Boston Dynamics, Tesla, Figure AI, Agility Robotics, and Unitree represent the most prominent players, the humanoid robotics ecosystem extends far broader with numerous companies pursuing specialised approaches, distinct market segments, or emerging technologies.

Apptronik, spun out from University of Texas at Austin's Human Centered Robotics Lab, has developed the Apollo platform emphasising modular design philosophy. The approach allows swappable end effectors for different applications—grippers for logistics, tool interfaces for manufacturing, gentle manipulation systems for healthcare—enabling a single base platform to serve diverse markets. Force-controlled actuators enable safe physical interaction with humans, critical for collaborative applications. The company has announced partnerships with NASA for potential space applications and Mercedes-Benz for automotive manufacturing, though deployment scale remains limited as of 2026 compared to leaders like Agility or Figure AI.

UBTECH Robotics, established in China in 2012 and initially focused on educational and entertainment robots, has evolved its Walker platform for service applications. The strategic differentiation lies in prioritising human interaction over pure task performance. Walker emphasises natural language conversation through integrated AI language models, facial recognition and emotion detection for personalised interaction, appealing aesthetic design targeting consumer acceptance, and applications in retail, hospitality, museums, and eventually home environments.

The massive investment UBTECH attracted—"a massive $1 billion strategic financing facility" in 2025 (Automate.org, 2025)—signals investor confidence in the service robot market segment. This contrasts with Western emphasis on industrial and logistics applications, reflecting different market opportunities in China's large service sector and different cultural comfort with service robots.

Emerging Chinese manufacturers represent diverse specialised approaches:

Fourier Intelligence focuses explicitly on medical and rehabilitation applications, developing robots for patient assistance, physical therapy support, and eldercare. This specialisation positions the company for regulated healthcare markets requiring stringent safety validation but offering potentially higher margins than commodity logistics applications.

EX Robots pursues perhaps the most distinctive strategy: hyper-realistic human appearance and emotional intelligence design. The robots feature remarkably lifelike faces, skin textures, and expressions, targeting applications emphasising emotional connection—companionship for elderly individuals, mental health support, hospitality services. This approach generates both interest and controversy regarding appropriate human–robot relationships.

Xiaomi, the consumer electronics giant, views humanoid robots as ecosystem components integrating with its vast smart home and Internet of Things (IoT) infrastructure rather than standalone products. The CyberOne platform demonstrated in 2022–2025 shows capabilities but Xiaomi has not announced mass production, suggesting strategic positioning for future market timing rather than immediate commercialisation.

AgiBot emerged as a significant new player, joining Unitree at CES 2026 to showcase Chinese humanoid capabilities. "At CES 2026, Unitree and AgiBot led China's showcase of humanoid robots" (TrendForce, 2026), indicating the company's rise to prominence and expanding competitive landscape.

The ecosystem diversity reveals several important patterns. First, the solution space remains large—no single design philosophy, target application, or business model has proven dominant. Companies pursuing athletic capability (Boston Dynamics), mass manufacturing (Tesla), AI-first approaches (Figure AI), application-specific optimisation (Agility), cost leadership (Unitree), and service/social focus (UBTECH, EX Robots) can all potentially succeed by capturing different market segments.

Second, significant capital is flowing to diverse approaches rather than concentrating on a single winner. The $1 billion UBTECH raised, the valuations achieved by Figure AI ($2.6B) and Agility ($1.2B), and the $7 billion Chinese sector funding indicate investor belief that multiple companies will capture substantial market share rather than winner-take-all dynamics.

Third, geographic and cultural factors shape strategic approaches. Chinese companies disproportionately emphasise cost optimisation and service applications reflecting domestic market characteristics. U.S. companies emphasise AI integration and industrial applications reflecting different technology ecosystems and labour economics. Japanese companies (Honda, Softbank) maintain presence but have lost competitive edge to Chinese manufacturers and U.S. AI leaders.

Fourth, vertical integration versus specialisation represents a strategic fault line. Tesla pursues extreme vertical integration designing and manufacturing most components in-house. Others specialise in system integration while sourcing components from specialised suppliers. The optimal approach remains unproven—vertical integration maximises optimisation but requires enormous capital; specialisation enables faster iteration but may sacrifice performance or cost optimisation.

The competitive landscape as of 2026 is dynamic, well-funded, and globally distributed. This positions the industry for rapid progress through competitive pressure while creating questions about standards, interoperability, safety frameworks, and geopolitical implications of concentrated Chinese manufacturing and installation leadership.

## 1.6 Challenges, Constraints, and Adoption Thresholds

Despite rapid progress, the central adoption threshold is whether humanoid robots can deliver safe, reliable, economically useful work over extended duty cycles in real operating environments. Understanding these constraints grounds realistic expectations for deployment timelines and appropriate applications.

### 1.6.1 Technical Bottlenecks

Power and energy: Battery technology limits operational duration. Most humanoid robots operate 2–5 h before requiring recharge, substantially less than 8-h human work shifts. “Robots still face technical limits such as short battery life” (International Federation of Robotics, 2026), and battery improvements remain incremental relative to the demands of full-shift humanoid operation. Revolutionary battery breakthroughs could accelerate deployment, absent such breakthroughs, operational limitations persist for years.

Perception in complex environments: “Many robots still struggle outdoors, on rough terrain, or in unpredictable conditions” as reported in the same paper. Cluttered environments, variable lighting, transparent or reflective surfaces, and novel objects challenge vision systems. Humans effortlessly navigate these situations; robots fail unpredictably.

Dexterous manipulation: While hand designs, tactile sensing, and manipulation policies continue to advance, reliable manipulation of deformable objects (cloth, food), delicate items (eggs, electronics), or objects requiring complex reorientation remains difficult.

Real-time control: Humanoid control requires millisecond-level response times for balance and manipulation. Integrating large AI models (which may require hundreds of milliseconds for inference) with real-time control creates architectural challenges. Hybrid approaches partitioning fast reactive control from slower deliberative planning represent current solutions but add complexity.

### 1.6.2 Safety and Reliability

Safety certification represents a major deployment barrier. Robotic systems must be engineered and validated against established safety standards, with clearly delineated liability structures in place. However, the existing regulatory landscape consists of fragmented standards that were never built with intelligent, autonomous machines in mind. Regulatory frameworks lag technological capabilities, creating legal uncertainty that conservative companies find paralysing.

Reliability in extended deployment is the most important near-term gating factor for broad adoption. Battery life, safety certification, autonomy, integration cost, and workforce acceptance all matter, but most customers will not scale humanoid fleets until robots can deliver useful work at predictable uptime with manageable maintenance. Figure AI's forearm failures during BMW deployment (Sinha, 2025) illustrate that even impressive demonstrations may mask reliability issues revealed only through extended use. Industrial applications typically require high availability over thousands of operating hours, most humanoid robots have not yet publicly demonstrated such reliability across diverse customer environments.

The workforce acceptance challenge compounds technical safety concerns. Beyond engineering reliability, the sector continues to grapple with intertwined obstacles spanning physical safety, cybersecurity vulnerabilities, and human willingness to work alongside automated systems. Even technically safe robots may face worker resistance due to job displacement fears or discomfort with robot colleagues.

### 1.6.3 Economic Viability Thresholds

Whether humanoid robots achieve widespread deployment depends critically on crossing economic viability thresholds where robot costs plus operating expenses fall below human labour costs for the same tasks.

For industrial applications in developed nations with high labour costs ($30-$50/hour), some use cases may approach economic viability, but the calculation depends on utilisation, reliability, integration cost, and whether operating-cost estimates include depreciation. If a $30,000 robot operates for five years at 2,000 h per year and incurs $5/hour in operating costs excluding purchase depreciation, the five-year total is $80,000, or $8/hour across 10,000 h. If the $5/hour figure already includes depreciation, the model should itemise depreciation, energy, maintenance, integration, downtime, and support separately before comparing robot labour with human wages, benefits, and overhead.

However, this calculation assumes the robot actually performs useful work for those hours rather than failing or standing idle. It assumes work can be reorganised around robot capabilities and limitations. It assumes regulatory approval, worker acceptance, and technical integration. Each assumption may not hold.

For developing nations with low labour costs ($2–$5/hour), economic viability requires even more dramatic cost reductions or identification of applications where robots provide value beyond cost (24/7 operation, extreme environments, superhuman strength or precision).

### 1.6.4 Skills Gap and Deployment Friction

A frequently underestimated challenge is the scarcity of expertise required to deploy, maintain, and operate advanced robotic systems. Recently, the binding bottleneck has shifted from hardware availability to human capital. Sectors that previously managed with a small group of automation engineers now depend on multidisciplinary teams encompassing roboticists, system integrators, safety specialists, and skilled technicians simply to maintain operational continuity. Purchasing the machinery is no longer the difficult part; assembling the talent needed to install, sustain, and properly oversee these systems is where organisations consistently fall short.

This skills gap creates deployment friction even when economic and technical viability exists. Small and medium enterprises particularly struggle, lacking resources to hire specialised personnel. This may limit initial deployment to large corporations with resources for expertise acquisition, potentially concentrating economic benefits.

Educational systems lag demand because robotics expertise requires hardware integration, control theory, safety engineering, and field maintenance skills beyond general AI literacy. Developing educational pipelines that produce sufficient qualified personnel will take years even with aggressive investment.

## 1.7 Book Structure and Reading Guide

This book explores humanoid robotics and physical AI across technical, economic, social, and philosophical dimensions. Understanding the complete picture requires engaging with multiple perspectives—engineering capabilities, business models, economic implications, safety frameworks, psychological dynamics, ethical considerations, and governance challenges.

### 1.7.1 Chapter Organisation

The book's structure progresses from technical foundations through applications and implications to future trajectories:

This chapter (current): Establishes fundamental concepts, taxonomy, current state of the art, and case studies of leading platforms.
Chapter 2: Hardware and embodiment technologies—actuation systems, structural materials, power systems, locomotion, and manipulation.
Chapter 3: Perception and interaction systems—computer vision, sensing modalities, sensor fusion, and human–robot interaction.
Chapter 4: Agentic AI architectures—foundation models for robotics, reinforcement learning, planning and reasoning, world models, and benchmarks.
Chapter 5: Integration and control systems—real-time control, whole-body coordination, simulation, and software development.
Chapter 6: Safety, security, and reliability—physical safety standards, cybersecurity, fail-safe mechanisms, and verification.
Chapter 7: Economic transformation and labour markets—industry disruption, cost analysis, productivity, workforce implications, and market projections.
Chapter 8: Governance, ethics, and philosophy—regulatory frameworks, liability, ethical design principles, rights, and existential questions.

Chapter 9: Psychology and human–robot coexistence—perception, trust, emotional bonding, mental health, developmental psychology, and cultural adaptation.

Chapter 10: Technology convergence and future of human purpose—Artificial General Intelligence (AGI) timelines, universal high income, ageing societies, longevity, meaning, and scenarios for 2030–2050.

### 1.7.2 Reading Paths for Different Audiences

For Technical Readers (Engineers, Researchers, Computer Scientists): Deep engagement with Chaps. 2–5 provides comprehensive technical foundation. This chapter and Chaps. 6 and 7 offer context on state of the art, safety requirements, and market dynamics. Chapters 8–10 can be skimmed for social and ethical context.

For Business Leaders and Investors: Focus on this chapter and Chaps. 7 and 10 for market analysis, deployment economics, and future scenarios. Chapters 2–3 provide necessary technical context without requiring deep expertise. Case studies throughout chapters illuminate competitive dynamics and strategic positioning.

For Policy Makers and Regulators: this chapter and Chaps. 6–8 provide essential foundation for informed policy: technical capabilities, safety challenges, economic impacts, and governance frameworks. Chapters 9 and 10 address social psychology and long-term implications relevant to policy planning.

For General Readers and Students: Sequential reading provides comprehensive understanding. Each chapter builds on earlier material while remaining accessible to non-specialists. Comprehension questions enable self-assessment. Readers can skip technical depth while still gaining valuable insights from surrounding context and case studies.

For Ethicists and Philosophers: this chapter and Chaps. 8–10 directly address philosophical questions about purpose, meaning, consciousness, and values in an age of capable machines. Chapters 2–7 provide concrete technical and economic grounding for abstract philosophical analysis.

### 1.7.3 Cross-References and Thematic Integration

Throughout the book, extensive cross-referencing highlights connections across topics:

- Hardware capabilities (Chaps. 2–3) enable or constrain applications (Chap. 7)
- AI architectures (Chap. 4) determine autonomy levels affecting economic viability (Chap. 7)
- Safety frameworks (Chap. 6) shape regulatory requirements (Chap. 8) influencing deployment pace

- Psychological factors (Chap. 9) interact with economic displacement (Chap. 7) determining social outcomes
- All preceding chapters inform future scenarios (Chap. 10)

These cross-references appear as explicit chapter and section citations throughout. Readers focusing on specific topics should follow references to develop comprehensive understanding across relevant dimensions.

**Comprehension Questions**

1. Historical Evolution: Trace the evolution of robotics from the Unimate industrial arm (1961) through the DARPA Robotics Challenge (2012–2015) to contemporary humanoid robots in 2026. What were the key technical limitations that prevented earlier humanoid robot development? Which limitations have been substantially overcome versus which remain significant challenges?
2. Embodied AI Concept: Explain the concept of "embodied AI" or "physical AI" and why it represents a qualitative shift from previous AI systems. Using the example of Figure AI's BMW deployment (90,000 parts loaded, 30,000 vehicles produced), analyse what capabilities embodied AI provides that disembodied AI cannot.
3. Agentic vs. Traditional AI: Distinguish between agentic AI and traditional AI systems using specific characteristics discussed in the chapter. Why does Gartner predict that "40% of enterprise applications will embed AI agents by the end of 2026, up from less than 5% in 2025"? What business value does agentic capability create?
4. Form Factor Justification: Analyse the engineering trade-offs involved in choosing humanoid form factor versus alternatives (wheeled, quadrupedal, specialised). Under what circumstances does the humanoid form's versatility justify its complexity? When would alternative form factors be superior?
5. Autonomy Levels Application: Using the five-level autonomy framework, classify Boston Dynamics Atlas, Tesla Optimus Gen 2, and Agility Robotics Digit as of 2026. Justify your classifications with specific capability examples. What would each platform need to achieve to advance one autonomy level?
6. Market Analysis: The chapter presents several dramatic market projections: growth from \$2.92B (2025) to \$15.26B (2030) to potentially \$5T (2050). What technical and economic assumptions underlie these projections? What factors could cause reality to substantially exceed or fall short of projections?
7. Cost Curve Implications: Analyse the significance of reported cost reductions and Unitree's \$5,900 R1 entry-level pricing. Using the economic viability threshold analysis from Sect. 1.6.3, calculate at what price point humanoid robots become economically competitive with human labour in various wage contexts (\$10/hour, \$30/hour, \$50/hour).
8. Geographic Competition: Compare and contrast the approaches of U.S. companies (Boston Dynamics, Tesla, Figure AI, Agility) versus Chinese manufacturers

(Unitree, UBTECH). How do different strengths—AI research versus manufacturing scale, venture capital versus government support—shape competitive dynamics? What are implications for global technology leadership?

9. Deployment Lessons: Figure AI's 11-month BMW deployment revealed that "the robot's forearm was the top hardware failure point." What does this teach about the gap between laboratory demonstrations and commercial deployment? Why do extended real-world trials reveal failure modes that testing cannot predict?
10. Critical Assessment: The chapter documents both extraordinary progress and persistent limitations (battery life, autonomy, reliability). Construct an argument for either optimistic (widespread deployment by 2028–2030) or cautious (limited deployment until 2032–2035+) timelines for mass adoption. Support your position with specific evidence from the chapter.

## References


Agility Robotics. (2025). *Digit moves over 100,000 totes in commercial deployment.* Agility Robotics News. https://www.agilityrobotics.com/content/digit-moves-over-100k-totes

Automate.org. (2025). *Industry insights: Humanoid robots in 2025: The next stage of evolution.* Association for Advancing Automation. https://www.automate.org/industry-insights/humanoid-robots-are-evolving

BMW Group. (2025). *Successful test of humanoid robots at BMW Group Plant Spartanburg.* BMW Press Release. https://www.press.bmwgroup.com/global/article/detail/T0444265EN/successful-test-of-humanoid-robots-at-bmw-group-plant-spartanburg?language=en

Boston Dynamics. (2026a). *Boston Dynamics unveils new Atlas robot to revolutionize industry.* Boston Dynamics Blog. https://bostondynamics.com/blog/boston-dynamics-unveils-new-atlas-robot-to-revolutionize-industry/

Boston Dynamics. (2026b). *Atlas: The world's most dynamic humanoid robot.* Boston Dynamics Products. https://bostondynamics.com/products/atlas/

Campbell, I. C. (2026). *Boston Dynamics unveils production-ready version of Atlas robot at CES 2026.* Engadget. https://www.engadget.com/big-tech/boston-dynamics-unveils-production-ready-version-of-atlas-robot-at-ces-2026-234047882.html

China Daily. (2025). *Zhejiang robots sweeping the nation.* China Daily Technology. https://www.chinadaily.com.cn/a/202504/03/WS67ede108a3104d9fd381d684.html

CGTN. (2026). *CES 2026: Is Boston Dynamics' Atlas a game changer in humanoid robots?* CGTN Technology News. https://news.cgtn.com/news/2026-01-09/CES-2026-Is-Boston-Dynamics-Atlas-a-game-changer-in-humanoid-robots--1JNuewVLmgM/p.html

Deloitte. (2026). *AI goes physical: Navigating the convergence of AI and robotics.* Deloitte Tech Trends 2026. https://www.deloitte.com/us/en/insights/topics/technology-management/tech-trends/2026/physical-ai-humanoid-robots.html

Digital Commerce 360. (2025). *Mercado Libre to test Agility Robotics' humanoid robots in fulfillment center.* Digital Commerce 360 News. https://www.digitalcommerce360.com/2025/12/10/mercado-libre-agility-robotics-texas-fulfillment-center/

Figure AI. (2025). *Figure 02 contributed to the production of 30,000 cars at BMW.* Figure AI News. https://www.figure.ai/news/production-at-bmw

Gartner. (2025). *Gartner predicts 40% of enterprise applications will feature task-specific AI agents by 2026, up from less than 5% in 2025.* Gartner Newsroom. https://www.gartner.com/en/newsroom/press-releases/2025-08-26-gartner-predicts-40-percent-of-enterprise-apps-will-feature-task-specific-ai-agents-by-2026-up-from-less-than-5-percent-in-2025
International Federation of Robotics. (2026). *Top 5 global robotics trends 2026.* IFR Press Release. Retrieved from https://ifr.org/ifr-press-releases/news/top-5-global-robotics-trends-2026
MarketsandMarkets. (2025). *Humanoid robot market size, share, industry report trends, 2025 to 2030* (MarketsandMarkets Industry Reports). https://www.marketsandmarkets.com/Market-Reports/humanoid-robot-market-99567653.html
Morgan Stanley. (2025). *Humanoid robot market expected to reach $5 trillion by 2050.* Morgan Stanley Research. https://www.morganstanley.com/insights/articles/humanoid-robot-market-5-trillion-by-2050
Nextgov. (2025). *2026 is set to be the year of agentic AI, industry predicts.* Nextgov Federal Technology. https://www.nextgov.com/artificial-intelligence/2025/12/2026-set-be-year-agentic-ai-industry-predicts/410324/
NVIDIA. (2026). *NVIDIA releases new physical AI models as global partners unveil next-generation robots.* NVIDIA Newsroom. https://nvidianews.nvidia.com/news/nvidia-releases-new-physical-ai-models-as-global-partners-unveil-next-generation-robots
Rai, A. (2025). *The great physical breakout: Why 2026 will be the year of embodied AI.* Medium. https://medium.com/@ashishrai12_23721/the-great-physical-breakout-why-2026-will-be-the-year-of-embodied-ai-d962ff8b2cf5
Sinha, S. (2025). *Figure humanoid robots retire bruised after 11 months of work at BMW.* Interesting Engineering. https://interestingengineering.com/ai-robotics/figure-humanoid-robots-retires-bmw
The Register. (2026). *Boston Dynamics beats Tesla to the humanoid robot punch.* The Register Technology News. https://www.theregister.com/2026/01/06/boston_dynamics_atlas_production/
TrendForce. (2026). *China to showcase humanoid robots at CES 2026 led by Unitree and AgiBot; AR glasses also in focus.* TrendForce News. https://www.trendforce.com/news/2026/01/06/news-china-to-showcase-humanoid-robots-at-ces-2026-led-by-unitree-and-agibot-ar-glasses-also-in-focus/
Unitree Robotics. (2026). *Universal humanoid robot H1.* Unitree Robotics Product Page. https://www.unitree.com/h1/

**Ken Huang** is a globally recognized expert in AI and Web3 security, a prolific author, and a leading figure in shaping industry standards. He is the CEO and Chief AI Officer (CAIO) of DistributedApps.ai, specializing in Generative AI training and consulting. Ken is deeply involved in driving the development of secure AI systems. He serves as a Research Fellow and Co-Chair of the AI Safety Working Groups at the Cloud Security Alliance (CSA), leading Agentic AI initiatives. He is also a Co-Chair of the AI STR Working Group at the World Digital Technology Academy under the UN Framework, and a core contributor to the OWASP GenAI project, focusing on Agentic AI security. His expertise extends to his contributions to OWASP's Top 10 Risks for LLM Applications and his active participation in the NIST Generative AI Public Working Group. He is also a member of Open AI forum.

**Grace Huang** is a seasoned product and technology professional with extensive experience at organizations including PIMCO (a global investment management firm managing over $2 trillion in assets), IBM, SoftBank, and DataRobot. Throughout her career, she has successfully launched multiple products and managed large-scale projects, leveraging her expertise in product management, market analysis, strategic planning, and cross-functional team leadership. Her work brings a finance and product strategy perspective to physical AI, particularly the operational, risk, and adoption considerations involved in bringing advanced robotics into enterprise environments. Always at the forefront of emerging technology, she has explored innovative tools including the implementation of Agentic AI powered solutions within the product management process—streamlining communication, improving decision-making, enhancing customer satisfaction, and driving business growth. In addition to her professional accomplishments, Grace holds a degree from Babson College, where she built a strong foundation in business management and entrepreneurship. She continues to lead at the intersection of product strategy and advanced technology, bringing a rare combination of financial rigor and deep technical acumen to complex, high-stakes initiatives.

Her work brings a finance and product strategy perspective to physical AI, particularly the operational, risk, and adoption considerations involved in bringing advanced robotics into enterprise environments. Always at the forefront of emerging technology, she has explored innovative tools including the implementation of Agentic AI powered solutions within the product management process—streamlining communication, improving decision-making, enhancing customer satisfaction, and driving business growth.

In addition to her professional accomplishments, Grace holds a degree from Babson College, where she built a strong foundation in business management and entrepreneurship. She continues to lead at the intersection of product strategy and advanced technology, bringing a rare combination of financial rigor and deep technical acumen to complex, high-stakes initiatives.

**Yuyan Duan** is a product leader and angel investor at the forefront of AI and robotics innovation. As an earlystage investor, she has backed a robotics startup supported by Salesforce CEO Marc Benioff, Skild's founder, and Nat Friedman, as well as a healthcare LLM venture backed by Anthropic's co-founder. Known for writing first checks, Lynn has already seen one portfolio company advance to Series B funding. Since 2019, Lynn has served as Product Leader at Treasure Data, a pre-IPO customer data platform company, where she has architected roughly one-third of the company's product suite. She pioneered the company's AI transformation by creating its first AI Agent Framework, which has generated 30+ enterprise-grade agents and driven 30+ new customer engagements. Her work spans enterprise big data, data security, and AI-driven marketing solutions.

In 2023, Lynn founded AI+, a thriving AI startup community with 50+ contributing members that has hosted 150+ events over two years, connecting 2,000+ founders annually with 200+ investors and corporate partners. She holds degrees from the University of Michigan (Psychology & Statistics, International Relations) and Georgia Tech (M.S. in Human-Computer Interaction with a Certificate in Management of Technology).

**Chunxiao Xing** is a Foreign Academician of the Russian Academy of Engineering. He currently serves as the Dean of the Tsinghua-Kelan Advanced Intelligent Database Research Institute, and Vice Dean of the Institute for **Medical-Robotic Engineering** Interdisciplinary Research at Tsinghua University. His research focuses on artificial intelligence, blockchain technology, big data, smart cities, and e-government. He has published over 380 academic papers, holds 43 invention patents and 33 software copyrights, and has led more than 40 major national research projects including National 973 Program, 863 Program, and National Key R&D projects. Professor Xing has received numerous honors including the National Science and Technology Progress Award (Second Class) and recognition as one of the "Metaverse 30" and "Blockchain 60" innovation leaders in China. He holds leadership positions in several professional organizations including the China Computer Federation, the Zhongguancun Blockchain Industry Alliance, and the Chinese Institute of Electronics Blockchain Branch.

# 2 Hardware and Embodiment Technologies

Grace Huang, Ken Huang, Bo Tian, and Yuyan Duan

## 2.1 Introduction

Humanoid robots represent the integration of numerous engineering disciplines—mechanical design, electrical engineering, materials science, control theory, and manufacturing engineering—into unified physical systems capable of operating in three-dimensional human environments. Unlike software systems where updates can be deployed instantly worldwide, or wheeled robots where mechanical simplicity dominates, humanoid platforms must simultaneously satisfy demanding and often conflicting requirements. They need sufficient strength to manipulate objects and support their own body weight during dynamic movement, adequate speed for practical task completion, appropriate compliance for safe physical interaction, acceptable energy efficiency for multi-hour operation, and achievable manufacturing costs to enable commercial deployment at scale.

These requirements create a complex, multi-dimensional optimization problem. In this chapter, "winning" does not mean maximizing any single specification; it means achieving useful work per hour at acceptable safety, reliability, maintenance burden, and total cost of ownership for a target application. A warehouse robot, a manufacturing robot, and a healthcare robot therefore optimize different

G. Huang
PIMCO, Austin, TX, USA

K. Huang (✉)
DistributedApps.ai, Fairfax, VA, USA
e-mail: ken@distributedapps.ai

B. Tian
Tesla, Palo Alto, CA, USA

Y. Duan
AI+ Founders Club, Silicon Valley, CA, USA



K. Huang et al. (eds.), *Humanoid Robots and Physical AI*, Advances in Data Analytics, AI, and Smart Systems, https://doi.org/10.1007/978-3-032-33669-9_2

points in the design space. High-power actuators that enable dynamic movement consume substantial energy, reducing runtime. Lightweight structures improve energy efficiency but may sacrifice strength and durability. Dexterous hands with numerous degrees of freedom enable sophisticated manipulation but increase complexity, cost, and control challenges. Every design choice involves trade-offs that fundamentally shape a robot's capabilities, suitable applications, and commercial viability.

Boston Dynamics Atlas, in its all-electric form presented as an enterprise industrial humanoid, prioritizes dynamic capability through 56 degrees of freedom (DoF), a 190 cm height, 90 kg mass, 30 kg sustained payload capacity, and self-swappable batteries for continuous operation (Boston Dynamics, 2026). The Tesla Optimus project emphasizes manufacturing scalability, leveraging vertically integrated electric actuation and automotive production techniques. Unitree H1 targets a balance of cost and performance, with manufacturer specifications reporting a 47 kg body weight, 0.864 kWh battery, and 3.3 m/s moving speed (Unitree Robotics, 2026). Meanwhile, Agility Robotics' Digit is optimized specifically for logistics applications, sacrificing some general capability for reliability and cost-effectiveness in its target domain.

These divergent approaches illuminate the multi-dimensional design space and the various paths being explored toward commercial viability. No single approach has proven dominant; instead, competition is driving innovation across multiple technical directions. This chapter examines the hardware technologies enabling these robots, organized around five technical subsystems and two system-level commercialization lenses. Section 2.2 analyzes actuation technologies. Section 2.3 explores structural design and materials. Section 2.4 examines power systems. Section 2.5 investigates locomotion systems. Section 2.6 analyzes manipulation systems. Section 2.7 provides a comparative specifications analysis across leading platforms, illuminating design trade-offs. Section 2.8 addresses manufacturing and scaling challenges, and we conclude by synthesizing insights about hardware evolution and its implications.

Figure 2.1 provides a high-level map of the hardware embodiment stack, showing how actuation, structure, power, locomotion, and manipulation converge into the chapter's central outcome: useful work per hour in human environments.

## 2.2 Actuation Technologies: The Muscles of Humanoid Robots

Actuators are the "muscles" of humanoid robots, converting stored energy into controlled mechanical motion. The choice of actuation technology fundamentally shapes a robot's capabilities across all performance dimensions: force, speed, controllability, compliance, energy efficiency, size, weight, cost, and reliability. As of 2026, while diverse approaches exist, electric motor systems have emerged as the clear dominant technology for commercial applications.

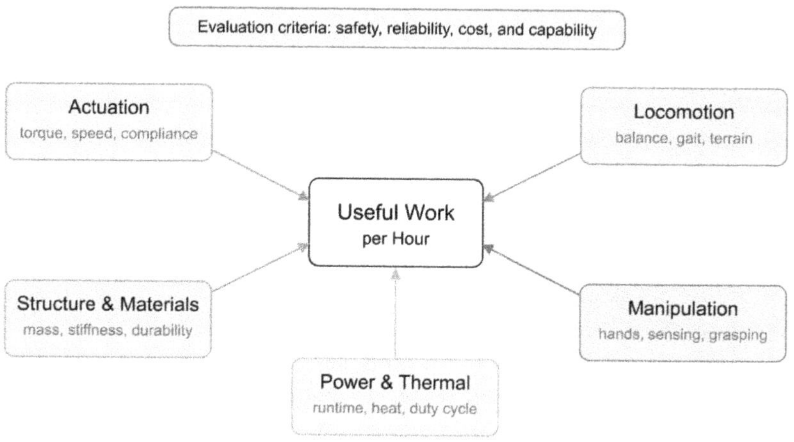


**Fig. 2.1** Hardware embodiment stack

### 2.2.1 Electric Motor Systems

Electric motors have become "the mainstream technological route for humanoid robots, thanks to their high precision, fast speed and reasonable price" (Bank of America, 2025). Modern humanoids primarily employ brushless DC (BLDC) motors combined with reduction gearing to achieve the required torque levels at the robot's joints.

**Box: Brushless DC motor (BLDC)**
A brushless DC motor is an electric motor that uses electronic commutation instead of mechanical brushes to switch current between stator windings, generating a rotating magnetic field that drives a permanent-magnet rotor. Unlike traditional brushed DC motors, BLDC motors eliminate physical brush contact, reducing friction, heat, and maintenance while improving efficiency, reliability, and lifespan. Speed and torque are controlled through an external motor controller that precisely times current delivery based on rotor position feedback, making BLDC motors widely used in robotics, drones, electric vehicles, and industrial automation.

The technical architecture of these electric actuators is sophisticated. As one industry analysis notes, "An electric actuator comprises a servo drive for motion control, a servo motor (usually frameless torque motor) for torque output, a transmission system (harmonic/planetary reducer for rotary actuator, and planetary roller screw for linear actuator) to convert the torque for desired use, and a sensory system (encoder and torque/force sensor) to collect system data" (Bank of America, 2025).

Frameless torque motors are particularly important for humanoid applications. These motors consist solely of a rotor and stator, without a frame, housing, or shaft, allowing for direct integration into the robot's joints. This design reduces weight, eliminates mechanical complexity, and enables the compact, anthropomorphic form factor required for humanoids (Articsledge, 2025).

The dominance of electric actuation reflects multiple practical advantages. Precise position and velocity control are straightforward. System efficiency, including gearing losses, is relatively high, typically in the range of 70–85%. The technology allows for compact packaging and benefits from mature manufacturing ecosystems in the automotive and industrial automation sectors, which helps to drive down costs.

Tesla Optimus exemplifies this modern approach. The company's expertise in electric motor design and manufacturing for its vehicles provides an advantage in optimizing actuators for performance, cost, and large-scale production.

However, electric motors are not without their challenges. They naturally produce high speeds and low torques, whereas robot joints require the opposite: high torques at moderate speeds. Reduction gears are used to bridge this gap, but they can introduce issues like backlash (reducing positional accuracy), friction losses, and reflected inertia. Contemporary designs mitigate these issues through high-ratio planetary gearboxes (with ratios often in the 50:1–200:1 range), harmonic drives that offer zero-backlash operation, and advanced motor control strategies.

### 2.2.2 The Hydraulic Legacy and Electric Transition

Hydraulic actuators offer exceptional power density, enabling feats of strength and dynamic agility that were long the hallmark of Boston Dynamics' research platforms. The classic hydraulic Atlas was famous for its ability to run, jump, and perform acrobatic maneuvers, a testament to the sheer force that hydraulic systems can deliver.

**Box: Hydraulic systems**

Hydraulic systems transmit power using pressurized fluid—typically oil—to generate force and motion. A pump converts mechanical energy into fluid pressure, which is directed through valves and hoses to actuators such as cylinders or hydraulic motors. Because liquids are nearly incompressible, hydraulic systems can deliver very high force and torque with smooth, precise control, making them well suited for heavy-duty applications like construction machinery, aircraft control surfaces, industrial presses, and some high-power robots. However, they tend to be heavier, noisier, and more maintenance-intensive than electric systems due to fluid leakage risks and the need for pumps and reservoirs.

However, Boston Dynamics' strategic pivot in 2024 underscores the commercial limitations of hydraulics. In April 2024, Boston Dynamics retired the hydraulic Atlas and unveiled a new, all-electric version designed for real-world applications. This new Atlas is described as "stronger, more dexterous, and more agile" than its predecessor, signaling that electric actuation technology has matured to the point where it can meet and exceed the performance of hydraulics in a commercially viable package.

The operational complexity of hydraulic systems—including high-pressure pumps, fluid reservoirs, thermal management, potential leak points, and contamination risks—presents significant challenges for widespread commercial deployment and maintenance. The transition to a fully electric design by the industry's most prominent developer of dynamic robots effectively signals a consensus: electric actuation is the path to commercial viability, even for the most demanding applications.

### 2.2.3 Series Elastic Actuators: Compliance and Force Control

Series Elastic Actuators (SEAs) represent an important innovation that addresses some of the limitations of conventional, rigid electric actuators. SEAs intentionally incorporate a compliant element, typically a spring, in series between the motor and the load. This design offers several key advantages:

- Accurate Force Sensing: By measuring the deflection of the spring, the actuator can very accurately determine the force it is applying.
- Shock Tolerance: The spring can absorb impacts, protecting the motor and gearbox from sudden shocks.
- Energy Storage: The spring can store and release energy, which can improve efficiency in dynamic movements like walking or running.
- Natural Compliance: The inherent springiness makes the robot safer for interaction with humans and delicate objects.

Recent research continues to advance SEA technology. A February 2025 publication, for example, presented a novel 3D-printed SEA for an elbow exoskeleton, making the design openly available to accelerate adoption (Frontiers, 2025). Advanced control strategies are also emerging, such as adaptive stiffness control that allows a robot to be stiff for precise tasks and compliant for safe interaction (Wiley, 2026).

Despite these advantages, SEAs have not become the dominant approach in mainstream humanoid robots. The primary drawback is a trade-off between compliance and control bandwidth. The compliant element that provides the benefits also reduces the actuator's ability to respond quickly to position commands, limiting its performance in rapid, precise movements. For many applications, designers have opted for stiff actuators paired with external force-torque sensors to achieve a balance of performance and force-sensing capability.

### 2.2.4 Manufacturing and Supply Chain Dynamics

The supply chain for actuators is an important factor in the commercial viability and scaling potential of humanoid robots. As a Bank of America report notes, component manufacturers are working to innovate and scale up production of the

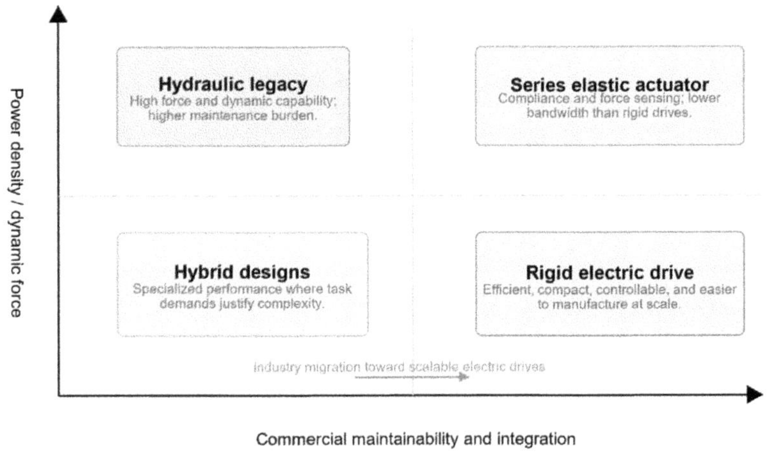


**Fig. 2.2** Actuation technology trade-offs

mechatronics—motors, sensors, and actuators—that meet the specific requirements of humanoids (Bank of America, 2025). (Note: How data from these sensors is processed by the robot's AI is covered in Chap. 3).

This dynamic is playing out differently across the globe. Chinese manufacturers benefit from a highly developed and cost-effective supply chain for electromechanical components, which has enabled companies like Unitree to offer robots at aggressive price points. Western manufacturers, on the other hand, are pursuing a mix of strategies. Tesla is leveraging its vertical integration to design and manufacture its own actuators, giving it tight control over optimization and cost. Other companies, like Figure AI and Agility Robotics, source actuators from specialized suppliers, allowing for faster iteration and more efficient use of capital.

The market for these components is projected to grow substantially. As humanoid robot production scales from thousands to millions of units in the coming years, the suppliers who can achieve high volumes and low costs for these critical components will be positioned for significant success.

Figure 2.2 compares major actuation approaches, clarifying why electric drives dominate commercial humanoids while hydraulic, series elastic, and hybrid designs remain relevant for specific performance requirements.

## 2.3 Structural Design and Materials

The mechanical structure of a humanoid robot must satisfy a difficult set of contradictory requirements: it must be strong enough to support its own weight plus any payload during dynamic movement, yet lightweight for energy efficiency and agility. It requires precise dimensions for reliable kinematics, effective thermal management for its high-power components, and a design that is manufacturable at a cost that enables commercial deployment. The selection of materials and the structural design are therefore intimately coupled with choices made in actuation, power systems, and overall performance goals.

### 2.3.1 Materials Selection: Aluminum, Carbon Fiber, and Advanced Polymers

Contemporary humanoid robots employ a diverse palette of materials, each selected to optimize for specific requirements and cost constraints. As one analysis notes, “Humanoid robots use lightweight materials—aluminum alloys, carbon fiber, and advanced plastics—to balance strength and weight” (Articsledge, 2025). Each of these material classes offers a distinct profile of advantages and limitations.

Aluminum alloys remain a dominant material in robot manufacturing, offering a favorable strength-to-weight ratio, excellent machinability for prototyping and production, good thermal conductivity for heat dissipation, and a relatively low cost compared to more advanced alternatives. High-strength alloys like 7075 aluminum, with a yield strength over 500 MPa, are often used for highly stressed components, while alloys like 6061 are used for more general structural parts (Custom Plastic Mold, 2025).

Carbon fiber composites are increasingly used to achieve further weight reduction, particularly in less-stressed structures. Robots such as Agility Robotics’ Digit illustrate this mixed-material strategy, using structural materials selected to balance weight, strength, durability, and cost. The market for carbon fiber in this sector is projected to grow significantly, with one report forecasting it to become a USD 710 million market by 2032, expanding at a remarkable 37.9% compound annual growth rate (CAGR) (Newstrail, 2025). Note: an independent forecast for the humanoid robot cooling-system market (Sect. 2.3.3) reports the same 37.9% headline figure for the same 2025–2032 window, we treat the coincidence as artefactual of the two publishers’ converging assumptions about humanoid-platform shipment growth, not as evidence of a single underlying source. This growth reflects the increasing priority placed on weight optimization as manufacturing costs for composites decline.

Advanced polymers, particularly PEEK (polyetheretherketone), have emerged as critical materials for modern humanoids. PEEK offers an exceptional combination of properties: high strength comparable to some metals, high thermal resistance with a continuous use temperature of up to 260 °C (500°F), excellent wear resistance for components like gears and joints, and chemical inertness. Tesla Optimus Gen 2 famously achieved a 10 kg weight reduction, a feat attributed in part to the strategic use of materials including PEEK composites and magnesium alloys (PEEK China, 2025). Industry sources estimate that a single robot might use a significant amount of PEEK, both in its pure form for gears and joints and as a carbon-fiber-reinforced composite (CF/PEEK) for limbs and skeletal structures.

However, the adoption of these advanced materials is still moderated by cost. As one analysis points out, "The application of carbon fiber, magnesium alloy and other composite materials is still constrained by application costs and technical levels" (Custom Plastic Mold, 2025). As production volumes scale up and material processing technologies mature, these advanced composites will likely see wider adoption, but for the near future, designs will continue to balance the performance gains of premium materials against the economic realities of the market.

### 2.3.2 Weight Optimization

Weight is a critical parameter in humanoid robot design, directly affecting multiple dimensions of performance. A lighter robot consumes less energy, enabling longer runtimes or smaller, lighter batteries. Lower inertia allows for faster acceleration and more agile movements, which is crucial for dynamic stability and performance. It also directly impacts the payload capacity relative to the robot's own body weight and reduces the stress on actuators and structural components.

The industry has made impressive strides in weight optimization. Tesla Optimus project serves as a clear illustration. The initial prototype presented in late 2022 weighed approximately 73 kg. The Gen 2 model, unveiled in December 2023, was significantly lighter, and the current target for the Gen 3 robot is a weight of only 57 kg (125 lbs) at a height of 173 cm. This represents a substantial weight reduction achieved across generations through refined design and advanced materials.

Unitree H1 platform offers another example of weight-focused design. At a height of about 180 cm, the H1 weighs about 47 kg and uses a 0.864 kWh battery, according to Unitree's published specifications (Unitree Robotics, 2026). This lightweight design supports the H1's high reported moving speed, but it may involve trade-offs in payload capacity, structural robustness, or endurance when compared with heavier platforms.

The challenge of weight optimization extends beyond total mass to include mass distribution. The position of the robot's center of mass is critical for balance and energy efficiency during walking. Concentrating mass in the torso reduces the inertia of the limbs, allowing for faster leg movements with less torque. This principle, however, can conflict with other design considerations, such as component packaging and thermal management.

Advanced design tools, including topology optimization, are now commonly used to find the most efficient distribution of material to meet strength requirements while minimizing weight. These computational methods can generate highly organic and efficient structures, but their outputs must always be tempered by the constraints of manufacturability. A theoretically optimal part is of little use if it cannot be produced, assembled, and inspected reliably and cost-effectively.

### 2.3.3 Thermal Management

High-power-density humanoid robots generate significant waste heat that must be effectively managed to prevent performance degradation and component failure. Although actuators and motor controllers contribute to thermal generation during high-torque operation, the onboard AI computing system is increasingly the dominant source of thermal load. Modern humanoids rely on high-performance GPUs and AI accelerators to perform real-time perception, planning, and control. Similar to data-center computing hardware, nearly all electrical power consumed by these processors is ultimately converted into heat. As a result, advanced AI modules can dissipate hundreds of watts during operation. When combined with heat from actuators, battery packs, and power electronics, the total thermal load of a humanoid robot can approach the kilowatt scale, making efficient thermal management a critical aspect of system design.

The growing importance of this challenge is reflected in market forecasts. Intel Market Research, a market-research publisher rather than Intel Corporation, projects the global market for humanoid robot cooling systems to grow from USD 91.4 million in 2025 to USD 623 million by 2032, exhibiting a 37.9% CAGR (Intel Market Research, 2025). This indicates that thermal management is evolving from an afterthought to a critical design consideration.

Contemporary approaches to thermal management span a range of passive and active strategies:

- Passive Cooling: This approach leverages the high thermal conductivity of the robot's metallic structure, typically aluminum. Heat is conducted away from components like motors and electronics through the frame to large surface areas where it can be dissipated into the environment through convection and radiation. This method is simple, silent, and reliable, but its cooling capacity is limited.
- Active Air Cooling: This method uses fans to circulate air through actuator compartments and over heat sinks. It is a lightweight and moderately effective solution, but it adds noise and complexity.
- Liquid Cooling: For the highest-performance platforms, liquid cooling offers superior heat removal capacity. These systems circulate a coolant through jackets or cold plates attached to heat-generating components, transferring the thermal energy to radiators where it is dissipated. While this is the most effective method, it also adds the most weight, complexity, and potential failure points (e.g., leaks).

The choice of cooling technology involves clear trade-offs. The design selection depends heavily on the robot's intended application. Industrial robots in controlled environments can accommodate the complexity of liquid cooling, while service robots intended for homes or public spaces may require quieter, less complex passive or air-cooled designs. Companies with experience in other domains, such as Tesla with its automotive background, can bring mature thermal management solutions and supply chains to bear on this problem, providing a potential advantage.

## 2.4 Power Systems and Energy Management

Energy storage remains one of the defining constraints limiting the widespread deployment of autonomous humanoid robots, but the bottleneck is best understood as uptime under real duty cycles rather than battery chemistry alone. Current humanoids commonly operate in the 2- to 4-h range, and increasing operating time depends on hot-swappable packs or higher-energy-density batteries (TrendForce, 2026). Operators can manage around this constraint with spare packs, opportunity charging, task scheduling, and battery swapping, so energy is not an absolute blocker for every pilot. It becomes a deployment bottleneck when runtime, recharge time, peak-current limits, thermal load, and battery mass reduce productive hours per day or require extra robots to cover one workflow. The ability to store sufficient energy in a lightweight, compact, and safe package therefore remains a paramount challenge in modern robotics.

### 2.4.1 Lithium-Ion Battery Technology and Evolution

Virtually all contemporary humanoid robots are powered by lithium-ion battery packs. The capacity of these packs typically falls in a range that reflects a tradeoff between runtime, weight, and cost. While some platforms may have larger custom packs, many current-generation humanoids appear to use batteries in the sub-3 kWh range. For example, Unitree lists the H1 with a 0.864 kWh pack, while public estimates for Tesla Optimus commonly cite a roughly 2.3 kWh pack (Robotics, 2026; Unitree LinkedIn, 2025). A larger capacity extends operational time but adds significant weight, which in turn increases energy consumption and reduces dynamic performance.

The specific chemistry of the lithium-ion cells has a major impact on performance characteristics:

- Lithium Iron Phosphate (LFP): This chemistry is known for its excellent safety profile and long cycle life, often rated for 2,000–5,000 cycles or more. However, it offers a lower energy density, typically in the range of 150–180 Wh/kg, though some manufacturers claim to be reaching up to 200 Wh/kg.
- Nickel Manganese Cobalt (NMC): NMC batteries provide a higher energy density, generally in the 200–250 Wh/kg range, which allows for more energy storage in the same mass. This comes at the cost of a shorter cycle life and more stringent thermal management requirements to ensure safety.

Emerging battery technologies promise transformative improvements. Grepow, a specialized battery manufacturer, offers semi-solid-state NMC batteries that achieve an energy density of up to 350 Wh/kg (CleanTechnica, 2025). Meanwhile, REPT Battery has announced its goal of producing all-solid-state batteries exceeding 400 Wh/kg, specifically targeting robotics applications. Perhaps most notably, CATL, the world's largest battery manufacturer, has unveiled a "condensed matter" battery with a claimed energy density of up to 500 Wh/kg. It is crucial, however, to interpret such announcements with precision. For instance, another of CATL's recent high-profile announcements for a new energy storage system touted a figure of 430, but this referred to a volumetric energy density of 430 Wh/L (Watt-hours per liter), not a gravimetric density of 430 Wh/kg. Confusing these two metrics can lead to a significant overestimation of a battery's weight-based performance.

If these advanced, high-density chemistries can be commercialized at scale, they could materially improve robot uptime. In a simplified pack-level comparison, doubling gravimetric energy density from 220 to 440 Wh/kg could roughly double stored energy for the same cell mass. In practice, runtime gains would be smaller because pack structure, power electronics, thermal limits, task intensity, and safety margins also matter. These chemistries still face challenges in manufacturability, cost, qualification, and long-term reliability before mass deployment.

### 2.4.2 Battery Integration and Packaging

Beyond the selection of cell chemistry, the design of the battery pack itself is a complex engineering discipline. The mechanical packaging must be robust enough to protect the cells from impacts and vibrations while minimizing weight and volume. A sophisticated Battery Management System (BMS) is required to monitor the voltage and temperature of each cell, balance the charge across the pack, and protect against overcharge, over-discharge, and thermal runaway conditions. An effective thermal management system is also critical to dissipate the heat generated during high-power charging and discharging, which is essential for both performance and safety.

The maturation of the humanoid robot market is evidenced by the fact that major battery manufacturers are now making significant investments in production infrastructure specifically for robotics. At the end of 2025, CATL announced that its "Zhongzhou" production base in Luoyang, China, had officially put into operation a production line for battery packs intended for humanoid robots (Tycorun, 2025). This move by the world's largest battery supplier signals a strong belief in the future scale of the humanoid robot market.

Battery costs represent a significant portion of a robot's total bill of materials. BloombergNEF's 2025 battery price survey reported average lithium-ion pack prices of approximately $108/kWh, with EV packs averaging about $99/kWh, but humanoid robot packs are likely to cost more at current production volumes because they are lower-volume, mechanically integrated, and safety-critical assemblies (Maisch, 2025). As robot production scales, companies that can reuse automotive battery supply chains, qualification processes, and pack-management electronics should be better positioned to reduce per-unit energy-storage costs.

### 2.4.3 Runtime Limitations and Operational Implications

The current reality for most humanoid robot platforms is a runtime of approximately 2–5 h per charge. This limitation, as noted in a 2025 analysis, substantially underperforms a typical 8-h human work shift and creates significant operational hurdles for commercial deployment (LinkedIn, 2025).

The economic implications are substantial. To cover a single 8-h shift with a robot that has a 4-h runtime and requires 2 h to recharge, a business would need at least two robots and a charging station. To cover two full shifts (16 h), the capital expenditure increases further. This reality has forced the industry to develop innovative operational models to bridge the gap until battery technology matures.

Several Strategies Are Being Employed to Address This Limitation:

- Multi-Robot Operations with Battery Swapping: This model involves robots working for their allotted runtime and then autonomously navigating to a station to have their depleted battery swapped for a fully charged one. UBTech's Walker S2 is a pioneer in this area, demonstrating the ability to autonomously swap its own battery in approximately 3 min. This capability dramatically reduces downtime and allows for near-continuous operation with a smaller fleet of robots.
- Task Optimization and Hybrid Operations: This approach focuses on deploying robots only for the highest-value tasks where they offer a clear advantage over humans—such as heavy lifting, highly repetitive actions, or work in hazardous environments. Humans then handle the remaining tasks that require more flexibility, judgment, or complex manipulation. This hybrid human-robot model can improve return on investment even when robot runtimes remain limited.
- Differentiated Deployment by Application: Not all applications have the same runtime requirements. A robot in a logistics warehouse might need to operate for a full 8-h shift, while a robot in a retail environment may only be needed during peak business hours. Matching the robot's capabilities to the specific demands of the application is key to successful deployment.
- Tethered Power: For stationary or near-stationary applications, such as a robot working in a fixed manufacturing work cell, a power tether can eliminate runtime constraints entirely. This trades mobility for unlimited operation, which is an acceptable compromise for certain industrial tasks.

The industry is pursuing a dual strategy of engineering operational models that work with current battery limitations while simultaneously investing in the research and development of next-generation battery technology. Projections suggest that gradual, incremental improvements in conventional battery chemistry (in the range of 5–8% annually) may extend runtimes to 6–7 h by the end of the decade. However, achieving a full 8+ hour runtime will likely require a breakthrough in a new chemistry, such as the commercialization of the high-density solid-state or condensed matter batteries currently in development.

### 2.4.4 Charging Infrastructure and Energy Distribution

Beyond the batteries themselves, the infrastructure required to charge a fleet of robots is a significant deployment consideration. The most common method is conductive charging, where robots autonomously dock with a station to make a physical electrical connection. This approach is simple, reliable, and allows for high-power charging. Inductive (wireless) charging is an alternative that eliminates physical contacts, potentially improving long-term reliability. However, it typically suffers from lower efficiency (with energy losses of 15–30% being common) and higher costs.

The infrastructure requirements extend beyond the charging stations to the facility's electrical distribution system. A facility deploying 100 robots, with perhaps 50 charging simultaneously at 2–3 kW each, would require an additional 100–150 kW of electrical capacity. This is a substantial load, comparable to the average power consumption of 80–125 typical U.S. homes, and may necessitate significant and costly upgrades to the building's electrical service.

As deployments scale into the hundreds or thousands of robots per facility, energy management becomes a major engineering challenge. This includes the physical layout of charging stations (distributed vs. centralized) and the development of sophisticated software to optimize charging schedules across the fleet. This software can help to balance electrical loads, avoid peak demand charges from utilities, and ensure that a sufficient number of charged robots are always available to meet operational needs. Standardizing these infrastructure components will be crucial for enabling the cost-effective deployment of humanoid robots at a massive scale.

Figure 2.3 illustrates the power and uptime loop, emphasizing that battery capacity, heat generation, charging infrastructure, and task scheduling jointly determine practical duty cycle.

## 2.5 Locomotion Systems: Enabling Bipedal Walking and Running

Bipedal locomotion is one of the most defining and challenging aspects of humanoid robotics. It requires a sophisticated interplay of real-time dynamic balance, trajectory planning, and force control. Humans walk and run with an effortless grace that belies the complexity of the underlying neuro-muscular control system, which has been honed through millions of years of evolution and years of individual learning. Replicating this capability in a machine requires solving a

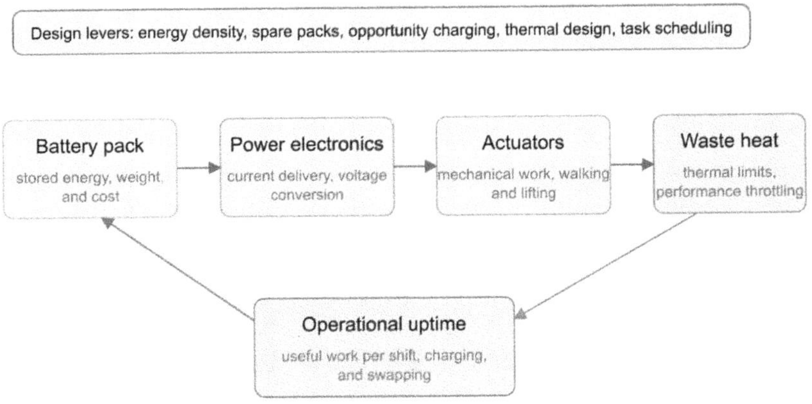


**Fig. 2.3** Power, thermal load, and uptime loop

series of fundamental control problems. A bipedal robot is an under-actuated system that is inherently unstable, requiring continuous, active control to prevent it from falling. The act of walking involves constantly changing contact states as feet strike and lift from the ground, creating discontinuities that the controller must manage. Furthermore, the real world is filled with unmodeled disturbances, from uneven terrain to external pushes, that demand robust recovery strategies.

### 2.5.1 Zero-Moment Point Theory and Application

The dominant theoretical framework for controlling the balance of bipedal robots for the past several decades has been the concept of the Zero-Moment Point (ZMP). The ZMP is defined as the point on the ground where the net moment of the inertial and gravitational forces has no horizontal component. In simpler terms, it is the point where the total tipping moment is zero. The core principle of ZMP-based control is that if the ZMP is kept within the boundaries of the support polygon—the area formed by the feet in contact with the ground—the robot will remain dynamically balanced (Articsledge, 2025).

This elegant formulation reduces the complex, high-dimensional problem of dynamic balance to a single, verifiable criterion. ZMP-based controllers work by planning a trajectory for the robot's center of mass and a sequence of foot placements that together ensure the ZMP remains safely within the support polygon at all times. This approach has proven to be remarkably effective and remains a foundational element in the control of most modern humanoid robots (Springer, 2025).

However, classic ZMP control has its limitations. It is most rigorously applicable to quasi-static walking where accelerations are moderate. For highly dynamic movements like running, jumping, or aggressive direction changes, the ZMP may intentionally move outside the support polygon for brief periods, leveraging dynamic effects for propulsion. Therefore, more advanced control strategies are needed to handle these high-performance regimes.

Recent research has explored various enhancements to this fundamental framework. One area of investigation involves the use of Control Moment Gyroscopes (CMGs) to generate internal torques that can help to stabilize the robot without requiring ground contact, potentially improving disturbance rejection and enabling more dynamic maneuvers (Springer, 2025). More significantly, data-driven approaches, particularly reinforcement learning (RL), are increasingly being used to complement or even replace traditional model-based controllers. An early 2025 paper, for instance, demonstrated a humanoid traversing extreme terrains using a whole-body locomotion algorithm based on dynamic balance and reinforcement learning, relying only on proprioceptive feedback (Xie et al., 2025). These learning-based controllers can discover novel, efficient gaits and naturally adapt to terrain variations, though ensuring their safety and reliability remains a significant challenge compared to traditional, mathematically verifiable control methods.

### 2.5.2 Gait Generation and Dynamic Walking

Generating a walking gait involves planning the coordinated movements of all the robot's joints—which can number from 30 to over 50 in contemporary humanoids—to achieve a desired body velocity while maintaining balance and respecting all physical constraints. This is a high-dimensional optimization problem that must be solved in real-time.

Model Predictive Control (MPC) has emerged as a powerful and widely used technique for gait generation. MPC works by using a simplified model of the robot's dynamics to predict its future states over a short time horizon. It then calculates the optimal sequence of control actions (e.g., foot placements and joint torques) that will achieve the desired motion while satisfying all constraints, such as keeping the ZMP within the support polygon and respecting joint limits. This optimization is performed repeatedly at a high frequency, allowing the robot to continuously re-plan its movements in response to new sensor data and disturbances.

This MPC-based planning is often combined with a whole-body controller that translates the high-level plan into low-level motor commands. A whole-body controller treats the robot as a single, unified dynamic system and works to satisfy multiple objectives simultaneously: maintaining balance, tracking the desired velocity, minimizing energy consumption, avoiding joint limits, and preventing self-collisions. This is typically formulated as a quadratic programming problem that can be solved efficiently in real-time.

The practical results of these advanced control techniques are impressive. Public demonstrations from companies like Unitree show their robots performing genuine dynamic locomotion, including sprinting, jumping, and maintaining balance even after being subjected to heavy pushes (WINS Solutions, 2025). Unitree H1, for example, has demonstrated a walking speed of 3.3 m/s (7.4 mph). While some claims, such as a potential 10–12 km/h walking speed for a future Tesla Optimus, remain speculative, the demonstrated capabilities of current-generation robots represent a significant leap forward from just a few years ago and are beginning to approach human-level agility.

### 2.5.3 Terrain Adaptation and Robustness

While walking on flat, level surfaces in a controlled environment is now a largely solved problem, this is insufficient for most real-world applications. The frontier of locomotion research lies in enabling robots to navigate unstructured terrain—the outdoor environments, cluttered warehouses, and unpredictable public spaces where they are ultimately intended to operate. As the International Federation of Robotics noted in its 2026 trends report, many robots still struggle when faced with rough terrain or unpredictable conditions (International Federation of Robotics, 2026).

True terrain adaptation requires the seamless integration of multiple complex capabilities:

- Perception: The robot must use its sensors (cameras, depth sensors, LiDAR) to build a model of the surrounding environment, identifying the geometry of the terrain, classifying surface types (e.g., concrete, gravel, ice), and detecting obstacles. (The computational architectures for this visual and depth processing are examined in detail in Chap. 3.)
- Planning: Based on this perceived model, the gait planner must adapt its strategy, adjusting foot placement to find stable footholds, varying step height to clear obstacles, and modulating walking speed and ground contact forces to suit the surface properties.
- Reactive Control: Perception and planning are never perfect. The robot must be able to react in real-time to unexpected events, such as a foot slipping or the ground yielding unexpectedly. This requires fast and robust reactive controllers that can detect instability using inertial sensors and compute and execute corrective motions within milliseconds to prevent a fall. The required response times are typically in the range of 50–100 ms.

Achieving the level of reliability required for commercial deployment in these unstructured environments remains a monumental challenge. A video of a robot successfully navigating a difficult obstacle course is an impressive research demonstration, but it does not guarantee reliable performance over thousands of hours of continuous operation. Consider the arithmetic: a warehouse robot taking two steps per second over an 8-h shift executes 57,600 steps. A failure rate of just 0.01% would still result in nearly six falls per shift, which is commercially unacceptable.

True viability likely requires a step success rate of 99.99% or better, where failures are not catastrophic falls but graceful stops where the robot can call for assistance.

Closing this gap between "usually works" in a controlled test and "reliably works" in a commercial deployment is the central challenge of locomotion research today, and it is the difference between an impressive technology demonstration and a transformative commercial product.

## 2.6 Manipulation and Dexterity

Manipulation—the ability to grasp, hold, and use objects and tools—is what distinguishes a truly useful humanoid robot from a simple mobile platform. The human hand is a marvel of biological engineering, with 27 bones, 27 joints, and 34 muscles working in concert to enable a remarkable range of dexterous behaviors, from the delicate precision of threading a needle to the powerful grip required to wield a hammer. Replicating even a fraction of this capability in a robotic end-effector represents a formidable engineering challenge.

### 2.6.1 The Dexterous Hand Revolution of 2025–2026

The period from 2025 to early 2026 witnessed an extraordinary acceleration in the development and commercialization of dexterous robotic hands. After years where most humanoid platforms relied on simple parallel-jaw grippers or hands with limited (3–5) degrees of freedom, the industry saw the emergence of multiple companies offering highly sophisticated, multi-DoF hands. The CES 2026 trade show, in particular, served as a major showcase for this trend.

ZWHAND, a Chinese robotics company, exemplified this revolution with the announcement of a full product line of dexterous hands. Their flagship B20 model boasts 20 active degrees of freedom (DoF), enabling it to replicate a wide range of human hand movements for complex grasping and fine manipulation. The company also introduced the A17, a 17-DoF model designed as a cost-effective workhorse for industrial and service applications, and the B06, a 6-DoF entry-level hand priced at just $699 for mass production, aimed at making basic dexterous grasping accessible for education and simple automation (PR Newswire, 2026). This stratified product line allows developers to match hand capability to specific application needs and budgets.

Sharpa Robotics also made a significant impact at CES 2026 with its SharpaWave hand. This advanced end-effector features 22 DoF and, critically, integrates a dense array of tactile sensors, with the company claiming more than 1,000 tactile pixels per fingertip (Kalil, 2026). This level of tactile feedback is crucial for enabling the fine control of grasp forces and the detection of slip, which are essential for robust manipulation of a wide variety of objects.

Another key player, DexRobot, showcased an interesting portfolio of manipulation solutions. Their flagship DexHand021 Pro features 22 DoF, while their mass-produced DexHand021 offers 19 DoF. The company also produces the three-fingered DexHand021 S, a pragmatic design that provides a significant dexterity advantage over simple grippers while avoiding the cost and complexity of a full five-fingered anthropomorphic hand (Robotics Tomorrow, 2026).

This rapid progression in hand dexterity is not limited to specialized component suppliers. Major robot developers are also pushing the envelope. Tesla, for example, has indicated that its Optimus Gen 3 will feature hands with 22 DoF, a major upgrade from the 11-DoF hands used on its Gen 2 prototype. This industry-wide push toward high-DoF, sensorized hands signals a clear recognition that sophisticated manipulation is a prerequisite for unlocking the most economically valuable applications for humanoid robots, moving beyond simple logistics to complex assembly, tool use, and human-collaborative tasks.

### 2.6.2 Actuation Approaches for Dexterous Hands

Packing the required number of actuators—often 15 to 22—into the constrained volume of a human-sized hand is a major design challenge. Several competing architectural approaches are used:

- Motors in Fingers: Placing small, direct-drive motors at each joint is the most straightforward control architecture. However, this adds mass at the fingertips, which can reduce dynamic performance and result in bulky, oversized fingers.
- Tendon-Driven Systems: This approach locates the motors in the forearm or palm and uses a system of cables (tendons) to transmit force to the finger joints. This concentrates the mass proximally, allowing for more slender, lightweight, and anthropomorphic finger designs. However, tendon-driven systems introduce challenges with friction, cable tensioning, and long-term durability.
- Hybrid Approaches: Many designs use a combination of these methods, optimizing the actuation for each joint independently. For example, the large, powerful knuckle joints might use direct-drive motors, while the smaller, more numerous fingertip joints are actuated by tendons.

The recent announcements from companies like ZWHAND, which describes its 20-DoF hand as being in the "tendon-driven class," suggest that this approach has matured to the point where it can deliver high performance at a commercial scale (PR Newswire, 2026).

### 2.6.3 Tactile Sensing Integration

As the SharpaWave hand demonstrates, the integration of tactile sensing is becoming a critical component of dexterous manipulation. The ability to "feel" an object is essential for adjusting grasp force, detecting slip, and identifying object properties. From a hardware perspective, integrating these sensors presents significant packaging challenges, including wiring routing, protecting the sensors within durable artificial skin or silicone fingertips, and managing the physical bulk at the end-effector. The algorithmic processing and transduction principles of these tactile sensors—such as capacitive, resistive, and optical arrays—are detailed in Chap. 3, Sect. 3.3.

To accelerate the development of manipulation skills, data collection systems are also becoming more sophisticated. DexRobot, for example, offers the DexCap Exoskeleton, a system capable of capturing human upper-limb motion at a rate of 1000 Hz across 65 DoF. This high-fidelity data can be used to train robot manipulation policies through imitation learning, allowing robots to learn complex tasks by observing human demonstrators.

### 2.6.4 Force Control and Impedance

Successful manipulation requires not just controlling the position of the fingers, but also regulating the forces they apply to objects. Applying too much force can crush a delicate object, while applying too little can result in a dropped tool. Humans perform this force modulation unconsciously, using a combination of tactile and proprioceptive feedback. Robots must be explicitly programmed with sophisticated force control strategies to achieve the same result.

Impedance control is a powerful framework for achieving force-regulated manipulation. Instead of commanding a specific position or force, an impedance controller specifies the desired dynamic behavior—the mechanical impedance (stiffness, damping, and inertia)—that the robot should exhibit. This allows the robot to behave like a programmable spring-damper system. It can be made very stiff to hold a precise position, or very compliant to safely interact with a person or a delicate object. This approach provides a natural and robust way to handle the uncertainties of physical interaction.

Implementing impedance control requires force feedback, which can be obtained through several methods. Wrist-mounted force-torque sensors can measure the overall forces and torques acting on the hand. Fingertip tactile sensors can provide local contact force information. It is also possible to estimate applied forces by sensing the current drawn by the joint motors, though this method is less accurate than direct sensing. Finally, as discussed previously, Series Elastic Actuators (SEAs) provide inherent force sensing through the measurement of their spring deflection. (How this kinesthetic data is fused with other sensor modalities is explored in Chap. 3.)

### 2.6.5 Learning Dexterous Manipulation

Historically, programming a robot to perform a manipulation task was a painstaking, manual process. A programmer would have to specify every detail of the task: grasp points, approach trajectories, force profiles, and error recovery strategies. This programming bottleneck severely limited the practical deployment of robots in tasks that involved any variability.

Contemporary approaches are increasingly relying on machine learning, particularly learning from demonstration and reinforcement learning, to dramatically accelerate this process. Using data from systems like the DexCap exoskeleton, robots can learn manipulation strategies by observing humans. In parallel, reinforcement learning in simulation allows a robot to practice a task for millions of simulated trials, learning from its own mistakes without risking physical hardware. Sim-to-real transfer techniques are then used to adapt the policies learned in simulation to the physical robot.

These learning-based approaches have enabled robots to acquire complex skills, such as in-hand object reorientation and tight-tolerance peg-in-hole insertion, that would be prohibitively difficult to program by hand. However, significant challenges remain. Learned policies can sometimes fail in unpredictable ways when faced with novel situations, and ensuring their safety and reliability is an active area of research. Despite these challenges, learning-based manipulation is likely to remain central to the industry because it offers a scalable path toward broader manipulation capability, especially when paired with rigorous validation and runtime safeguards.

## 2.7 Comparative Specifications Analysis

Detailed technical specifications provide a quantitative lens through which to examine the design trade-offs and performance envelopes of contemporary humanoid platforms. Table 2.1 presents a comparative analysis of leading systems as of early 2026, compiled from manufacturer specifications, technical reports, and other verified sources. It is important to note that many specifications, particularly for platforms in early development, are subject to change and may be based on estimates or targets rather than final production figures.

Several key patterns and design philosophies emerge from this comparative analysis:

- Weight-Performance Correlation: A clear trade-off exists between weight, speed, and payload. The exceptionally lightweight Unitree H1 (47 kg) achieves the highest confirmed walking speed at 3.3 m/s. In contrast, heavier platforms like the 90 kg Atlas and 70 kg Figure 02, while potentially more robust and capable of handling larger payloads (30 and 20 kg respectively), exhibit more moderate locomotion speeds. This highlights a fundamental principle: lighter limbs require less torque to accelerate, enabling a faster and more dynamic gait.
- DoF Allocation Philosophy: The total number of degrees of freedom (DoF) is a strong indicator of a robot's intended versatility. The Boston Dynamics Atlas sits at the high end with 56 DoF, a design choice that prioritizes maximum

**Table 2.1** Comparative specifications of leading humanoid robots (as of early 2026)

| Platform | Height (cm) | Weight (kg) | Total DoF | Hand DoF | Payload (kg) | Speed (m/s) | Battery/capacity | Runtime (hr) | Price (USD) |
|---|---|---|---|---|---|---|---|---|---|
| Boston Dynamics Atlas (Electric) | 190 | 90 | 56 | Not specified | 30 (sustained) | Unconfirmed | Not specified | ~4 | Not Disclosed |
| Tesla Optimus (Gen 3 Target) | 173 | 57 | Unconfirmed | 22 per hand | 20 (est.) | Unconfirmed | ~2.3 kWh | Unconfirmed | $20,000–$30,000 (target) |
| Unitree H1 | 180 | 47 | 27 | 3 per hand | Unspecified | 3.3 | 0.864 | Unconfirmed | ~$90,000 |
| Unitree G1 | 132 | 35 | 23–43 | 0 base; +7 + 2 wrist per arm (opt. hand) | ~2 (per arm) | >2 | 9 Ah pack | ~2 | ~$13,500+ |
| Agility Robotics Digit | 175 | 65 | 28 | Simple Gripper | 16 | ~1.4 | Unconfirmed | 2–4 (heavy use; ~8 light) | ~$250,000 (RaaS ~$30,000/year) |
| Figure 02 | 168 | 70 | 28 | 16 per hand | 20 | 1.2 | Unconfirmed | 5 | Not Disclosed |

*Source* Compiled from manufacturer websites, technical reports, and verified sources cited throughout the chapter, including current manufacturer pages for Atlas and Unitree platforms where available (Boston Dynamics, 2026; Unitree Robotics, 2026). Many figures remain estimates, targets, or configuration-dependent values and are subject to change

dynamic capability and a wide range of motion. At the other end of the spectrum, Agility Robotics' Digit, with 28 DoF and simple grippers, is a minimalist design highly optimized for its specific logistics tasks, trading broad versatility for increased reliability and cost-effectiveness. Most other platforms, such as the Figure 02 and Unitree G1, occupy a middle ground, balancing capability against complexity.
- Hand Dexterity as a Key Differentiator: The rapid progression in hand dexterity is one of the most significant trends. The contrast between the simple grippers on the logistics-focused Digit and the sophisticated, 16-DoF and 22-DoF hands on the Figure 02 and Tesla Optimus, respectively, is stark. This reflects a clear industry recognition that manipulation capability is a critical factor in determining the total addressable market for humanoid robots. While simple grippers are sufficient for pick-and-place tasks in a warehouse, the ability to perform complex assembly, use human tools, or provide caregiving assistance requires near-human levels of dexterity.
- Runtime and Energy Efficiency: The data reveals significant variation in energy efficiency. The Figure 02, for example, achieves a 5-h runtime on its 2.25 kWh battery, while the Unitree G1 is rated for only about 2 h. Agility Robotics' Digit operates roughly 2–4 h under heavy duty cycles, extending toward 8 h under lighter loads, per Agility's own statements. These differences are a function of not only battery size but also the efficiency of the actuators, the robot's weight, and the optimization of its walking gait. A robot like Digit, designed for the relatively flat and predictable environment of a warehouse, can have its gait highly optimized for energy efficiency, whereas a more general-purpose robot designed for varied terrain may have to compromise efficiency for robustness.
- Cost-Capability Trade-offs: The price points, where available, reflect vastly different market strategies. The Unitree G1, with a starting price around $13,500, is positioned to be a highly accessible platform for research and development. In contrast, the Unitree H1 and Agility Robotics Digit are priced in the tens to hundreds of thousands of dollars, targeting commercial customers who require a higher degree of reliability and support. Tesla's ambitious target price of $20,000–$30,000 for Optimus, if achieved, would be a disruptive force, predicated on leveraging automotive-scale manufacturing to dramatically reduce costs.

This analysis makes it clear that there is no single "optimal" humanoid robot. The ideal design is highly dependent on the target application. A robot for healthcare will prioritize safety and compliance, while one for manufacturing will prioritize precision and reliability, and one for logistics will optimize for cost and runtime. This application-specific optimization explains the diversity of designs seen across the industry and suggests that the market may support both specialized, niche platforms and more expensive, general-purpose robots for the foreseeable future.

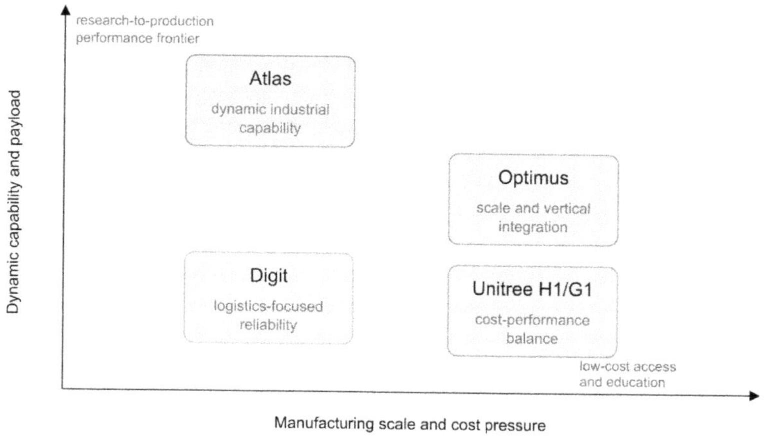


**Fig. 2.4** Platform design trade-space

Figure 2.4 summarizes how leading humanoid platforms occupy different regions of the hardware design trade-space, showing why the chapter treats capability, cost, reliability, and specialization as interacting design choices rather than a single ranking.

## 2.8 Manufacturing and Scaling Challenges

The specifications detailed in product announcements and research papers represent engineering targets; actually manufacturing thousands or millions of these complex machines reliably and economically poses a completely different set of challenges. As the humanoid robotics industry transitions from building research prototypes to commercial products, the principles of manufacturing are becoming a primary driver of design.

### 2.8.1 Design for Manufacturability

A design that is optimized for pure performance without regard for how it will be built can be prohibitively expensive or even impossible to produce at scale. Design for Manufacturability (DFM) is a critical discipline that involves designing products in a way that makes them easy to manufacture. This includes principles like minimizing part count, standardizing components, using materials and geometries that are compatible with high-volume processes like injection molding and die casting, and designing for automated assembly.

Tesla's entry into the humanoid robot space is particularly notable in this context. The company's deep experience in manufacturing electric vehicles at a scale of millions of units per year provides it with a significant advantage. The expertise in automation, supply chain management, and quality control developed for the automotive industry is directly applicable to the challenge of mass-producing robots. However, the translation is not one-to-one. Humanoid robots have a much higher degree of mechanical complexity than a car, with 30–50 actuated degrees of freedom compared to a handful in a vehicle. The precision required for joint alignment and calibration also exceeds typical automotive tolerances. Therefore, while automotive expertise is a powerful asset, it must be adapted to the unique challenges of robotic manufacturing.

### 2.8.2 Supply Chain and Component Sourcing

A robust supply chain is essential for scaling production. Key components like actuators, sensors, batteries, and computing hardware must be sourced, qualified, and integrated. Companies are pursuing different strategies to manage this. A vertically integrated company like Tesla designs and manufactures many of its own components, giving it tight control over the design and cost. Other companies choose to rely on a network of specialized suppliers, which can allow for faster development and lower capital investment.

The development of this supplier ecosystem is critical for the health of the entire industry. As a 2025 Bank of America report noted, component manufacturers are now actively working to innovate and scale up their production to meet the specific demands of humanoid robot OEMs (Bank of America, 2025). The maturity of the electromechanical component supply chain in China, for example, has been a key enabler for Chinese robot manufacturers to achieve aggressive price points.

### 2.8.3 Quality Control and Reliability Engineering

Ensuring that every robot that comes off the assembly line meets its specifications and will operate reliably for thousands of hours is a monumental task. The challenge arguably exceeds that of automotive production due to the higher complexity, tighter tolerances, and the deep integration of hardware and AI-driven software (the latter explored in Chap. 3).

The real-world deployment of Figure AI's robots at a BMW manufacturing plant provided a stark lesson in reliability engineering. Despite extensive testing, the deployment revealed that the robot's forearm was a top hardware failure point, leading to a redesign of the wrist electronics in the next-generation Figure 03 to improve reliability and thermal management (Interesting Engineering, 2025). This experience highlights the gap that often exists between performance in a controlled laboratory environment and reliability in a real-world industrial setting.

Bridging this gap requires a rigorous approach to reliability engineering, including:

- Failure Mode and Effects Analysis (FMEA): A systematic process for identifying potential failures and designing mitigations.
- Accelerated Life Testing: Subjecting components to extreme stress to identify and predict failures before deployment.
- Redundancy and Fault Tolerance: Designing critical systems with backups to allow for continued operation even if a component fails.
- Graceful Degradation: Ensuring that when failures do occur, the system fails in a safe and predictable way.
- Predictive Maintenance: Using sensor data to monitor the health of components and predict failures before they happen.

The reliability of the AI systems that control these robots is a particularly challenging new frontier. Unlike deterministic software, the behavior of a learned AI model can be difficult to predict and verify. The industry is actively working to develop new reliability frameworks that combine traditional engineering discipline with AI-appropriate methods like adversarial testing and runtime monitoring to ensure that these intelligent machines can be deployed safely and reliably at scale.

## 2.9 Conclusion

The hardware technologies that form the physical basis of modern humanoid robots are the product of a sophisticated integration of mechanical, electrical, materials, and manufacturing engineering. As of early 2026, the field is characterized by a healthy diversity of approaches, with different companies placing different bets on the optimal path to commercial viability.

Electric actuation has become the dominant commercial direction, a trend strengthened by Boston Dynamics' 2024 transition from hydraulic to electric systems for Atlas. This shift prioritizes controllability, efficiency, maintainability, and integration with commercial workflows over the raw power density of hydraulics.

In structural design, aluminum alloys remain the workhorse material, but advanced composites like carbon fiber and high-performance polymers like PEEK are enabling significant weight reductions, which in turn improve energy efficiency and dynamic performance. However, the high cost of these advanced materials means that their adoption is still limited.

Power systems remain a defining hardware constraint, but the bottleneck is better understood as uptime under real duty cycles rather than battery chemistry alone. Current humanoids commonly operate for roughly 2–4 h unless supported by hot-swappable packs or higher-energy-density cells (TrendForce, 2026). Operators can partly manage this through spare batteries, opportunity charging, task scheduling, and autonomous battery swapping, so energy is not an absolute blocker for every pilot. It becomes a scaling bottleneck when short runtime, charging downtime, peak-current limits, and thermal constraints reduce the useful work per day below what industrial customers need.

Locomotion capabilities have advanced to the point where robots can walk and run at near-human speeds in controlled environments. However, achieving this same level of reliability and robustness in the unstructured and unpredictable real world remains a major open challenge.

Manipulation is another major frontier. The emergence of commercially available, high-DoF, sensorized hands in the 2025–2026 timeframe marks a substantial capability gain. The shift from simple grippers toward dexterous hands expands the addressable market for humanoids beyond logistics into more complex tasks such as assembly, tool use, and assisted-care workflows.

Looking forward, the evolution of humanoid robot hardware will be driven by incremental improvements in existing technologies and selective breakthroughs in areas such as battery chemistry, actuator efficiency, thermal design, and manufacturable lightweight structures. The compound effects of these improvements will produce robots that are more capable, more reliable, and less expensive than the current generation. Critical milestones to watch include commercially demonstrated 8+hour uptime under real duty cycles, 99.99%+ reliability in field deployments, and price reductions that make fleet-scale adoption economically defensible.

The hardware foundation described in this chapter both enables and constrains the intelligence of the software. The most sophisticated AI in the world cannot overcome fundamental physical limits such as battery energy density, actuator torque, sensor noise, thermal headroom, or mechanical backlash. At the same time, modern software increasingly tolerates imperfect hardware: learned controllers compensate for compliance and drift, perception systems work around noisy sensors, and planning systems adapt to limited dexterity or runtime. Conversely, capable hardware without intelligent software is just a tele-operated machine. The central tension in humanoid robotics is therefore not whether hardware enables software or software tolerates hardware, but how tightly the two co-evolve into a reliable cyber-physical system.

## Questions

1. Actuation Technology Evolution: Explain why Boston Dynamics transitioned its Atlas platform from hydraulic to electric actuation in 2024. What does this industry-leading move signal about the priorities for commercial viability versus pure research performance?
2. Materials Trade-offs: Compare and contrast the use of aluminum alloys, carbon fiber composites, and PEEK polymers in humanoid robot structures. Why do aluminum alloys remain a dominant material despite the performance advantages of advanced composites? Discuss how a combination of materials contributed to the significant weight reduction seen in platforms like the Tesla Optimus.
3. Battery Bottleneck Analysis: The Figure 02 robot has an estimated 5-h runtime on its 2.25 kWh torso-integrated battery (Figure AI, 2024), calculate its average power consumption. What operational strategies can companies use to cope with

the 2- to 4-h runtime limitation common across many current platforms? Clarify the critical difference between gravimetric (Wh/kg) and volumetric (Wh/L) energy density, using CATL's recent announcements as an example.

4. Series Elastic Actuators: Explain the core principle of a Series Elastic Actuator (SEA) and the main advantages it offers for robotics (force sensing, shock tolerance, safety). What is the primary performance trade-off associated with SEAs, and why has this limited their adoption in mainstream humanoid platforms?
5. Dexterous Hand Comparison: Compare the three ZWHAND models (the 20-DoF B20, the 17-DoF A17, and the 6-DoF B06 priced at $699). For each of the following applications, justify which hand would be the most appropriate choice, considering the balance between capability, cost, and complexity: (a) logistics box sorting, (b) delicate electronics assembly, (c) academic research on tactile sensing, (d) a university-level educational robot.
6. ZMP Control: Explain the Zero-Moment Point (ZMP) concept and its role in maintaining dynamic balance for bipedal robots. Why is this concept less applicable to highly dynamic motions like running and jumping, and what control strategies (like reinforcement learning) are being explored to overcome these limitations?
7. Thermal Management: The humanoid robot cooling system market is projected to grow at a 37.9% CAGR through 2032. Why is thermal management becoming such a critical design consideration? If an actuator produces 1 kW of mechanical power at 90% efficiency, calculate the thermal power (in watts) that must be dissipated as waste heat.
8. Manufacturing Scaling: Discuss the primary challenges involved in scaling up the production of a complex humanoid robot from hundreds of units to hundreds of thousands of units. What aspects of automotive manufacturing expertise provide an advantage, and what aspects of robotic manufacturing are fundamentally different and more challenging?

## References


Articsledge. (2025). *AI humanoid robots 2025: Technology, builders & future.* Articsledge Technology Analysis. Retrieved from https://www.articsledge.com/post/ai-humanoid-robots

Bank of America. (2025). *Humanoid robots 101* Bank of America Institute Report. Retrieved from https://institute.bankofamerica.com/content/dam/transformation/humanoid-robots.pdf

Boston Dynamics. (2026). *Atlas.* Boston Dynamics. Retrieved from https://bostondynamics.com/products/atlas/

CleanTechnica. (2025). *Humanoid robot battery production: CATL achieves a world first.* CleanTechnica. Retrieved from https://cleantechnica.com/2025/12/24/humanoid-robot-battery-production-catl-achieves-a-world-first/

Custom Plastic Mold. (2025). *Why are carbon fiber composites plastic resin so suitable for use in humanoid robots?* Industry Knowledge Article. Retrieved from https://www.custom-plastic-mold.com/info/why-are-carbon-fiber-composites-plastic-resin-103054230.html

Figure AI (2024). Figure 02: The new humanoid robot. IEEE Spectrum. Retrieved from https://spectrum.ieee.org/figure-new-humanoid-robot

Frontiers. (2025). *OpenSEA: A 3D printed planetary gear series elastic actuator for a compliant elbow joint exoskeleton.* Frontiers in Robotics and AI, 12. Retrieved from https://doi.org/10.3389/frobt.2025.1528266/full

Intel Market Research. (2025). *Humanoid robot cooling system market outlook 2025–2032.* Intel Market Research Reports. Retrieved from https://www.intelmarketresearch.com/humanoid-robot-cooling-system-market-7280

International Federation of Robotics. (2026). *Top 5 global robotics trends 2026.* IFR Press Release. Retrieved from https://ifr.org/ifr-press-releases/news/top-5-global-robotics-trends-2026

Interesting Engineering. (2025). *Figure humanoid robots retire bruised after 11 months of work at BMW.* Interesting Engineering Robotics. Retrieved from https://interestingengineering.com/ai-robotics/figure-humanoid-robots-retires-bmw

Kalil, M. (2026). *Sharpa debuts North humanoid at CES 2026.* Mike Kalil Technology Analysis. Retrieved from https://mikekalil.com/blog/sharpa-north-ces-2026/

LinkedIn. (2025). *Battery performance challenges for humanoid robots.* Hermitage Capital Analysis. Retrieved from https://www.linkedin.com/pulse/battery-performance-challenges-humanoid-robots-hermitagecapital-4mwxc

Newstrail. (2025). *Humanoid robot carbon fiber market to hit USD 710 million by 2032 at 37.9% CAGR.* Newstrail Industry News. Retrieved from https://www.newstrail.com/humanoid-robot-carbon-fiber-market/

PEEK China. (2025). *Why PEEK is the key material for Tesla's humanoid robot Optimus-Gen2?* PEEK China Technical Blog. Retrieved from https://www.peekchina.com/blog/key-material-for-humanoid-robots-peek-polymer.html

PR Newswire. (2026). *ZWHAND shines at CES 2026: Redefining robotic hands for mass embodied AI.* PR Newswire Press Release. Retrieved from https://www.prnewswire.com/news-releases/zwhand-shines-at-ces-2026-redefining-robotic-hands-for-mass-embodied-ai-302659035.html

Maisch, M. (2025). *Global lithium-ion battery pack prices fall to $108/kWh, says BNEF.* pv magazine International. Retrieved from https://www.pv-magazine.com/2025/12/09/global-lithium-ion-battery-pack-prices-fall-to-108-kwh-says-bnef/

Robotics Tomorrow. (2026). *DexRobot shines at CES 2026, showcasing its flagship DexHand021 Pro and closed-loop dexterous solution.* Robotics Tomorrow News. Retrieved from https://www.roboticstomorrow.com/news/2026/01/07/dexrobot-shines-at-ces-2026-showcasing-its-flagship-dexhand021-pro-and-closed-loop-dexterous-solution/25973/

Springer. (2025). Experimental research of control moment gyroscope effect on bipedal robot walking trajectory. *Journal of the Brazilian Society of Mechanical Sciences and Engineering, 47.* Retrieved from https://doi.org/10.1007/s40430-025-05477-1

TrendForce. (2026). *Humanoid robots move toward commercialisation; solid-state batteries set to break through power bottlenecks.* TrendForce Press Centre. Retrieved from https://www.trendforce.com/presscenter/news/20260128-12902.html

Tycorun. (2025). *Powering the future: How the humanoid robot boom is reshaping the lithium battery industry.* Tycorun Battery Technology. Retrieved from https://batteryswapcabinet.com/powering-the-future-how-the-humanoid-robot-boom-is-reshaping-the-lithium-battery-industry/

Unitree Robotics. (2026). *Unitree G1 and H1 product specifications.* Unitree Robotics. Retrieved from https://shop.unitree.com/products/unitree-g1 and https://www.unitree.com/h1/

Wiley. (2026). Adaptive stiffness control for series elastic actuators in robotic systems using dynamic systems. *Computational Intelligence, 43.* Retrieved from https://doi.org/10.1111/coin.70163

WINS Solutions. (2025). *Innovative humanoid robots in 2025–2026: Reality or hype?* WINS Solutions Industry Analysis. Retrieved from https://www.winssolutions.org/humanoid-robots-2025-2026-reality-hype/

Xie, W., Bai, C., Shi, J., Yang, J., Ge, Y., Zhang, W., & Li, X. (2025). *Humanoid whole-body locomotion on narrow terrain via dynamic balance and reinforcement learning.* arXiv. https://arxiv.org/abs/2502.17219

**Grace Huang** is a seasoned product and technology professional with extensive experience at organizations including PIMCO (a global investment management firm managing over $2 trillion in assets), IBM, SoftBank, and DataRobot. Throughout her career, she has successfully launched multiple products and managed large-scale projects, leveraging her expertise in product management, market analysis, strategic planning, and cross-functional team leadership. Her work brings a finance and product strategy perspective to physical AI, particularly the operational, risk, and adoption considerations involved in bringing advanced robotics into enterprise environments. Always at the forefront of emerging technology, she has explored innovative tools including the implementation of Agentic AI powered solutions within the product management process—streamlining communication, improving decision-making, enhancing customer satisfaction, and driving business growth. In addition to her professional accomplishments, Grace holds a degree from Babson College, where she built a strong foundation in business management and entrepreneurship. She continues to lead at the intersection of product strategy and advanced technology, bringing a rare combination of financial rigor and deep technical acumen to complex, high-stakes initiatives.

Her work brings a finance and product strategy perspective to physical AI, particularly the operational, risk, and adoption considerations involved in bringing advanced robotics into enterprise environments. Always at the forefront of emerging technology, she has explored innovative tools including the implementation of Agentic AI powered solutions within the product management process—streamlining communication, improving decision-making, enhancing customer satisfaction, and driving business growth.

In addition to her professional accomplishments, Grace holds a degree from Babson College, where she built a strong foundation in business management and entrepreneurship. She continues to lead at the intersection of product strategy and advanced technology, bringing a rare combination of financial rigor and deep technical acumen to complex, high-stakes initiatives.

**Ken Huang** is a globally recognized expert in AI and Web3 security, a prolific author, and a leading figure in shaping industry standards. He is the CEO and Chief AI Officer (CAIO) of DistributedApps.ai, specializing in Generative AI training and consulting. Ken is deeply involved in driving the development of secure AI systems. He serves as a Research Fellow and Co-Chair of the AI Safety Working Groups at the Cloud Security Alliance (CSA), leading Agentic AI initiatives. He is also a Co-Chair of the AI STR Working Group at the World Digital Technology Academy under the UN Framework, and a core contributor to the OWASP GenAI project, focusing on Agentic AI security. His expertise extends to his contributions to OWASP's Top 10 Risks for LLM Applications and his active participation in the NIST Generative AI Public Working Group. He is also a member of Open AI forum.

**Bo Tian** is a Sr Staff Hardware Engineer at Tesla, where he manages an electrical engineering team focused on the board-level architecture and design for the Optimus AI computer. Operating at the intersection of hardware and software, his work applies knowledge of transformer architecture tradeoffs to board architecture decisions, striving for optimal AI inferencing engines. Prior to his work on humanoid robotics and AI inferencing clusters at Tesla, he served as a Senior Hardware Engineer at Apple. He holds a Master of Science in Electrical and Electronics Engineering from Texas A&M University and a Bachelor of Engineering from The Hong Kong Polytechnic University.

**Yuyan Duan** is a product leader and angel investor at the forefront of AI and robotics innovation. As an earlystage investor, she has backed a robotics startup supported by Salesforce CEO Marc Benioff, Skild's founder, and Nat Friedman, as well as a healthcare LLM venture backed by Anthropic's co-founder. Known for writing first checks, Lynn has already seen one portfolio company advance to Series B funding.Since 2019, Lynn has served as Product Leader at Treasure Data, a pre-IPO customer data platform company, where she has architected roughly one-third of the company's product suite. She pioneered the company's AI transformation by creating its first AI Agent Framework, which has generated 30+ enterprise-grade agents and driven 30+ new customer engagements. Her work spans enterprise big data, data security, and AI-driven marketing solutions.In 2023, Lynn founded AI+, a thriving AI startup community with 50+ contributing members that has hosted 150+ events over two years, connecting 2,000+ founders annually with 200+ investors and corporate partners. She holds degrees from the University of Michigan (Psychology & Statistics, International Relations) and Georgia Tech (M.S. in Human-Computer Interaction with a Certificate in Management of Technology).

# Perception and Interaction Systems

3

Ken Huang, Grace Huang, Bo Tian, and Manish Bhatt

## 3.1 Computer Vision Systems

Computer vision serves as the primary perceptual modality for most humanoid robots, providing rich information about the environment at relatively low cost and power consumption. Modern humanoid platforms employ sophisticated camera arrays and deep learning-based perception pipelines to enable object recognition, scene understanding, and navigation in complex environments. These pipelines are inseparable from the onboard AI inference computer: camera streams must pass through serialisers, image signal processors, memory, GPUs, NPUs or other accelerators, and real-time middleware within strict latency and thermal budgets. NVIDIA Jetson Thor is a prominent current example for physical AI and humanoid robotics, but the engineering problem is broader than choosing a single vendor module; it includes sensor I/O bandwidth, memory capacity, model compression, power delivery, cooling, and fail-safe scheduling (NVIDIA, 2025). The fundamental challenge lies in extracting actionable information from visual data in real time, under varying lighting conditions, and in the presence of occlusions and motion blur that are inherent to dynamic humanoid operation. Motion blur is often harder for humanoids than for road vehicles because cameras move with the head, torso,

K. Huang (✉)
DistributedApps.ai, Fairfax, VA, USA
e-mail: ken@distributedapps.ai

G. Huang
PIMCO, Austin, TX, USA

B. Tian
Tesla, Palo Alto, CA, USA

M. Bhatt
Amazon Kuiper Security/OWASP, Seattle, WA, USA



K. Huang et al. (eds.), *Humanoid Robots and Physical AI*, Advances in Data Analytics, AI, and Smart Systems, https://doi.org/10.1007/978-3-032-33669-9_3

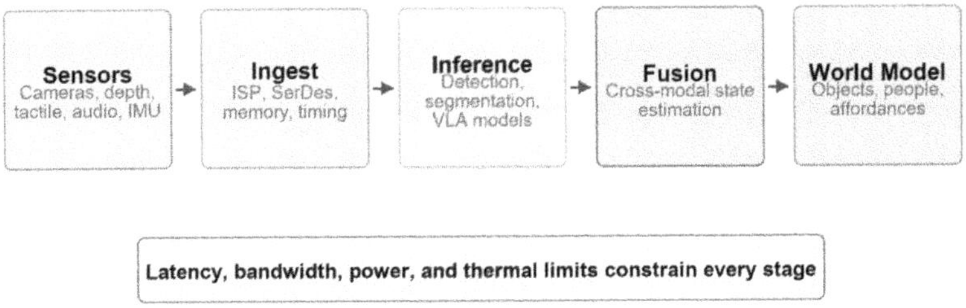


**Fig. 3.1** Humanoid perception pipeline

arms, and walking impacts rather than with a comparatively rigid chassis; mitigation therefore begins with locomotion control, mechanical isolation, exposure control, and camera placement before image processing. Consumer-style digital stabilisation is useful for recorded video, but ISP-based stabilisation can require frame buffering and introduce latency that is unacceptable for closed-loop vision pipelines.

The evolution of computer vision in robotics has been remarkable over the past decade. Early systems relied heavily on hand-crafted features and classical computer vision algorithms, such as edge detection, corner detection, and optical flow. These approaches required significant domain expertise to tune and often failed in real-world conditions where lighting, viewpoint, and occlusions varied unpredictably. The advent of deep learning, first through convolutional neural networks and later through Vision Transformers (ViTs), revolutionised robotic perception by enabling learned feature extraction from raw pixel data to semantic understanding (Dosovitskiy et al., 2021). This shift has enabled humanoid robots to achieve unprecedented levels of scene understanding and object recognition accuracy.

Figure 3.1 summarises the perception pipeline from raw sensors to a task-ready world model, emphasising that inference, fusion, latency, bandwidth, and thermal limits must be engineered together.

### 3.1.1 Camera Configurations and Hardware

Humanoid robots typically employ multiple cameras with complementary characteristics to achieve comprehensive visual coverage. The most common configurations include stereo camera pairs for depth estimation, RGB-D cameras that directly measure distance, fisheye lenses for wide field-of-view coverage, and task-specific close-range cameras for manipulation. The placement of cameras on humanoid platforms involves careful engineering trade-offs between field of view, resolution, computational load, focal distance, mechanical protection, cable routing, thermal packaging, and aesthetics. Because cameras shape both perception quality and the robot's apparent gaze, their placement also affects human trust and social readability during interaction.

The choice between global shutter and rolling shutter sensors represents a fundamental trade-off in camera selection for humanoid robotics. Global shutter sensors capture all pixels simultaneously, eliminating motion artefacts during rapid movements—critical for humanoid robots executing dynamic manoeuvres such as walking, running, or manipulation tasks. When a robot moves quickly, a rolling shutter sensor captures different rows of the image at different times, resulting in geometric distortions that can significantly degrade downstream perception. Global shutter designs also push more simultaneous data off the sensor, increasing demands on camera links, serialiser-deserialiser hardware, cabling, and compute ingest. However, global shutter sensors typically exhibit lower light sensitivity and higher system cost compared to rolling shutter alternatives, which capture pixels sequentially but often offer superior low-light performance and higher effective resolution per dollar.

The sensitivity difference between global and rolling shutter sensors arises from their underlying pixel architectures. Global shutter pixels require additional transistors to store charge during readout, reducing the photosensitive area and thus the light-gathering capability. A related trade-off exists between resolution and low-light performance: for a fixed sensor size, higher resolution usually pushes pixel pitch downward, reducing each pixel's photon-collection area and signal-to-noise ratio even as spatial acuity improves. This is a sensor-design constraint rather than a purely software problem; optics, exposure time, gain, illumination, and image processing can compensate only within limits (Chen et al., 2000). For humanoid robots operating primarily in well-lit industrial environments, these penalties may be acceptable given the benefits of motion artefact elimination. However, robots designed for homes, warehouses, and mixed indoor-outdoor spaces must balance acuity for manipulation against versatility across dim or uneven lighting conditions.

High Dynamic Range (HDR) imaging presents another critical consideration for humanoid robot vision systems. Real-world environments often contain extreme brightness variations—a robot looking from a dimly lit interior toward a bright window, for example, must handle illuminance differences spanning several orders of magnitude. Standard cameras with linear response functions cannot capture such scenes without saturating highlights or crushing shadows. HDR cameras address this through multiple exposure fusion or logarithmic response curves, enabling perception across challenging lighting conditions. Tesla's Optimus, designed for factory environments with variable lighting, incorporates HDR cameras to maintain consistent perception performance throughout industrial facilities.

Tesla's Optimus robot represents a distinctive approach by adopting a vision-only perception system derived from Tesla's Full Self-Driving (FSD) technology. While Tesla's autonomous vehicles utilise eight external cameras for 360° coverage, the Optimus platform adapts this neural architecture to a humanoid form factor by concentrating high-definition cameras primarily within the robot's head to simulate human binocular vision. This configuration deliberately eschews LiDAR or structured light depth sensors, reflecting Tesla's design philosophy that neural networks can learn robust 3D scene understanding from 2D images alone. Tesla has reported that the cameras run at approximately 36 frames per second, inherited from the FSD camera stack (Tesla, 2023), feeding into neural networks originally trained on Tesla's extensive autonomous driving dataset and fine-tuned for embodied bipedal applications.

The transfer of perception technology from autonomous vehicles to humanoid robots represents both an opportunity and a challenge. Vehicle perception systems benefit from massive training datasets collected from millions of miles of driving, robust camera mounts on stable platforms, and relatively predictable motion patterns. Humanoid robots face more varied motion profiles, closer interaction distances with objects, and different perceptual requirements for manipulation versus navigation. Focal length is a particularly important transfer issue: vehicle cameras can often be fixed-focus for mid- to long-range acuity, while humanoids need high acuity at short manipulation distances. Practical designs may use overlapping fixed-focus cameras with different focal ranges, actuated lenses for autofocus, or coordinated head and hand cameras to preserve close-range detail without sacrificing latency. Tesla's approach involves adapting vehicle-trained networks to humanoid contexts while collecting new training data specific to bipedal robot operation, progressively improving perception capabilities through fleet learning.

Figure AI's Figure 02 humanoid, unveiled in August 2024, employs a multi-camera array consisting of six RGB cameras positioned for comprehensive environmental awareness (Figure AI, 2024). This configuration includes both high-resolution RGB cameras for detailed object recognition and dedicated depth-sensing capabilities. The platform's head-mounted cameras provide a human-like viewpoint for manipulation tasks, while additional cameras offer coverage during dexterous manipulation. The integration with Figure's proprietary Helix model (Figure AI, 2025)—a vision-language-action (VLA) model, discussed in detail in Sect. 3.6.4 and revisited in Chap. 4—represents a significant advancement over traditional modular perception pipelines, enabling end-to-end learning from pixels to actions without explicit depth estimation or object detection modules.

The camera positioning on Figure 02 reflects careful consideration of humanoid-specific perception requirements. Unlike autonomous vehicles where cameras are distributed around a fixed chassis, humanoid robots can orient their "head" to direct visual attention toward regions of interest. This active vision capability reduces the number of cameras needed for comprehensive coverage while enabling more efficient allocation of computational resources toward relevant scene regions (Bohg et al., 2017). However, this approach requires sophisticated attention mechanisms that can quickly identify where to look based on task requirements and environmental context.

Engineered Arts' Ameca robot features eye-mounted cameras built into physically moving mechanical eyeballs, enabling highly natural gaze behaviour and human-like attention mechanisms. This anthropomorphic design facilitates more intuitive human–robot interaction by allowing the robot to physically pan and tilt its cameras to visibly "look" at objects of interest, providing nonverbal communication cues that humans naturally understand. The mechanical eye implementation requires precise coordination between camera movement, image stabilisation, and

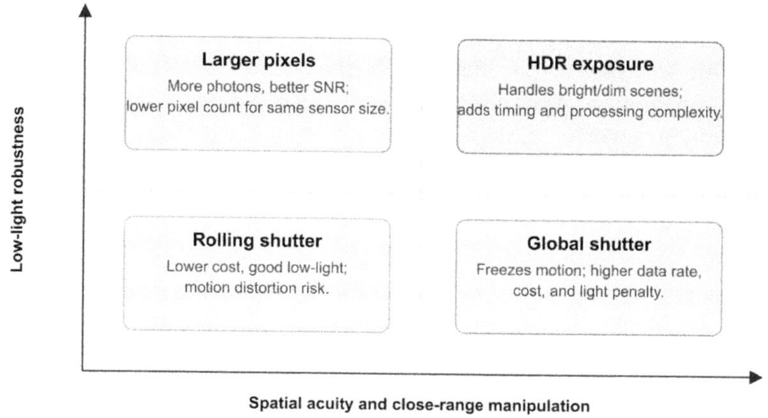


**Fig. 3.2** Camera hardware trade-offs

gaze tracking to maintain perceptual accuracy while conveying natural social signals. For manipulation, head-centric vision alone is insufficient: arms, tools, and carried objects frequently occlude head-mounted cameras near the workspace. Eye-in-hand or other distal cameras mounted near wrists, hands, or end-effectors provide local views for visual servoing, closed-loop grasp correction, and contact-rich manipulation when global head cameras lose line of sight (Chaumette & Hutchinson, 2006).

Figure 3.2 shows the main camera-hardware trade-offs for humanoid robots, including shutter type, pixel size, low-light robustness, acuity, and close-range manipulation requirements.

### 3.1.2 Object Detection and Segmentation

Modern humanoid robots rely heavily on deep learning-based object detection and segmentation algorithms to identify and localise objects in their environment. The YOLO (You Only Look Once) family of detectors has become particularly popular for real-time robotics applications due to its favourable speed-accuracy trade-off. More recent versions like YOLOv8 and YOLOv9 achieve detection speeds exceeding 100 frames per second on modern GPUs while maintaining high accuracy across diverse object categories. These single-stage detectors predict bounding boxes and class probabilities directly from image features in a single forward pass, enabling real-time performance essential for dynamic humanoid operation.

For manipulation tasks requiring precise object boundaries, instance segmentation networks provide pixel-level object delineation. The Mask R-CNN architecture extends object detection by adding a segmentation branch that predicts a binary mask for each detected instance. This detailed boundary information is essential for grasp planning, where the robot must understand object geometry to select appropriate grip points. More recent architectures like the Segment Anything Model (SAM) from Meta have demonstrated remarkable zero-shot segmentation capabilities, segmenting arbitrary objects without task-specific training. SAM's ability to generate masks for any object in an image based on point prompts or automatic detection makes it particularly valuable for humanoid robots operating in unstructured environments where the specific objects encountered cannot be predicted in advance.

The integration of segmentation models into robotic perception pipelines involves careful consideration of computational requirements and latency constraints. High-resolution images typical of modern cameras can result in significant processing delays when running sophisticated segmentation models, potentially introducing unacceptable latency for real-time control. Approaches to mitigate this include running segmentation on region proposals from faster object detectors, using lower-resolution inputs for segmentation while maintaining full resolution for detection, and employing hardware acceleration through dedicated neural network processors.

Semantic segmentation, which assigns a class label to every pixel in an image, provides complementary information to instance segmentation. While instance segmentation identifies individual objects with separate masks, semantic segmentation provides dense scene labelling useful for navigation and terrain understanding. For a humanoid robot navigating through an environment, semantic segmentation can identify walkable surfaces, obstacles, and regions of interest without requiring explicit object detection. Modern architectures like SegFormer and Mask2Former have achieved impressive results by combining transformer-based feature extraction with efficient decoding strategies, enabling real-time semantic segmentation on edge computing hardware.

### 3.1.3 3D Reconstruction and SLAM

Simultaneous Localisation and Mapping (SLAM) enables robots to build maps of unknown environments while tracking their own position within those maps. Visual SLAM systems use camera images as the primary input, making them attractive for cost-sensitive humanoid platforms. The fundamental challenge of SLAM involves maintaining a consistent map while tracking robot motion, solving what is essentially a chicken-and-egg problem: accurate localisation requires a good map, but building a good map requires accurate localisation (Cadena et al., 2016).

The mathematical foundations of modern SLAM can be categorised into feature-based methods and direct methods, each with distinct advantages for humanoid applications. Feature-based SLAM systems detect distinctive visual features—corners, edges, or blobs—and track these features across frames to estimate camera motion and build sparse 3D maps. The key advantage of feature-based approaches is their computational efficiency: by processing only distinctive image regions rather than every pixel, these systems can achieve real-time performance on modest hardware while maintaining robustness to illumination changes and viewpoint variations.

Building upon earlier versions like ORB-SLAM2 (Mur-Artal & Tardós, 2017), ORB-SLAM3 (Campos et al., 2021) represents the current state-of-the-art in feature-based visual SLAM, supporting monocular, stereo, and RGB-D cameras while handling dynamic environments. The system builds sparse 3D maps using Oriented FAST and Rotated BRIEF (ORB) features that are invariant to rotation and scale changes. The feature extraction process identifies corners in images that can be reliably detected across different viewpoints and lighting conditions. Feature-based methods excel in computational efficiency and robustness to illumination changes but struggle in textureless environments common in industrial settings such as blank walls, uniform floors, and monochrome machinery.

In contrast, direct methods like LSD-SLAM (Large-Scale Direct SLAM) estimate motion by minimisng photometric error over entire image regions rather than relying on sparse feature points (Engel et al., 2014). By using all image pixels with sufficient gradient, direct methods can operate in environments where feature extraction would fail due to lack of distinctive corners. The photometric error measures the difference in pixel intensities between corresponding regions across frames, and minimisng this error yields estimates of camera motion and scene depth. Direct methods produce dense or semi-dense reconstructions valuable for manipulation planning but require significant computational resources and are sensitive to exposure changes between frames.

The choice between feature-based and direct SLAM for humanoid robots depends on the target operating environment and computational budget. Industrial environments with textured surfaces and controlled lighting may benefit from feature-based approaches with their lower computational requirements. Environments with poor texture but significant intensity gradients, such as corridors with uniform walls, may require direct methods to maintain tracking. Hybrid approaches that switch between methods based on scene characteristics have shown promise for robust operation across diverse environments.

Neural SLAM methods, particularly Neural Radiance Fields (NeRF) (Mildenhall et al., 2020) and more recently 3D Gaussian Splatting (3DGS) representations (Kerbl et al., 2023), have emerged as powerful tools for dense 3D reconstruction. In robotics, the important shift is not only novel-view rendering but online tracking and mapping: systems such as SplaTAM and Gaussian Splatting SLAM use 3D Gaussians as the map representation for camera pose estimation, dense reconstruction, and fast rendering (Keetha et al., 2024; Matsuki et al., 2024). Unlike traditional SLAM methods that produce sparse point clouds or mesh representations, these approaches learn continuous volumetric scene representations that can be rendered from arbitrary viewpoints. NeRF represents a scene as a neural network that maps 3D coordinates and viewing direction to colour and density values. Training this network on a collection of images enables photorealistic novel-view synthesis, and the learned representation can be queried for 3D geometry. Recent work has demonstrated real-time Gaussian Splatting SLAM achieving over 150

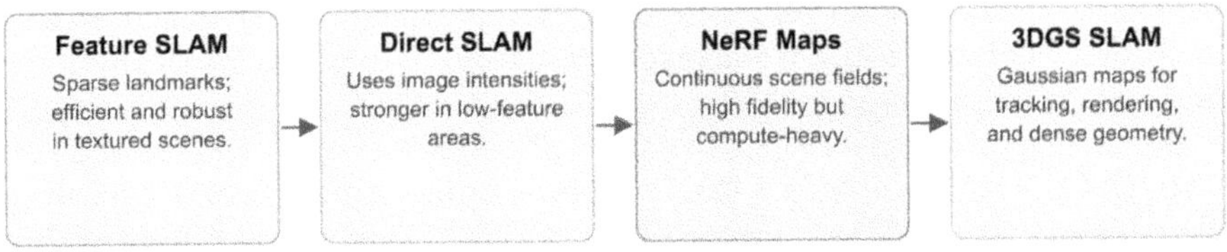


**Fig. 3.3** SLAM and 3D reconstruction options

frames per second for high-fidelity reconstruction, making it viable for online robotics applications.

The application of neural rendering to robotics perception offers several advantages over traditional representations. Neural representations can fill in gaps in observations through learned priors, enabling complete scene reconstruction from partial views. The continuous nature of these representations supports queries at arbitrary resolution, useful for manipulation tasks requiring detailed geometric information. However, the computational requirements of neural rendering have historically limited real-time applications, and ongoing research focuses on efficient architectures that can operate within the power and latency constraints of mobile humanoid platforms.

Figure 3.3 compares SLAM and neural reconstruction approaches, showing how map representations vary from sparse landmarks to dense neural or Gaussian scene models.

Boston Dynamics' electric Atlas, unveiled in April 2024 to replace the retired hydraulic version (Ackerman, 2024; Boston Dynamics, 2024), employs advanced SLAM algorithms to maintain accurate localisation during dynamic manoeuvres like parkour, where rapid motion and occlusions challenge standard tracking approaches. The system must handle extreme viewpoint changes, motion blur, and temporary occlusions of tracked features during acrobatic movements. This requires robust data association mechanisms that can correctly match features across frames even when the robot's motion dramatically changes the observed scene between consecutive camera frames.

### 3.1.4 Visual Odometry and Ego-Motion Estimation

Visual odometry estimates a robot's motion by tracking features across consecutive image frames (Scaramuzza & Fraundorfer, 2011). This provides critical feedback for control systems and complements inertial measurements for state estimation. The incremental nature of visual odometry—estimating frame-to-frame motion and accumulating these estimates—makes it susceptible to drift over time, but the rich visual information provides valuable constraints on robot motion that inertial sensors alone cannot offer.

Learning-based visual odometry has shown promising results, with networks trained end-to-end to predict motion from image sequences. These approaches can potentially handle challenging scenarios like motion blur and low texture that confound traditional geometric methods. The key advantage of learned methods is their ability to leverage large datasets to develop robustness to challenging conditions that would cause classical algorithms to fail. However, learned methods may also exhibit failure modes that are difficult to predict or characterise, requiring careful validation before deployment in safety-critical applications.

Tesla's Optimus system employs neural network-based odometry estimation trained on diverse driving and manipulation scenarios, enabling robust motion estimation across varied visual conditions. The system benefits from Tesla's extensive data collection infrastructure, with millions of miles of driving data providing diverse training examples. Transfer learning approaches adapt these vehicle-trained models to humanoid locomotion, with additional fine-tuning on humanoid-specific motion patterns and visual perspectives.

### 3.1.5 Low-Light and High Dynamic Range Performance

Humanoid robots must operate across varied lighting conditions, from bright outdoor environments to dimly lit interiors. This requires cameras with high dynamic range (HDR) capabilities and sophisticated exposure control strategies. Standard cameras with linear response can capture approximately 8–10 stops of dynamic range, while real-world scenes may span 20 or more stops from deepest shadow to brightest highlight. HDR cameras address this through multiple exposure fusion, capturing several images at different exposures and combining them to extend the effective dynamic range.

The challenge of HDR imaging for robotics extends beyond capture to the downstream perception algorithms. Object detection and segmentation models trained on standard imagery may fail when presented with HDR images containing unusual exposure distributions. Training data augmentation and specialised HDR-aware architectures help address this challenge, but the fundamental tension remains between capturing scene information and maintaining algorithm compatibility.

Event cameras represent an emerging technology that addresses both dynamic range and temporal resolution limitations by asynchronously reporting per-pixel brightness changes with microsecond resolution rather than fixed-rate frames. Their high dynamic range and low-latency characteristics are examined in detail in Sect. 3.8.1.

## 3.2 LiDAR and Depth Sensing

While computer vision provides rich semantic information, direct depth measurement through LiDAR and structured light sensors offer complementary geometric data that enhances robustness, particularly in textureless environments or poor lighting conditions. The strategic choice between vision-only and LiDAR-enhanced perception represents a fundamental architectural decision in humanoid robot design, with implications for cost, power consumption, reliability, and operational envelope.

### 3.2.1 LiDAR Principles and Technologies

Light Detection and Ranging (LiDAR) sensors measure distance by timing the round-trip of laser pulses reflected from surfaces. Time-of-flight (ToF) LiDAR systems emit short laser pulses and measure the time until reflected light returns, enabling range measurements up to hundreds of metres. The range resolution depends on the timing precision of the detection electronics, with modern systems achieving centimeter-scale accuracy at useful ranges. The angular resolution depends on the beam divergence and scanning mechanism, determining the point density achieved at different distances (Li & Ibanez-Guzman, 2020).

Traditional mechanical scanning LiDAR systems use rotating mirrors or rotating sensor assemblies to sweep laser beams across the field of view. These systems provide excellent angular resolution and range but suffer from moving parts that limit durability and increase cost. The mechanical complexity of scanning systems introduces points of failure and requires careful calibration to maintain accuracy over time. Despite these limitations, mechanical scanning LiDAR remains popular for high-end applications where performance justifies cost and complexity.

Solid-state LiDAR alternatives using optical phased arrays or MEMS (Micro-Electro-Mechanical Systems) mirrors achieve beam steering without macro-mechanical rotation. Optical phased arrays manipulate light using arrays of tunable elements that control the phase of transmitted light, electronically steering the beam without moving parts. MEMS-based systems use tiny mirrors that can be tilted at high speeds to scan the laser beam. Both approaches offer improved reliability and potentially lower manufacturing costs compared to mechanical systems, though achieving the same angular resolution and range remains challenging.

Flash LiDAR represents another solid-state approach, illuminating the entire scene with a single laser pulse and using a detector array to capture range information simultaneously across all directions. This approach eliminates scanning entirely, capturing depth images at high frame rates. However, the limited optical power per direction reduces effective range compared to scanning systems that concentrate laser energy in a narrow beam. For close-range applications typical of humanoid manipulation tasks, flash LiDAR may provide an attractive trade-off between range, resolution, and update rate.

As noted in Chap. 2's specification analysis, budget-friendly platforms like the Unitree G1 (Unitree, 2024) demonstrate that meaningful LiDAR integration need not be limited to flagship platforms. The G1 integrates both a LIVOX-MID-360 LiDAR (360° by 59° field of view, 40 m range, plus or minus 3 cm accuracy) and an Intel RealSense D435 depth camera (1280 by 720 resolution at 30 frames per second) for robust navigation in complex environments. This dual-sensor approach provides reliable obstacle detection for path planning while vision handles manipulation and object recognition tasks. The Intel RealSense D435i variant adds an IMU for improved motion tracking, with depth accuracy of approximately 2 percent at 4 m distance.

### 3.2.2 Structured Light and Time-of-Flight Depth Cameras

Structured light depth cameras project a known pattern onto the scene and analyse pattern deformation to estimate depth. The original Microsoft Kinect popularised this approach for consumer applications, using an infrared projector to cast a speckle pattern and an infrared camera to observe pattern distortion. The displacement of observed pattern points from their expected positions encodes depth information. Structured light systems work well at close range, making them suitable for manipulation tasks, but suffer in outdoor environments where sunlight overwhelms the projected pattern.

Time-of-flight (ToF) cameras measure depth directly by timing the return of modulated infrared light. Continuous-wave ToF systems measure the phase shift between transmitted and received light, which encodes distance (Horaud et al., 2016). Pulsed ToF systems directly measure the round-trip time of short light pulses. ToF cameras provide dense depth maps at high frame rates but have limited range compared to scanning LiDAR and may suffer from multipath interference in complex geometries where light reflects multiple times before returning to the sensor.

The Intel RealSense family includes both structured light and stereoscopic depth-sensing approaches. The D400 series uses stereo vision with active infrared illumination to improve matching in low-texture regions. This hybrid approach combines the advantages of passive stereo vision—which works at any range and in any lighting where stereo matching succeeds—with active illumination to handle textureless surfaces. The integration of depth cameras with RGB sensors in the same package simplifies depth-colour alignment, providing registered depth and colour images for downstream processing.

### 3.2.3 Vision-Only Versus LiDAR-Enhanced Approaches

The choice between vision-only and LiDAR-enhanced perception represents a fundamental trade-off in humanoid robot design. Vision-only approaches, exemplified by Tesla's Optimus, offer lower hardware costs, reduced power consumption, and no moving parts. Proponents argue that neural networks can learn robust depth estimation from monocular or stereo vision, as demonstrated by autonomous vehicle perception systems. The massive scale of data collection from Tesla's vehicle fleet enables training of depth estimation networks that generalise across diverse environments.

However, vision-only approaches face inherent limitations. Monocular depth estimation remains fundamentally ambiguous without strong priors—a small object nearby and a large object far away can produce identical images. Stereo vision resolves this ambiguity through triangulation but fails in textureless regions where corresponding points cannot be reliably matched. The accuracy of stereo depth degrades with distance, as the baseline between cameras becomes insignificant compared to the object distance. For manipulation tasks requiring millimeter-scale precision at close range, stereo vision may suffice, but navigation and long-range perception benefit significantly from active depth sensing.

Boston Dynamics' electric Atlas maintains a multi-sensor approach combining LiDAR with stereo vision. This architectural decision reflects the company's emphasis on maximum robustness for industrial applications. The LiDAR provides absolute distance measurements that help correct visual odometry drift, while vision fills in geometric detail between LiDAR scan lines. This combination enables Atlas to execute precision tasks with centimeter-level accuracy across varied lighting conditions—from bright sunlight to complete darkness where vision-only systems fail.

The redundancy provided by multiple sensing modalities offers fault tolerance critical for industrial deployment. If LiDAR fails due to a dirty lens or hardware malfunction, the vision system can maintain basic navigation capabilities. If vision is degraded by poor lighting, LiDAR continues to provide reliable geometric information. This graceful degradation is essential for robots operating in safety–critical environments where perception failure could result in property damage or human injury.

The optimal choice depends on application requirements and constraints. Cost-sensitive consumer robots may favour vision-only approaches, accepting limitations in challenging conditions to achieve accessible pricing. Industrial and research platforms prioritise the robustness and precision that LiDAR provides, justifying the additional cost and complexity through expanded operational capability. For safety-critical applications, the redundancy of independent sensing modalities increases overall system reliability—a principle codified in ISO 13849 performance level requirements for safety-related control systems.

### 3.2.4 Point Cloud Processing and Scene Understanding

LiDAR sensors generate point clouds—sets of 3D points representing surface measurements in the environment. Processing these point clouds efficiently requires specialised algorithms for filtering, segmentation, and feature extraction. The Point Cloud Library (PCL) provides widely-used implementations of standard algorithms for robotics applications, including filtering, feature estimation, and surface reconstruction.

Ground plane removal represents a fundamental preprocessing step for mobile robot navigation, separating ground surfaces from obstacles and objects of interest. RANSAC (Random Sample Consensus) based plane fitting efficiently identifies the dominant ground plane by repeatedly sampling small sets of points, fitting planes to these samples, and counting inliers that lie close to the fitted plane. The plane with the most inliers is selected as the ground plane, and points near this plane are removed from further processing. This approach handles noisy measurements and partial occlusions but may fail in environments with significant terrain variations.

Point cloud segmentation groups points into semantically meaningful clusters representing individual objects or surfaces. Euclidean clustering groups spatially proximate points based on distance thresholds, assuming that points belonging to the same object are close together. More sophisticated approaches incorporate surface normals and colour information from aligned RGB cameras, enabling separation of objects that are spatially adjacent but differ in orientation or appearance.

Deep learning methods like PointNet and PointNet++ process point clouds directly, learning hierarchical feature representations for classification and segmentation. Unlike image-based networks that operate on regular grids, PointNet processes unordered point sets using symmetric functions that are invariant to point ordering. PointNet++ extends this with local feature learning that captures fine geometric structures. These architectures enable end-to-end learning on point cloud data without requiring conversion to intermediate representations like voxels or meshes.

## 3.3 Tactile and Haptic Sensing

While vision and depth-sensing enable robots to perceive the environment at a distance, tactile and haptic feedback provides critical information during physical contact with objects and surfaces. The development of tactile sensing capabilities for humanoid robots has drawn inspiration from human touch, which provides rich information about contact forces, object properties, and surface textures through mechanoreceptors distributed throughout the skin (Dahiya et al., 2010).

### 3.3.1 Transduction Principles and Sensor Technologies

Tactile sensors convert mechanical stimuli into electrical signals through various transduction principles. Resistive sensors, including force-sensing resistors (FSRs), change electrical resistance under applied pressure. They offer simple construction and low cost, but their nonlinear response, hysteresis, and calibration drift make them better suited to contact detection or coarse force estimation than to high-fidelity manipulation. Capacitive arrays improve sensitivity and spatial resolution, while optical tactile sensors such as GelSight-style devices use an internal camera and illumination system to image deformation of a soft elastomer surface, enabling high-resolution estimates of contact geometry, slip, and local force patterns (Agarwal et al., 2025; Yuan et al., 2017). This shift from scalar force sensing toward vision-based tactile sensing is especially important for dexterous hands,

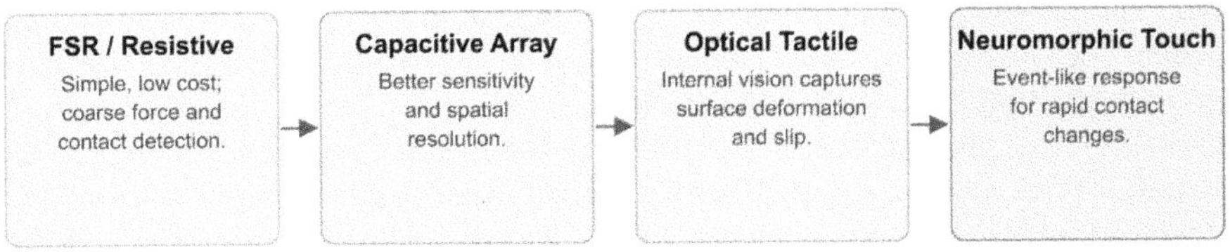


**Fig. 3.4** Tactile sensing for manipulation

where local surface geometry and incipient slip often matter more than normal force alone.

Figure 3.4 situates tactile sensing technologies along the path from coarse contact detection toward high-resolution contact geometry, slip detection, and rapid event-based touch.

Capacitive sensors detect changes in capacitance caused by surface deformation. When force is applied to a capacitive tactile sensor, the distance between conductive plates changes, altering the capacitance. The capacitance is proportional to the overlap area of the plates divided by the separation distance. Capacitive approaches provide higher sensitivity and faster response times than resistive sensors, making them preferred for high-resolution tactile arrays. However, capacitive sensors are susceptible to electromagnetic interference and require careful shielding in electrically noisy environments.

Piezoelectric sensors generate electrical charge in response to mechanical stress, providing direct force measurement without requiring external excitation. This self-generating property simplifies sensor design but limits low-frequency response: piezoelectric sensors respond to changing forces but cannot measure static loads. For manipulation tasks involving dynamic contact, piezoelectric sensors provide valuable information about contact events and force transients, but static grasp force estimation requires complementary sensing technologies.

Optical tactile sensors use cameras or photodetectors to observe deformation of a compliant sensing surface. The GelSight sensor, developed at MIT, uses a camera observing an illuminated elastomer to capture detailed surface geometry and texture. When an object contacts the elastomer, it deforms, and the camera captures this deformation at high resolution. Optical approaches can achieve very high spatial resolution limited only by camera resolution and optics, and can capture texture information invisible to conventional tactile sensors. However, the requirement for internal illumination and imaging increases sensor bulk and complexity.

Tesla's Optimus features dense tactile sensors integrated into its highly articulated fingers and palms (whose mechanical design is detailed in Chap. 2), providing feedback during object manipulation (Tesla, 2023). These tactile sensors enable the robot to detect contact with objects, estimate grasp forces, and adjust grip strength to prevent damage to delicate items. The integration of tactile sensing with vision-based grasping provides confirmation of successful contact and enables closed-loop force control that adapts to object properties during manipulation.

### 3.3.2 Distributed Tactile Arrays and Artificial Skin

Distributed tactile sensing arrays cover robot surfaces with multiple sensing elements, enabling spatial resolution of contact location and pressure distribution. These artificial skin technologies provide whole-body contact awareness essential for safe human–robot interaction and manipulation. The design of tactile arrays involves trade-offs between spatial resolution, sensitivity, durability, and cost that vary with application requirements.

High-density tactile arrays can resolve fine spatial features, enabling detection of small objects, edges, and surface textures. The spatial resolution required depends on the task: whole-body collision detection may only require resolution of several centimetres, while fingertip manipulation may require millimetre-scale resolution to provide useful feedback. The trade-off between resolution and channel count affects wiring complexity and data bandwidth, motivating multiplexing approaches that reduce the number of independent sensing channels.

Flexible printed circuit board technology enables fabrication of tactile arrays on conformal substrates that can cover curved robot surfaces. These arrays integrate sensing elements, interconnects, and sometimes local signal conditioning on flexible substrates that bend without damage. The mechanical properties of the substrate affect both durability and tactile perception: stiffer substrates provide better protection but may reduce sensitivity to small forces, while softer substrates conform better to surfaces but may be more susceptible to damage.

Temperature sensing integrated with tactile arrays provides additional information about contact properties. The thermal conductivity of contacted objects affects the rate of heat transfer between the sensor and the object, providing cues about material properties. Metal objects feel cold to the touch because they conduct heat away from the skin rapidly, while insulating materials like plastic feel closer to skin temperature. Human touch incorporates thermal sensing alongside mechanical sensing, and robots with similar capabilities can better estimate object properties during manipulation.

### 3.3.3 Neuromorphic Tactile Sensing

Neuromorphic tactile sensors inspired by human mechanoreceptors offer event-based output with microsecond temporal resolution. Unlike conventional tactile arrays that continuously sample at fixed rates, neuromorphic sensors asynchronously report contact events—spikes—when pressure changes exceed thresholds, mimicking the behaviour of biological mechanoreceptors (Osborn et al., 2018). This event-driven architecture reduces data bandwidth by orders of magnitude while enabling rapid response to dynamic contact (Liu et al., 2025).

The latency benefits of neuromorphic sensing are substantial. Conventional tactile sampling at 100 Hz implies a worst-case latency of 10 ms between contact and detection—unacceptable for rapid reflex responses. Neuromorphic sensors can detect contact onset within microseconds, enabling reflex responses that approach biological speeds. For manipulation tasks requiring rapid adjustment to unexpected contact events, this latency improvement can mean the difference between successful grasp and object damage.

Spiking neural networks (SNNs) process event streams from neuromorphic sensors in a biologically-inspired manner, where information is encoded in spike timing rather than activation magnitudes. The temporal precision of spike timing enables sophisticated temporal pattern recognition, and the sparse, event-driven nature of SNN computation offers potential energy efficiency advantages over conventional neural networks. Research has demonstrated that pairing neuromorphic tactile sensors with SNNs enables robust texture classification invariant to movement speed—a challenging problem for conventional approaches that require controlled contact velocity.

While still primarily in research settings, neuromorphic tactile sensors represent a promising path toward more natural and responsive manipulation in future humanoid platforms. The integration of neuromorphic sensing with conventional tactile arrays offers complementary capabilities: neuromorphic sensing provides rapid detection of contact events and dynamic changes, while conventional sensing provides absolute force estimates suitable for static grasp control.

### 3.3.4 Proprioception and Kinesthetic Feedback

Proprioception—the sense of body position and movement—relies on sensors distributed throughout the robot's kinematic structure. Joint encoders measure angular position with high precision, typically using optical or magnetic encoding methods. Combined with velocity estimates from encoder differentiation or tachometers, proprioceptive sensing enables accurate state estimation for control. The precision requirements for joint position sensing depend on the joint's role: leg joints for locomotion require encoder resolution of $0.1^\circ$ or better for precise foot placement, while arm joints for manipulation may require even finer resolution, particularly for distal joints where small angular errors translate to large endpoint deviations.

Joint torque sensing provides information about forces experienced by robot limbs, enabling compliant control and collision detection. As discussed in Chap. 2, Series Elastic Actuators (SEAs) provide built-in force estimates via their internal spring deflection. In the context of perception, this kinesthetic data must be fused with IMU and visual data at high frequencies to maintain a reliable state estimation, though the mechanical compliance introduced by SEAs must be accounted for in the overall control bandwidth.

Boston Dynamics' Atlas achieves whole-body state estimation at 1 kHz update rates by fusing proprioceptive sensing from joint encoders with inertial measurements and visual odometry. This high-frequency estimation enables the rapid control responses required for dynamic balance and agile locomotion. The integration of multiple proprioceptive modalities—position, velocity, and torque—provides redundancy that improves estimation accuracy and robustness to individual sensor failures.

## 3.4 Audio Processing and Speech Recognition

Audio perception enables humanoid robots to understand spoken commands, localise sound sources, and engage in natural dialogue with human users. For home and care settings, audio may sometimes be the only human input channel available in the moment, making emergency keywords and stop commands safety-relevant rather than merely conversational. Modern speech recognition systems based on deep learning have achieved strong benchmark performance, but safety–critical voice commands require redundant microphones, local keyword spotting that does not depend on cloud connectivity, noise-robust detection, cybersecurity protections against spoofing or malicious audio, and integration with safety-rated stop functions rather than replacement of physical emergency stops (ISO, 2025). The integration of audio perception with language understanding and robot control therefore represents both an HRI advance and a safety engineering challenge.

### 3.4.1 Microphone Arrays and Beamforming

Microphone arrays employ multiple spatially distributed microphones to enhance desired audio signals while suppressing noise and reverberation. Beamforming algorithms combine signals from array elements with appropriate delays and weights to synthesise directional sensitivity patterns (Benesty et al., 2008). The array can electronically "steer" its sensitivity toward a desired sound source while attenuating sounds from other directions, dramatically improving speech recognition accuracy in noisy environments.

The delay-and-sum beamformer represents the simplest beamforming approach, applying delays to each microphone signal to compensate for propagation delays from a target direction. When properly aligned, signals from the target direction add constructively while signals from other directions add incoherently, providing directional gain. More sophisticated adaptive beamformers like Minimum Variance Distortionless Response (MVDR) optimise the array weights to minimise output power while maintaining unity gain toward the target direction, effectively suppressing interfering sources.

Sound source localisation using Time Difference of Arrival (TDOA) methods estimates source location from inter-microphone timing differences. The Generalised Cross-Correlation with Phase Transform (GCC-PHAT) algorithm provides robust TDOA estimation in reverberant environments by emphasising phase information over magnitude. Cross-correlation of microphone pairs identifies the time lag that maximises correlation, corresponding to the propagation delay difference between microphones. Achieving localisation accuracy within 5–10° requires careful array geometry design and calibration, with larger arrays providing finer angular resolution.

The geometry of microphone arrays for humanoid robots involves trade-offs between localisation accuracy, array size, and aesthetic considerations. Linear arrays provide good directivity in the plane of the array but limited elevation resolution. Circular and spherical arrays offer more uniform spatial coverage at increased complexity. The physical placement of microphones on a humanoid head must also consider protection from damage, acoustic isolation from robot-internal noise sources, and aesthetic integration with the robot's appearance.

### 3.4.2 Speech Recognition in Noisy Environments

Humanoid robots must recognise speech in challenging acoustic environments with background noise, reverberation, and multiple simultaneous speakers. The performance of speech recognition systems degrades significantly in such conditions, as the acoustic models trained on clean speech fail to generalise to distorted inputs. Deep learning-based speech recognition systems have dramatically improved robustness through data augmentation, multi-condition training, and specialised architectures designed for noisy input.

End-to-end speech recognition systems like OpenAI's Whisper model (Radford et al., 2022) process raw audio waveforms directly, eliminating the need for handcrafted acoustic features. Whisper was trained on 680,000 h of multilingual web audio data, providing remarkable robustness to accents, background noise, and far-field recording conditions. The model's training data includes diverse audio quality, enabling it to handle challenging conditions that would defeat systems trained on cleaner data. For humanoid robots operating in real-world environments, this robustness is essential for reliable speech interaction.

Noise reduction and speech enhancement preprocessing can improve recognition accuracy in extremely noisy conditions. Classical approaches include spectral subtraction, which estimates noise characteristics during speech pauses and subtracts the noise estimate from the speech spectrum (Lee et al., 1990). Deep learning approaches learn mapping functions from noisy to clean speech, training on paired examples or through unsupervised methods that don't require clean reference signals. The effectiveness of enhancement preprocessing depends on the nature of the noise and the robustness of the downstream recognition system.

### 3.4.3 LLM Integration and Natural Language Understanding

Modern humanoid robots increasingly integrate large language models (LLMs) that provide sophisticated language understanding and generation capabilities. The architecture of LLM integration for robotics typically involves cascaded processing: automatic speech recognition converts audio to text, language models interpret intent and generate action specifications, and robot control policies translate specifications to motor commands. This modular approach allows independent improvement of each component but may introduce latency and error propagation between stages.

Figure AI initially partnered with OpenAI to integrate GPT-4-class language models into their humanoid robot control stack (Heater, 2025). However, in early

2025, Figure AI terminated this partnership after developing their own Vision-Language-Action model called Helix, which provides full-body control directly from pixels and language input without relying on external API access. This strategic shift reflects the importance of onboard, low-latency AI processing for practical robotics applications. Cloud-based LLM APIs introduce network latency that may be unacceptable for real-time robot interaction, and reliance on external services creates availability concerns for deployed systems.

The grounding of language in robotic perception remains a significant challenge. Language models trained on text data understand abstract concepts but lack direct connection to the physical world. Vision-language models such as Contrastive Language–Image Pre-training (CLIP) (Radford et al., 2021) address this by learning aligned visual–textual representations that enable robots to identify objects described in language without explicit programming; the underlying transformer-based mechanism that produces these aligned embeddings is examined in Sect. 3.6.3.

Speech synthesis for robot responses has similarly benefited from deep learning advances. Neural text-to-speech systems generate natural-sounding speech with appropriate prosody and emotional colouring. This enables humanoid robots to communicate more effectively with human users through natural spoken language, supporting collaborative tasks where verbal communication enhances coordination.

## 3.5 Inertial and Proprioceptive Sensing

Inertial and proprioceptive sensors provide robots with awareness of their own body configuration and motion, essential for balance, coordination, and precise movement control. These interoceptive sensors complement exteroceptive modalities like vision and tactile sensing, providing rapid feedback that enables stable locomotion and dexterous manipulation.

### 3.5.1 Inertial Measurement Units

Inertial Measurement Units (IMUs) combine accelerometers, gyroscopes, and often magnetometers to measure linear acceleration, angular velocity, and magnetic field orientation. MEMS-based IMUs integrate these sensors in compact packages suitable for mobile robotics, with costs decreasing and performance improving over time. The fundamental challenge in IMU-based state estimation arises from the need to integrate acceleration signals to obtain velocity and position—any errors in acceleration measurement accumulate over time, leading to unbounded drift in position estimates (Woodman, 2007).

Accelerometers measure specific force—the acceleration relative to free-fall. In the absence of other accelerations, a stationary accelerometer measures the upward acceleration required to counteract gravity, providing orientation information relative to the vertical direction. However, when the robot is moving, the measured acceleration includes both gravitational and kinematic components that cannot be directly separated without additional information.

Gyroscopes measure angular velocity about three orthogonal axes. Integration of gyroscope outputs yields orientation estimates, but accumulated errors from sensor noise and bias cause unbounded drift over time. The gyroscope bias—a non-zero output when stationary—represents a particularly troublesome error source as it integrates to orientation drift that grows linearly with time. Temperature variations affect bias, requiring continuous calibration or temperature compensation.

Magnetometers measure the local magnetic field vector, ideally dominated by Earth's magnetic field to provide absolute heading reference. However, magnetic disturbances from ferromagnetic materials and electrical currents often corrupt magnetometer measurements in indoor environments, limiting their utility for precision robotics. The calibration of magnetometers in environments with magnetic anomalies remains challenging, and many robotic systems rely primarily on gyroscopes and accelerometers for orientation estimation.

### 3.5.2 Orientation Estimation and Sensor Fusion

The fundamental challenge in IMU-based orientation estimation arises from the complementary error characteristics of accelerometers and gyroscopes. Accelerometers provide accurate low-frequency orientation information relative to gravity but suffer from noise and motion-induced errors during dynamic manoeuvres. When the robot accelerates, the accelerometer measures both gravitational and kinematic acceleration, and the tilt estimate derived from accelerometer data becomes corrupted. Gyroscopes offer precise high-frequency angular rate measurement but integrate sensor bias to unbounded orientation drift.

Complementary filtering exploits these complementary characteristics through frequency-domain separation. The filter combines gyroscope-integrated orientation for high-frequency content with accelerometer-derived tilt for low-frequency correction. The filter coefficient determines the crossover frequency below which accelerometer estimates are trusted and above which gyroscope estimates are trusted. This simple approach works well for moderate dynamics but requires careful tuning and may fail during extended periods of high acceleration.

The Extended Kalman Filter (EKF) provides a more principled framework for sensor fusion, recursively updating state estimates based on noisy measurements while explicitly modelling uncertainty through covariance matrices. For orientation estimation, the EKF fuses gyroscope, accelerometer, and magnetometer measurements while accounting for their respective noise characteristics. The filter maintains not only the current orientation estimate but also the uncertainty in that estimate, enabling optimal weighting of new measurements.

Unitree's H1 humanoid employs high-frequency IMU sampling at 1000 Hz combined with joint encoder measurements for robust balance control. The rapid update rate enables detection of incipient falls and generation of stabilising responses within the critical time window before unrecoverable imbalance develops. The integration of IMU data with proprioceptive joint measurements provides redundancy that maintains balance estimation accuracy even when individual sensors are temporarily corrupted by impacts or vibration. A key limitation is that many estimators assume an effectively rigid kinematic chain; under high torque, impacts, or lightweight structural design, limbs and joints can flex enough that encoder-derived posture and visual pose disagree. Multi-IMU proprioceptive estimation and deformation-aware filters can reduce this desynchronisation by tracking multiple bodies within the robot rather than treating the whole structure as perfectly rigid (Xavier et al., 2023).

## 3.6 Multimodal Sensor Fusion

Individual sensor modalities provide partial and often ambiguous information about the robot's state and environment. Multimodal sensor fusion combines complementary measurements to achieve more complete, accurate, and robust perception than any single modality can provide. The challenge of fusion lies in handling different data formats, update rates, coordinate systems, and reliability characteristics across modalities (Khaleghi et al., 2013).

### 3.6.1 Classical Fusion: Kalman Filtering

The Kalman filter provides the classical approach to optimal sensor fusion, recursively estimating system state from noisy measurements while explicitly modelling uncertainty (Kalman, 1960). The filter maintains a probability distribution over the state, represented by a mean estimate and covariance matrix. Prediction steps propagate the state forward in time using the system's dynamic model, while update steps incorporate new measurements to refine estimates.

For nonlinear systems typical in robotics, the Extended Kalman Filter (EKF) linearises dynamics around the current estimate, enabling application of Kalman filtering principles. The Unscented Kalman Filter (UKF) offers improved handling of nonlinearity by deterministically sampling the state distribution and propagating these samples through nonlinear transformations, capturing nonlinear effects that the EKF's linearisation ignores (Simon, 2006).

Particle filters represent the state distribution using a set of weighted samples, enabling representation of multi-modal and non-Gaussian distributions. Each particle represents a hypothesis about the system state, with particle weights reflecting the likelihood of that hypothesis given sensor measurements. Resampling steps eliminate low-weight particles and replicate high-weight ones, concentrating computational effort on probable state regions.

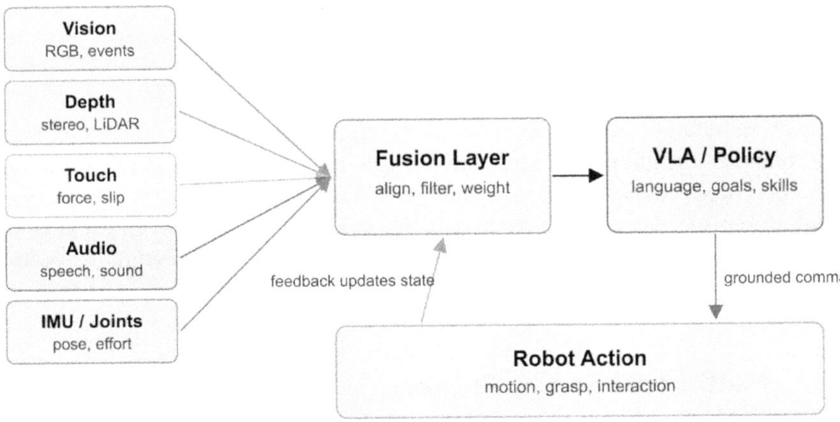


**Fig. 3.5** Multimodal fusion and VLA control

### 3.6.2 Deep Learning Fusion Architectures

Neural networks offer an alternative paradigm for sensor fusion, learning optimal combination of modalities from data rather than hand-crafted probabilistic models. Early fusion approaches concatenate raw sensor data before processing, allowing the network to learn correlations at the lowest level. Late fusion processes each modality independently before combining high-level features, offering modularity but potentially missing low-level correlations. Intermediate fusion provides a compromise, combining features at multiple hierarchical levels.

Attention mechanisms enable networks to dynamically weight different modalities based on context and reliability. In multimodal robotics perception, attention can down-weight corrupted or unreliable sensor streams while emphasising trustworthy modalities. Cross-modal attention explicitly models relationships between modalities, enabling features from one stream to guide processing in another. This learned attention provides flexibility that fixed fusion weights cannot achieve.

Figure 3.5 illustrates how multimodal sensor streams feed fusion layers and VLA policies, while robot actions create feedback that updates the perceptual state.

### 3.6.3 Transformer-Based Multimodal Integration

Transformer architectures have proven highly effective for multimodal perception in robotics (Vaswani et al., 2017). The self-attention mechanism models long-range dependencies within each modality, while cross-attention enables sophisticated integration across modalities. Given a query from one modality and keys and values from another, cross-attention computes weighted combinations that reveal cross-modal relationships.

This mechanism enables vision-language models like CLIP (Radford et al., 2021) to learn aligned representations where semantically similar visual and textual content maps to nearby points in a shared embedding space. For robotics, such alignments enable grounding of natural language commands in visual perception. When a human says "pick up the red block," the robot can identify the corresponding object through learned visual-linguistic associations without requiring explicit object models or hand-specified features.

Vision transformers (ViT) divide images into patches processed as sequences, enabling application of transformer architectures to visual perception. The self-attention mechanism captures relationships between image patches, building global understanding from local observations. Subsequent work demonstrated that self-supervised pre-training of vision transformers on unlabeled images yields rich visual representations useful for downstream robotics tasks, reducing the need for task-specific labeled training data.

### 3.6.4 Case Study: Vision-Language-Action Models

Vision-Language-Action (VLA) models represent a paradigm shift in robotic perception by jointly processing visual observations, natural language instructions, and robot actions within unified neural architectures. Unlike traditional perception pipelines that separate detection, segmentation, planning, and control into distinct modules, VLA models learn end-to-end mappings from sensory observations and language instructions to continuous actions.

OpenVLA (Kim et al., 2024), a 7-billion parameter open-source VLA model, was trained on 970,000 robot demonstrations from the Open X-Embodiment dataset (Padalkar et al., 2023), enabling cross-embodiment transfer where skills learned on one robot platform can be applied to others. The model processes camera images and language instructions through a transformer encoder, outputting predicted continuous actions. This unified architecture eliminates the need for explicit intermediate representations like object detections or scene graphs, instead learning task-relevant features directly from demonstration data.

Figure AI's Helix model, introduced in February 2025, represents the state-of-the-art in VLA architecture for humanoid robots. Helix unifies perception, language understanding, and learned control in a single neural network capable of controlling two robots simultaneously on shared, long-horizon manipulation tasks (Ackerman, 2025; Figure AI, 2025). The system can observe its surroundings, respond to natural language commands, and execute complex manipulation tasks like folding laundry—all without task-specific training. This capability demonstrates the power of foundation models pre-trained on diverse data to generalise to novel situations.

Earlier models like RT-1 (Brohan et al., 2022) laid the groundwork, and the RT-2 model from Google DeepMind (Brohan et al., 2023) demonstrated that vision-language models pre-trained on web-scale data can transfer semantic knowledge to robotic manipulation. The model can respond to abstract commands like "pick up the extinct animal" by leveraging semantic knowledge that dinosaurs are extinct to identify a toy dinosaur among other objects. This transfer of knowledge from internet-scale pre-training to robotic control represents a significant advancement, reducing the amount of robot-specific training data required for general-purpose manipulation. However, semantic recognition is not equivalent to physical affordance perception: identifying an object as a glass does not reveal its friction coefficient, fill level, centre of mass, fragility, or grasp stability. Practical VLA systems therefore still need embodiment-specific sensing, tactile feedback, force estimation, and interaction data to convert labels into physically grounded manipulation decisions.

### 3.6.5 Comparison of Fusion Strategies

Different humanoid platforms adopt varying sensor fusion philosophies reflecting distinct design priorities:

- Tesla Optimus employs vision-only fusion, leveraging massive neural networks trained on driving and humanoid data to extract depth and semantics from camera images alone. The approach prioritises cost efficiency and data scalability, using Tesla's fleet learning infrastructure to continuously improve perception from deployed robots. Limitations include reduced robustness in challenging lighting and textureless environments.
- Boston Dynamics Atlas uses multi-sensor fusion combining LiDAR, stereo vision, and IMU data through model-based fusion algorithms. The approach maximises robustness and precision across environmental conditions, accepting higher hardware costs and complexity. The redundancy of independent sensing modalities provides fault tolerance essential for industrial deployment.
- Figure 02 implements VLA-based fusion that integrates visual, linguistic, and proprioceptive information through learned attention mechanisms. The end-to-end learned architecture enables natural language interaction but requires significant computational resources for real-time inference.
- Unitree G1 adopts hybrid cost-optimised fusion, combining affordable LiDAR with depth cameras for navigation while relying primarily on vision for manipulation. This approach achieves practical capability at accessible price points, suitable for educational and research applications where moderate performance suffices.

## 3.7 Human–Robot Interaction Sensing

Safe and effective human–robot collaboration requires specialised perception capabilities beyond general environment sensing. Humanoid robots must detect human presence, interpret nonverbal communication, predict human intent, and maintain appropriate social behaviours. The design of human–robot interaction sensing systems draws from psychology, social science, and human factors research to create robots that interact naturally and safely with people (Goodrich & Schultz, 2008).

### 3.7.1 Gesture Recognition and Pose Estimation

Human pose estimation localises anatomical keypoints—joints, head, torso—in images or depth data, enabling robots to track human body configuration and motion (Zheng et al., 2023). Deep learning-based pose estimators like OpenPose (Cao et al., 2019) detect multiple people simultaneously in real-time, providing 2D keypoint locations for each person. The architecture uses part affinity fields that encode associations between body parts, enabling bottom-up parsing that groups detected keypoints into individual skeletons.

Three-dimensional pose estimation extends this to recover 3D body configuration, typically by combining multi-view images or by integrating 2D pose estimates with depth measurements. Parametric body models like SMPL represent human body shape and pose through learned statistical models, enabling recovery of complete 3D body meshes from images. These models can estimate detailed body shape, including limb lengths and body proportions, which may be useful for personalised interaction.

Gesture recognition interprets meaningful human motions—waves, pointing, hand signals—enabling intuitive robot control without speech. Hand pose estimation specifically tracks finger configurations, essential for understanding fine-grained gestures and sign language. MediaPipe Hands provides real-time hand tracking with 21 keypoints per hand, enabling precise gesture recognition on mobile and embedded platforms. The system tracks hand skeleton positions at video rates, supporting applications from sign language interpretation to virtual reality interaction.

### 3.7.2 Facial Expression and Emotion Recognition

Facial expression provides rich information about human emotional states and intentions. Automated facial expression recognition systems typically detect faces, localise facial landmarks, and classify expressions into discrete emotion categories or continuous affect dimensions. The Facial Action Coding System (FACS) decomposes facial expressions into component muscular activations called action units, providing fine-grained analysis of facial movements.

Deep learning systems trained on large datasets of labelled facial expressions can recognise basic emotions—happiness, sadness, anger, fear, surprise, disgust—with high accuracy. More sophisticated systems estimate continuous affect dimensions like valence (positive versus negative) and arousal (calm versus excited), providing nuanced understanding of emotional states beyond discrete categories.

Privacy concerns arise with emotion recognition systems, particularly regarding consent and potential misuse. Humanoid robots collecting facial data must implement appropriate privacy safeguards, including secure data handling, user consent mechanisms, and limitations on data retention. The ethical implications of emotion recognition technology continue to be debated, and responsible deployment requires careful consideration of potential harms as well as benefits.

### 3.7.3 Proxemics and Personal Space Detection

Proxemics, the Study of Spatial Relationships Between Individuals During Social Interaction, Was Pioneered by Anthropologist Edward T. Hall (1966). Hall Delineated Four Zones of Interpersonal Distance that Characterise Western Culture: Intimate Distance (0 m up to and including 0.45 m) Reserved for Close Relationships; Personal Distance (greater than 0.45 m up to and including 1.2 m) for Friends and Family; Social distance (greater than 1.2 m up to and including 3.6 m) for Casual Interactions and Colleagues; and Public Distance (greater than 3.6 m) for Formal Interactions and Presentations.

Robots must respect these social conventions to enable comfortable human–robot interaction. Implementation approaches include potential field methods that treat personal space as repulsive forces around people, creating virtual barriers that the robot's path planning avoids, finding paths that achieve navigation goals while maintaining appropriate distances from people (Sisbot et al., 2007). Optimisation approaches balance task requirements against social constraints.

Gaussian-based proxemic zone modelling provides smooth, differentiable representations amenable to gradient-based motion planning. The personal space around each person is modelled as a Gaussian distribution centred on the person, with variance that may depend on facing direction, motion velocity, or social context. The robot's navigation cost function includes penalties for entering high-cost regions of the proxemic field, encouraging socially appropriate paths.

The robot must also consider approach trajectory, as frontal approaches are generally more acceptable than approaches from behind that can startle humans. The facing direction of both the human and the robot affects appropriate approach distance, with face-to-face interactions permitting closer approach than situations where the human's back is turned (Walters et al., 2009). These contextual factors complicate the implementation of proxemic awareness but are essential for natural human–robot interaction.

### 3.7.4 Intent Prediction and Human Motion Forecasting

Predicting human intentions enables robots to anticipate and prepare for human actions, improving collaboration fluency (Mainprice et al., 2015). Social-LSTM, introduced by Alahi and colleagues, demonstrated that recurrent neural networks trained on human trajectory data can predict future paths considering social context (Alahi et al., 2016). The model encodes each person's motion history in an LSTM (Long-Short-Term Memory network) and shares hidden states between nearby people through a "social pooling" layer, enabling prediction of socially aware trajectories.

Gaze tracking provides powerful cues about human attention and intent. Humans typically fixate objects 0.5–1.0 s before reaching for them (Land & Tatler, 2009), allowing robots to predict grasp intentions from gaze patterns. Combined with pointing gesture recognition, this allows robots to resolve ambiguous references: when a human says "that one" while pointing, the robot determines the indicated object through visual attention analysis.

Theory of mind—the ability to reason about others' mental states including their beliefs, knowledge, and intentions—represents an advanced capability that sophisticated perception enables. By tracking what humans can see from their viewpoint, robots can infer what knowledge humans likely possess. This enables helpful behaviours like repositioning objects to make them visible to collaborators or providing information that fills gaps in others' knowledge.

Apptronik's Apollo robot emphasises worker safety in industrial environments by continuously monitoring human workers' positions and predicted trajectories, proactively modifying its path to maintain safe separation distances. The integration of human motion prediction with robot motion planning enables anticipatory collision avoidance rather than purely reactive safety responses (Kröger, 2010).

## 3.8 Emerging Research Directions

The field of humanoid robot perception is rapidly evolving, driven by advances in sensor technology, machine learning algorithms, and computing hardware. Several emerging technologies promise to significantly enhance perception capabilities beyond current state-of-the-art systems, addressing fundamental limitations in latency, power efficiency, and generalisation to novel environments. These advances span new sensor modalities that capture information previously inaccessible to robots, novel learning paradigms that reduce the need for task-specific training data, and architectural innovations that enable more efficient processing of perceptual information. While many of these technologies remain primarily in research settings, their potential impact on future humanoid platforms warrants careful examination of their principles, current capabilities, and remaining challenges for practical deployment.

### 3.8.1 Event Cameras for Ultra-Low Latency Perception

Event cameras represent a paradigm shift from conventional frame-based imaging to asynchronous event-based sensing. Each pixel operates independently, generating an event when brightness change exceeds a threshold. This event-driven operation provides microsecond temporal resolution, high dynamic range exceeding 120 dB compared to 60 dB for conventional cameras, and low data bandwidth as only changing pixels generate events (Gallego et al., 2020).

Applications in humanoid robotics include high-speed obstacle avoidance, slip detection during manipulation, and rapid visual stabilisation. Research prototypes have demonstrated event camera-based visual odometry with latencies under 1 ms, enabling tighter perception-action coupling than possible with conventional vision (Rebecq et al., 2017). The asynchronous nature of event data requires specialised processing algorithms, and integration with standard frame-based perception remains an active research area.

The Dynamic Vision Sensor (DVS) pioneered event camera technology for robotics applications. Modern sensors like the Prophesee evaluation kits offer improved resolution and sensitivity, making event cameras increasingly practical for real-world deployment. The combination of event cameras with standard cameras in hybrid systems provides both high-temporal-resolution event streams and rich texture information from conventional frames.

### 3.8.2 Foundation Models and Embodied AI

Foundation models pre-trained on massive datasets provide general-purpose perception capabilities that transfer to robotic tasks with minimal fine-tuning (Bommasani et al., 2021). These models learn rich representations from internet-scale data that capture semantic concepts, visual features, and language understanding useful for robotic perception and control.

A canonical example is the RT-2 model from Google DeepMind (Brohan et al., 2023), which demonstrated that vision-language models pre-trained on web-scale data can transfer semantic knowledge to robotic manipulation; this case is examined in detail in Sect. 3.6.4.

The Physical Intelligence $\pi 0$ (pi-zero) model, introduced in 2024, employs flow matching for action generation, achieving state-of-the-art performance on diverse manipulation tasks. The model learns a flow-based policy that can generate action trajectories conditioned on visual observations and language instructions, enabling zero-shot generalisation to novel tasks within the distribution of training data.

The Open X-Embodiment project aggregates robotics datasets from multiple institutions and robot platforms, enabling training of more general-purpose policies (Padalkar et al., 2023). The diversity of embodied experiences in this dataset—from industrial arms to mobile manipulators to humanoid robots—provides broader coverage of robotic skills than any single platform could accumulate. Models trained on this diverse data demonstrate improved generalisation across embodiments and tasks.

### 3.8.3 Self-Supervised Learning for Robotic Perception

Self-supervised learning trains models using supervisory signals derived from the data itself rather than manual labels. For robotics, this enables learning from vast amounts of unlabelled sensory data collected during robot operation. Common self-supervised tasks include predicting future frames from past observations, reconstructing masked image regions, and contrasting different augmentations of the same scene.

Contrastive learning methods like SimCLR and MoCo learn visual representations by maximising agreement between differently augmented views of the same image while minimising agreement with other images. These approaches have achieved impressive results on image classification benchmarks, often approaching supervised learning performance without requiring labelled data.

For robotics-specific self-supervision, temporal consistency provides powerful learning signals. Consecutive observations should correspond to smooth motion through the environment, enabling prediction of future states from current observations. Cross-modal prediction—predicting visual observations from proprioceptive state or vice versa—provides additional self-supervised objectives that learn the relationship between robot actions and sensory consequences.

## 3.9 Safety Considerations in Perception Systems

Perception failures can lead to dangerous robot behaviours, making robustness critical for practical deployment. The consequences of perception failure range from task failure (dropping a grasped object) to property damage (colliding with obstacles) to human injury (striking a nearby person). The acceptable failure rates and required safety measures depend on the application context and associated risks.

ISO 10218-1:2025 and ISO 10218-2:2025 specify safety requirements for industrial robots, including provisions for perception-based safety functions (ISO, 2025). These standards define requirements for safety-related control systems that monitor robot operation and initiate protective actions when hazardous conditions arise. Perception-based safety functions must achieve appropriate performance levels through measures including redundancy, diversity, and systematic failure analysis.

ISO 13849-1:2023 defines performance levels (PL a through e) for safety-related parts of control systems, with PL d typically required for hazardous robot applications (ISO, 2023). Achieving a given performance level requires satisfying requirements for hardware reliability (Mean Time to Dangerous Failure, as discussed in Chap. 2), diagnostic coverage (ability to detect failures), and resistance to common-cause failures. Perception-based safety functions present unique challenges for performance level verification, as the failure modes of learned perception systems are less well-characterised than those of traditional hardware components.

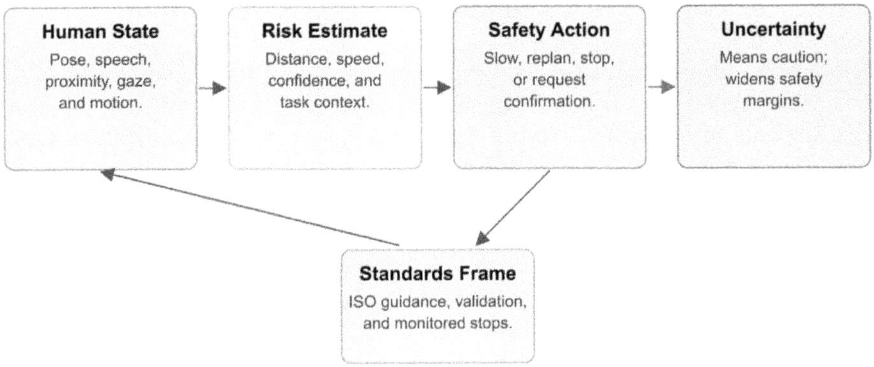


**Fig. 3.6** Perception for safe human–robot interaction

Multi-sensor redundancy provides fault tolerance, with the system remaining functional even if individual sensors fail. For perception systems, redundancy may involve multiple sensors of the same type (such as multiple cameras with overlapping fields of view) or diverse sensors that provide overlapping information through different physical principles (such as vision and LiDAR for obstacle detection). Diverse redundancy is more robust to common-cause failures that might affect all sensors of the same type.

Conservative uncertainty handling—rejecting or flagging low-confidence detections rather than acting on potentially erroneous information—is essential for safety–critical applications. Bayesian approaches to perception naturally provide uncertainty estimates, while neural networks require additional mechanisms such as Monte Carlo dropout or ensemble disagreement to estimate confidence. The integration of uncertainty-aware perception with risk-aware decision-making enables robots to take appropriate precautions when perception confidence is low.

Figure 3.6 summarises the safety loop connecting human-state perception, risk estimation, conservative action, and standards-based validation for shared human–robot spaces.

The integration of perception with safety-rated monitored stop functions enables robots to operate safely in shared workspaces with human workers while maintaining productivity. For service and personal-care robots, ISO 13482 provides a complementary safety baseline covering inherently safe design, protective measures, and information for use outside traditional industrial robot settings (ISO, 2014). When perception detects a human entering a potentially hazardous zone, the robot can reduce speed, modify its trajectory, or come to a complete stop before a collision becomes inevitable. These collaborative robot capabilities have enabled new applications where humans and robots work in close proximity, combining human flexibility and judgement with robotic precision and endurance.

## 3.10 Conclusion

Perception systems form the sensory foundation enabling humanoid robots to understand and interact with their environment. This chapter has examined diverse sensor modalities—vision, depth sensing, tactile feedback, audio processing, and proprioception—and explored how leading platforms integrate these modalities through sophisticated fusion architectures. Several key insights emerge from this analysis.

Architectural diversity persists across the humanoid robotics landscape. Tesla's vision-only approach prioritises cost and scalability, Boston Dynamics' multi-sensor systems maximise robustness, and Figure AI's language-grounded perception emphasises semantic understanding. These distinct philosophies reflect different application priorities and design constraints, with no single optimal architecture. The choice of perception architecture involves fundamental trade-offs between cost, capability, robustness, and operational envelope that different platforms resolve differently based on their target applications.

Foundation model integration is rapidly transforming perception capabilities, with Vision-Language-Action models enabling end-to-end learning from pixels to actions. Models like OpenVLA and Helix demonstrate that pre-trained systems can generalise across embodiments and tasks, reducing the engineering effort required for new applications. The transfer of knowledge from internet-scale pre-training to robotic control represents a significant advancement that may ultimately enable robots to benefit from the vast knowledge encoded in foundation models trained on text, images, and video from across the internet.

Real-time constraints demand careful optimisation, as dynamic balance requires sub-100 ms latencies. Edge AI computers and specialised accelerators now sit at the centre of humanoid perception design, where model inference, sensor ingest, memory bandwidth, power draw, thermal limits, and real-time scheduling must be engineered together. The tension between model complexity and inference speed continues to drive innovation in efficient architectures, model compression, and hardware acceleration tailored for robotic perception workloads.

Human–robot interaction capabilities—gesture recognition, emotion analysis, proxemics—enable more natural and safe collaboration. Social perception capabilities influence human trust and acceptance, connecting technical perception design to the social and psychological aspects of human–robot interaction. As humanoid robots move from controlled industrial environments into spaces shared with untrained humans, these interaction capabilities become increasingly important for safe and effective deployment.

Safety and reliability through multi-sensor redundancy remain paramount for practical deployment. ISO standards provide frameworks for ensuring perception-based safety functions meet required performance levels. The certification of perception systems for safety–critical applications remains challenging, particularly for learned systems whose failure modes are difficult to fully characterise, but ongoing work in verification and validation is establishing methods for assuring perception system safety.

Looking forward, perception capabilities will continue advancing through progress in sensors, algorithms, and computing hardware. Event cameras and neuromorphic processing promise ultra-low-latency perception with minimal power consumption. Foundation models trained on internet-scale data provide increasingly sophisticated semantic understanding. Active perception and learned attention enable efficient deployment of computational resources to task-relevant information. The perception systems described in this chapter enable the AI architectures and control systems examined in subsequent chapters, connecting sensing to understanding to action in the ongoing development of capable humanoid robots.

## Review Questions

1. Compare global shutter and rolling shutter camera sensors for humanoid robot applications. Explain why global shutter sensors are preferred for dynamic manoeuvres despite their lower light sensitivity, and describe the HDR challenges humanoid robots face when operating in real-world industrial environments with extreme brightness variations.
2. Contrast the operational principles of feature-based SLAM (exemplified by ORB-SLAM3) and direct SLAM (exemplified by LSD-SLAM). Under what environmental conditions would each approach excel or struggle, and why might a humanoid robot operating in an industrial facility with both textured and textureless surfaces benefit from hybrid approaches?
3. Analyse the architectural decision between vision-only perception (Tesla Optimus) and multi-sensor LiDAR-enhanced perception (Boston Dynamics Atlas). Beyond cost considerations, what specific operational scenarios and failure modes favour each approach, and how does sensor redundancy contribute to the ISO 13849 safety requirements for industrial robots?
4. Compare resistive and capacitive tactile sensing technologies in terms of sensitivity, response time, and calibration complexity. How do neuromorphic tactile sensors fundamentally differ from conventional tactile arrays in their data output, and what advantages does event-based sensing provide for rapid manipulation tasks?
5. Explain why a gyroscope alone cannot provide reliable orientation estimates over extended time periods. How does complementary filtering combine gyroscope and accelerometer measurements to leverage their complementary error characteristics, and under what motion conditions would this fusion approach become unreliable?
6. Describe how Vision-Language-Action (VLA) models like OpenVLA and Figure's Helix differ from traditional modular perception pipelines. What capabilities does the unified architecture enable that would be difficult to achieve with separate object detection, segmentation, and planning modules?
7. Define proxemics in the context of human–robot interaction and explain Hall's four interpersonal distance zones. How can Gaussian-based proxemic field modeling be integrated into robot navigation cost functions, and what contextual

factors beyond simple distance must be considered for socially appropriate robot approach behaviour?

8. Explain how event cameras fundamentally differ from conventional frame-based cameras in their data acquisition paradigm. What specific advantages do event cameras offer for humanoid robot applications requiring ultra-low-latency perception, and what challenges arise in processing asynchronous event streams compared to traditional frame-based computer vision algorithms?

## References


Ackerman, E. (2024). *Boston Dynamics research director on new Atlas robot video.* IEEE Spectrum. https://spectrum.ieee.org/boston-dynamics-new-atlas

Ackerman, E. (2025). *Video Friday: Helix from figure.* IEEE Spectrum. https://spectrum.ieee.org/video-friday-helix

Agarwal, A., Mirzaee, M. A., Sun, X., et al. (2025). A modularized design approach for GelSight family of vision-based tactile sensors. *The International Journal of Robotics Research, 44*(13). https://doi.org/10.1177/02783649251339680

Alahi, A., Goel, K., Ramanathan, V., Robicquet, A., Fei-Fei, L., & Savarese, S. (2016). Social LSTM: Human trajectory prediction in crowded spaces. In *Proceedings of the IEEE Conference on Computer Vision and Pattern Recognition* (pp. 961–971).

Benesty, J., Chen, J., & Huang, Y. (2008). *Microphone array signal processing.* Springer Science & Business Media.

Bohg, J., Hausman, K., Sankaran, B., Brock, O., Kragic, D., Schaal, S., & Sukhatme, G. S. (2017). Interactive perception: Leveraging action in perception. *Autonomous Robots, 41*(6), 1273–1292.

Bommasani, R., Hudson, D. A., Adeli, E., et al. (2021). *On the opportunities and risks of foundation models.* arXiv preprint arXiv:2108.07258.

Boston Dynamics. (2024). *An electric new era for Atlas.* https://bostondynamics.com/blog/electric-new-era-for-atlas

Brohan, A., Brown, N., Carbajal, J., et al. (2022). *RT-1: Robotics transformer for real-world control at scale.* arXiv preprint arXiv:2212.06817.

Brohan, A., Brown, N., Carbajal, J., et al. (2023). *RT-2: Vision-language-action models transfer web knowledge to robotic control.* arXiv preprint arXiv:2307.15818.

Cadena, C., Carlone, L., Carrillo, H., et al. (2016). Past, present, and future of simultaneous localization and mapping. *IEEE Transactions on Robotics, 32*(6), 1309–1332.

Campos, C., Elvira, R., Rodríguez, J. J. G., Montiel, J. M., & Tardós, J. D. (2021). ORB-SLAM3: An accurate open-source library for visual, visual-inertial, and multimap SLAM. *IEEE Transactions on Robotics, 37*(6), 1874–1890.

Cao, Z., Hidalgo, G., Simon, T., Wei, S. E., & Sheikh, Y. (2019). OpenPose: Realtime multi-person 2D pose estimation using part affinity fields. *IEEE Transactions on Pattern Analysis and Machine Intelligence, 43*(1), 172–186.

Chaumette, F., & Hutchinson, S. (2006). Visual servo control. I. Basic approaches. *IEEE Robotics & Automation Magazine, 13*(4), 82–90. https://doi.org/10.1109/MRA.2006.250573

Chen, T., Catrysse, P. B., El Gamal, A., & Wandell, B. A. (2000). How small should pixel size be? In *Proceedings of SPIE—The International Society for Optical Engineering*, Vol. 3965. https://doi.org/10.1117/12.385463

Dahiya, R. S., Metta, G., Valle, M., & Sandini, G. (2010). Tactile sensing—From humans to humanoids. *IEEE Transactions on Robotics, 26*(1), 1–20.

Dosovitskiy, A., Beyer, L., Kolesnikov, A., et al. (2021). An image is worth 16x16 words: Transformers for image recognition at scale. In *International Conference on Learning Representations*. https://iclr.cc/virtual/2021/oral/3458

Engel, J., Schöps, T., & Cremers, D. (2014). LSD-SLAM: Large-scale direct monocular SLAM. In *European Conference on Computer Vision* (pp. 834–849). Springer.

Figure AI. (2024). *Figure 02: General purpose humanoid robot*. https://figure.ai

Figure AI. (2025). *Helix: A vision-language-action model for generalist humanoid control*. https://www.figure.ai/news/helix

Gallego, G., Delbrück, T., Orchard, G., et al. (2020). Event-based vision: A survey. *IEEE Transactions on Pattern Analysis and Machine Intelligence, 44*(1), 154–180.

Goodrich, M. A., & Schultz, A. C. (2008). Human-robot interaction: A survey. *Foundations and Trends in Human-Computer Interaction, 1*(3), 203–275.

Hall, E. T. (1966). *The hidden dimension*. Doubleday.

Heater, B. (2025). *Figure will start alpha testing its humanoid robot in the home in 2025*. TechCrunch. https://techcrunch.com/2025/02/27/figure-will-start-alpha-testing-its-humanoid-robot-in-the-home-in-2025/

Horaud, R., Hansard, M., Evangelidis, G., & Ménier, C. (2016). An overview of depth cameras and range scanners based on time-of-flight technologies. *Machine Vision and Applications, 27*(7), 1005–1020.

ISO. (2014). *ISO 13482: Robots and robotic devices—Safety requirements for personal care robots*. International Organization for Standardization. https://www.iso.org/standard/53820.html

ISO. (2023). *ISO 13849-1:2023: Safety of machinery—Safety-related parts of control systems—Part 1: General principles for design*. International Organization for Standardization.

ISO. (2025). *ISO 10218-1:2025 and ISO 10218-2:2025: Robotics—Safety requirements*. International Organization for Standardization.

Kalman, R. E. (1960). A new approach to linear filtering and prediction problems. *Journal of Basic Engineering, 82*(1), 35–45.

Keetha, N., Karhade, J., Jatavallabhula, K. M., Yang, G., Scherer, S., Ramanan, D., & Luiten, J. (2024). SplaTAM: Splat, Track & Map 3D Gaussians for Dense RGB-D SLAM. In *Proceedings of the IEEE/CVF Conference on Computer Vision and Pattern Recognition*.

Kerbl, B., Kopanas, G., Leimkühler, T., & Drettakis, G. (2023). 3D Gaussian splatting for real-time radiance field rendering. *ACM Transactions on Graphics, 42*(4), 1–14.

Khaleghi, B., Khamis, A., Karray, F. O., & Razavi, S. N. (2013). Multisensor data fusion: A review of the state-of-the-art. *Information Fusion, 14*(1), 28–44.

Kim, S., Brohan, A., Ichter, B., et al. (2024). *OpenVLA: An open-source vision-language-action model*. arXiv preprint arXiv:2406.09246.

Kröger, T. (2010). *On-line trajectory generation in robotic systems: Basic concepts for instantaneous reactions to unforeseen (sensor) events*. Springer Science & Business Media.

Land, M. F., & Tatler, B. W. (2009). *Looking and acting: Vision and eye movements in natural behaviour*. Oxford University Press.

Lee, K. F., Hon, H. W., & Reddy, R. (1990). An overview of the SPHINX speech recognition system. *IEEE Transactions on Acoustics, Speech, and Signal Processing, 38*(1), 35–45.

Li, Y., & Ibanez-Guzman, J. (2020). Lidar for autonomous driving: The principles, challenges, and trends. *IEEE Signal Processing Magazine, 37*(4), 50–61.

Liu, T., Brayshaw, G., Li, A., Xu, X., & Ward-Cherrier, B. (2025). Neuromorphic touch for robotics—A review. *Neuromorphic Computing and Engineering, 5*(3), 032001. https://doi.org/10.1088/2634-4386/adf091

Mainprice, J., Hayne, R., & Berenson, D. (2015). Predicting human reaching motion in collaborative tasks using inverse optimal control. In *IEEE International Conference on Robotics and Automation* (pp. 885–892).

Matsuki, H., Murai, R., Kelly, P. H. J., & Davison, A. J. (2024). Gaussian splatting SLAM. In *Proceedings of the IEEE/CVF Conference on Computer Vision and Pattern Recognition* (pp. 18039–18048).

Mildenhall, B., Srinivasan, P. P., Tancik, M., et al. (2020). NeRF: Representing scenes as neural radiance fields. In *European Conference on Computer Vision* (pp. 405–421).
Mur-Artal, R., & Tardós, J. D. (2017). ORB-SLAM2: An open-source SLAM system for monocular, stereo, and RGB-D cameras. *IEEE Transactions on Robotics, 33*(5), 1255–1262.
NVIDIA. (2025). *NVIDIA Jetson Thor: Advanced AI for physical robotics.* https://www.nvidia.com/en-us/autonomous-machines/embedded-systems/jetson-thor/
Osborn, L. E., Dragomir, A., Betthauser, J. L., et al. (2018). Prosthesis with neuromorphic multilayered e-dermis perceives touch and pain. *Science Robotics, 3*(19), eaat3818.
Padalkar, A., Pooley, A., Jain, A., et al. (2023). *Open X-Embodiment: Robotic learning datasets and RT-X models.* arXiv preprint arXiv:2310.08864.
Radford, A., Kim, J. W., Hallacy, C., et al. (2021). Learning transferable visual models from natural language supervision. In *International Conference on Machine Learning* (pp. 8748–8763).
Radford, A., Kim, J. W., Xu, T., et al. (2022). *Robust speech recognition via large-scale weak supervision.* arXiv preprint arXiv:2212.04356.
Rebecq, H., Horstschaefer, T., & Scaramuzza, D. (2017). Real-time visual-inertial odometry for event cameras using keyframe-based nonlinear optimisation. In *British Machine Vision Conference.*
Scaramuzza, D., & Fraundorfer, F. (2011). Visual odometry: Part I—The first 30 years and fundamentals. *IEEE Robotics & Automation Magazine, 18*(4), 80–92.
Simon, D. (2006). *Optimal state estimation: Kalman, H infinity, and nonlinear approaches.* Wiley.
Sisbot, E. A., Marin-Urias, L. F., Alami, R., & Simeon, T. (2007). A human aware mobile robot motion planner. *IEEE Transactions on Robotics, 23*(5), 874–883.
Tesla. (2023). *Tesla Optimus Gen 2 update.* Official Company Release.
Unitree. (2024). *Unitree G1 humanoid robot specifications.* https://www.unitree.com
Vaswani, A., Shazeer, N., Parmar, N., et al. (2017). Attention is all you need. In *Advances in neural information processing systems* (pp. 5998–6008).
Walters, M. L., Dautenhahn, K., Te Boekhorst, R., et al. (2009). An empirical framework for human-robot proxemics. In *Proceedings of New Frontiers in Human-Robot Interaction* (pp. 144–149).
Woodman, O. J. (2007). *An introduction to inertial navigation.* (Computer Laboratory, Technical Report). University of Cambridge.
Xavier, F. E., Burger, G., Petriaux, M., Deschaud, J. E., & Goulette, F. (2023). Multi-IMU proprioceptive state estimator for humanoid robots. In *IEEE/RSJ International Conference on Intelligent Robots and Systems* (pp. 10880–10887). https://doi.org/10.1109/IROS55552.2023.10341849
Yuan, W., Dong, S., & Adelson, E. H. (2017). GelSight: High-resolution robot tactile sensors for estimating geometry and force. *Sensors, 17*(12), 2762. https://doi.org/10.3390/s17122762
Zheng, C., Wu, W., Chen, C., et al. (2023). Deep learning-based human pose estimation: A survey. *ACM Computing Surveys, 56*(1), 1–37.

**Ken Huang** is a globally recognized expert in AI and Web3 security, a prolific author, and a leading figure in shaping industry standards. He is the CEO and Chief AI Officer (CAIO) of DistributedApps.ai, specializing in Generative AI training and consulting. Ken is deeply involved in driving the development of secure AI systems. He serves as a Research Fellow and Co-Chair of the AI Safety Working Groups at the Cloud Security Alliance (CSA), leading Agentic AI initiatives. He is also a Co-Chair of the AI STR Working Group at the World Digital Technology Academy under the UN Framework, and a core contributor to the OWASP GenAI project, focusing on Agentic AI security. His expertise extends to his contributions to OWASP's Top 10 Risks for LLM Applications and his active participation in the NIST Generative AI Public Working Group. He is also a member of Open AI forum.

**Grace Huang** is a seasoned product and technology professional with extensive experience at organizations including PIMCO (a global investment management firm managing over $2 trillion in assets), IBM, SoftBank, and DataRobot. Throughout her career, she has successfully launched multiple products and managed large-scale projects, leveraging her expertise in product management, market analysis, strategic planning, and cross-functional team leadership.

Her work brings a finance and product strategy perspective to physical AI, particularly the operational, risk, and adoption considerations involved in bringing advanced robotics into enterprise environments. Always at the forefront of emerging technology, she has explored innovative tools including the implementation of Agentic AI powered solutions within the product management process—streamlining communication, improving decision-making, enhancing customer satisfaction, and driving business growth.

In addition to her professional accomplishments, Grace holds a degree from Babson College, where she built a strong foundation in business management and entrepreneurship. She continues to lead at the intersection of product strategy and advanced technology, bringing a rare combination of financial rigor and deep technical acumen to complex, high-stakes initiatives.

**Bo Tian** is a Sr Staff Hardware Engineer at Tesla, where he manages an electrical engineering team focused on the board-level architecture and design for the Optimus AI computer. Operating at the intersection of hardware and software, his work applies knowledge of transformer architecture tradeoffs to board architecture decisions, striving for optimal AI inferencing engines. Prior to his work on humanoid robotics and AI inferencing clusters at Tesla, he served as a Senior Hardware Engineer at Apple. He holds a Master of Science in Electrical and Electronics Engineering from Texas A&M University and a Bachelor of Engineering from The Hong Kong Polytechnic University.

**Manish Bhatt** Based in Silicon Valley, Manish Bhatt is a technical leader dedicated to making autonomous systems both safe and secure. He currently serves as the Application Security Tech Lead for AI & Crypto at Amazon Leo. With over 15 years in the field, Manish specializes in the critical space where complex hardware meets intelligent software.

He has been at the forefront of AI safety as a founding engineer of Purple Llama and the lead behind the CyberSecEval series, which helped set the standard for evaluating risks in frontier models. His research is widely recognized, covering everything from kernel rootkits to the safety of Large Language Models, and has earned over 20,000 citations.

As a founding member of the OWASP AI Security Verification Standard (AISVS), Manish is now focused on building the Artificial Intelligence Vulnerability Scoring System (AIVSS). His goal is to better understand risk in Agentic AI, ensuring that as AI technology moves into humanoid robotics, these systems remain resilient and trustworthy for everyone.

# Agentic AI Architectures and Models for Humanoid Robots

# 4

Grace Huang, Ken Huang, Tongzhou Mu, and Manish Bhatt

## 4.1 Introduction

The humanoid robot is universally regarded by researchers as the most demanding test case for artificial intelligence. For decades, roboticists have grappled with Moravec's paradox: the observation that high-level reasoning requires very little computation, while low-level sensorimotor skills demand substantial computational resources. Unlike a language model that operates within the bounded, discrete world of text, or a vision system that classifies static, two-dimensional images, a humanoid robot must continually perceive a chaotic, dynamic, three-dimensional physical environment. It must formulate intentions, decompose complex abstract goals into actionable steps, control dozens of independent degrees of freedom in real time, and execute all of this reliably in unstructured spaces designed specifically for human occupancy.

For the better part of fifty years, each of these major problems was addressed by separate engineering disciplines. Perception teams utilized classical computer vision or isolated convolutional neural networks; planning teams relied on search algorithms like A* or kinematic trajectory generators; and control teams utilized proportional-integral-derivative (PID) controllers or Model Predictive Control

G. Huang
PIMCO, Austin, TX, USA

K. Huang (✉)
DistributedApps.ai, Fairfax, VA, USA
e-mail: ken@distributedapps.ai

T. Mu
Rhoda AI, Mountain View, CA, USA

M. Bhatt
Amazon Kuiper Security/OWASP, Seattle, WA, USA

 
K. Huang et al. (eds.), *Humanoid Robots and Physical AI*, Advances in Data Analytics, AI, and Smart Systems, https://doi.org/10.1007/978-3-032-33669-9_4

(MPC). These distinct modules were then stitched together by hand-crafted, rule-based interfaces that worked only in controlled conditions. In these legacy architectures, a single unexpected object, a slightly unfamiliar instruction, or a minor change in ambient lighting could cause the entire pipeline to fail.

The past two years have seen that fragile, modular paradigm evolve rather than disappear. The convergence of large foundation models with advanced mechatronic hardware has produced a new class of system: the agentic humanoid robot. In this paradigm, complex behavior increasingly emerges from the combination of learned neural models, structured planners, skill libraries, and classical safety/control layers. These systems can receive a complex natural language instruction, such as "sort these fragile packages by size, stack the heavier ones on the bottom shelf, and leave the damaged ones on the table," decompose that instruction into logical sub-goals, reason about the objects and constraints in the scene, and execute the plan through learned policies and conventional controllers. The result is not the disappearance of traditional robotics code, but a shift in where that code sits: from hand-written task scripts toward reusable control, safety, orchestration, and verification infrastructure.

The scale of the industrial and commercial commitment to this vision is now unambiguous and global. Samsung officially announced in March 2026 its comprehensive plan to transform all global production facilities into "AI-driven factories" by 2030 (TechBuzz AI, 2026). This initiative aims to deploy agentic AI frameworks that independently plan and execute tasks to meet top-level corporate production goals, working alongside dexterous humanoid robots designed for complex, flexible operations previously reserved for human workers. Correspondingly, the International Federation of Robotics (2026) identified agentic AI as the top trend for 2026, describing it as a vital hybrid approach. This approach amalgamates analytical AI—for structured, rule-compliant decision-making—and generative AI, fostering greater adaptability and making robots capable of working independently in unpredictable, complex environments. Furthermore, Deloitte's (2025) extensive analysis of the decade ahead concludes that the intersection of digital agentic AI systems with physical AI robotic systems will result in robots whose central "brains" are agentic. These machines are increasingly able to adapt to novel environments, plan deep multi-step tasks, recover autonomously from physical failures, and operate safely under severe uncertainty.

This chapter maps the technical architecture that makes these capabilities possible. Section 4.2 examines the philosophical, mathematical, cognitive, and failure-handling design principles behind modern agentic robot systems. Section 4.3 introduces foundation models as the broader architectural family and then narrows to vision-language-action (VLA) models as one important subset for robot control. Section 4.4 analyzes leading industrial humanoid foundation-model systems, including GR00T N1, Gemini Robotics, and Helix. Section 4.5 addresses hierarchical task planning, classical constraints, and the role of large language models in long-horizon goal decomposition. Section 4.6 investigates multi-agent coordination frameworks for heterogeneous, multi-robot fleets. Section 4.7 treats simulation-to-reality transfer as an enabling methodology for training, testing, and validating agentic architectures rather than as a property unique to agents. Section 4.8 defines world models as predictive transition models and distinguishes them from memory systems and other spatial representations. Finally, Sect. 4.9 addresses the geopolitical dimension of agentic humanoid robotics, examining the widening open-source ecosystem divergence between China and the United States that is reshaping the competitive landscape for these architectures.

## 4.2 Design Principles of Agentic Robot Architectures

Agentic architectures represent a seismic shift from passive, reactive machinery to active, goal-oriented synthetic entities. To understand how humanoid robots achieved this level of autonomy, we must examine the core philosophical and cognitive design principles that structure their underlying software. This section dissects the transition from traditional pipelines to autonomous agents, the incorporation of dual-system cognitive frameworks inspired by human psychology, the necessity of modular skill composition, and the failure-mode analysis needed before these systems can be trusted in physical environments.

### 4.2.1 From Pipelines to Agents

The traditional robotic control stack was built on a rigid sense-plan-act pipeline. A perception module (often utilizing LiDAR or basic RGB cameras) would identify objects, generate point clouds, and estimate 6D poses; a deterministic motion planner would compute an obstacle-free kinematic sequence; and a low-level controller would generate the electrical currents to actuate motors and execute the movement. Each module was optimized separately, meticulously hand-tuned for a specific, static domain, and connected to the subsequent module through well-defined data structures. This architecture produced reliable performance in well-defined settings—such as automotive chassis welding lines—but it lacked the capacity to generalize. Any object falling slightly outside the perception model's training distribution, any user instruction not anticipated by the logic tree, or any environmental perturbation the controller had not been mathematically tuned to reject, would produce a total system failure.

The limitation was not an absence of planning or abstract reasoning. Classical stacks often included planning modules capable of symbolic search, trajectory optimization, model predictive control, and other forms of abstract reasoning. Their weakness was narrower: those reasoning processes were usually hand-specified, separated from perception and language understanding, and constrained by explicit state representations and brittle interfaces. They could reason effectively within a modeled domain, but they struggled to generalize when the world, task semantics, or sensor inputs diverged from the assumptions encoded by engineers.

Agentic architectures are better understood as systems that integrate perception, semantic interpretation, planning, action selection, execution, and feedback under persistent goal-directed control. Some implementations use reward functions or model-based reinforcement learning, but agentic robotics is broader than that single formalism. An agent may combine a foundation model, symbolic planner, learned policy, memory system, safety monitor, and classical controller. Its defining feature is not merely that it maintains a latent model or maximizes reward, but that it can interpret goals, select tools or skills, monitor progress, and replan when the task or environment changes. Closed-loop feedback is therefore necessary but not unique to agents; traditional closed-loop pipelines also monitor outcomes. What distinguishes modern agentic architectures is the integration of closed-loop control with semantic reasoning, flexible task decomposition, memory, and cross-task generalization.

The enabling technology for this deep, multi-layered integration has been the large foundation model, pre-trained on internet-scale multimodal data and subsequently fine-tuned via reinforcement learning for specific robotic control tasks. These models inherently carry vast semantic knowledge of the human world—they understand what a fragile wine glass looks like, how delicate raw eggs are, and what the abstract concept of "sort by colour" implies—and can spontaneously bring that knowledge to bear on novel physical tasks they have not explicitly practiced. Rather than relying exclusively on monolithic end-to-end models, the leading agentic frameworks treat robotic intelligence as something that emerges from deep reasoning, logical composition, and episodic experience.

### 4.2.2 The Dual-System Cognitive Architecture

The most influential and widely adopted design pattern to emerge in modern humanoid robot architectures is what practitioners colloquially refer to as the "dual-system" or "System 1/System 2" architecture—an explicit and deliberate borrowing from human cognitive science. In Nobel laureate Daniel Kahneman's famous framing, human thought operates in two distinct modes: fast, automatic, intuitive responses (System 1) and slow, deliberate, effortful reasoning (System 2). Modern humanoid robots face an exact, analogous computational challenge. Low-level motor control—dynamically balancing a 70 kg bipedal chassis on one foot, applying the correct friction to grip a smooth sphere, or instantly recovering from a sudden lateral push—requires continuous responses at millisecond timescales (often 500–1000 Hz) that cannot wait for the multi-second inference latency of a parameter language model. Conversely, high-level task planning—such as deducing how to prepare a complex meal given an incomplete and randomized set of ingredients in an unfamiliar kitchen—requires the kind of flexible, context-sensitive semantic reasoning that large foundation models excel at, but at a temporal frequency of perhaps 1 Hz or less.

The dual-system architecture elegantly separates these two computational regimes. System 2 is typically a vision-language model (VLM) possessing tens or hundreds of billions of parameters. It processes the current high-resolution visual scene, interprets complex natural language instructions, reasons about the multi-step task, and produces a structured plan, a set of intermediate physical goals, or a dense latent embedding. System 1 is a smaller, faster visuomotor policy—frequently utilizing a diffusion transformer or a flow-matching neural architecture. This fast system takes the System 2 latent representation as its conditioning input and rapidly generates continuous, fine-grained motor commands at the high frequencies required for stable physical control.

This architecture has become an influential design pattern because it maps well onto the computational hardware constraints of physical robotic deployment. Large vision-language models cannot reliably run at high control frequencies on a robot's onboard embedded computer systems, while smaller parameter-efficient diffusion or flow-matching policies can operate closer to the timing requirements of physical execution. Conversely, small policies lack the world knowledge to generalize to novel semantic instructions or visually novel environments; large foundation models possess this in abundance. The temporal boundary itself introduces a control problem: a slow System 2 update can create discontinuous target changes for a 500–1000 Hz controller, producing jerk, vibration, or unnecessary safety stops if passed directly to the motor layer. Practical dual-system stacks therefore insert target smoothing, interpolation, trajectory retiming, rate limiters, and safety filters between reasoning outputs and high-rate control loops so semantic goals become dynamically feasible setpoints rather than abrupt step changes. By decomposing robotic cognition along this temporal and computational boundary, dual-system architectures can combine semantic generalization with reliable physical execution. Some recent agentic frameworks (see Sect. 4.10) extend this two-layer model with a third "System 3" orchestration layer for multi-agent coordination, fleet-wide memory, and external API access; this third layer sits above the System 2 reasoner and does not replace the dual-system division of labor.

Figure 4.1 summarizes the dual-system cognitive architecture that has become the dominant design pattern for modern humanoid robots, separating slow VLM-based reasoning from a fast diffusion-based action policy with a safety-filtering buffer between them (Table 4.1).

### 4.2.3 Modularity and Skill Composition

A second key design principle that defines agentic robot architectures is skill compositionality. Rather than attempting the computationally intractable feat of training a single, monolithic end-to-end model to handle every possible task in every possible environment, these advanced systems construct a vast, modular library of "primitive skills." These are specialized, rigorously trained local policies for actions such as picking, placing, navigating, dynamic grasping, pushing, and visual inspection. The overarching agentic reasoning layer (System 2) is then tasked with intelligently sequencing and composing these discrete skills into novel, complex behaviors based on the user's prompt.

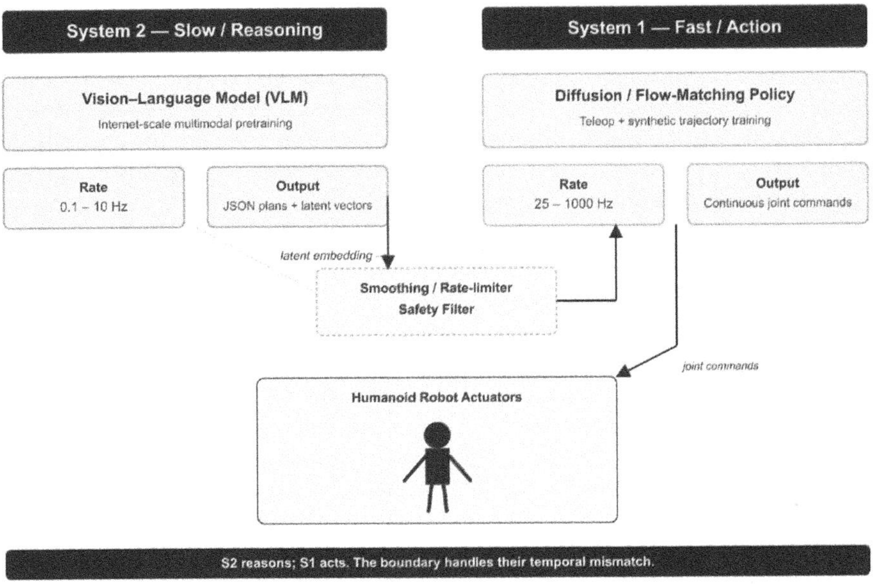


**Fig. 4.1** Dual-system cognitive architecture (System 1/System 2)

**Table 4.1** Characteristics of System 1 and System 2 components in humanoid robot architectures

| Characteristic | System 1 (action model) | System 2 (reasoning model) |
|---|---|---|
| Primary role | Low-level continuous motor execution | High-level semantic reasoning and planning |
| Typical model type | Diffusion transformer/ flow-matching policy | Vision-language model (VLM)/ Large Language Model (LLM) |
| Inference frequency | 25–1000 Hz | 0.1–10 Hz |
| Latency requirement | <20 ms | 100 ms–5 s |
| Input | Proprioceptive state, Joint positions, VLM latent vectors | High-res Images, Audio, Natural language, Scene context |
| Output | Continuous joint positions, velocities, or torques | JSON plans, intermediate coordinates, VLM latents |
| Training data | Teleoperated robot demonstrations, synthetic sim trajectories | Internet-scale multimodal data + vast robotic datasets |
| Generalization | Task-specific, Embodiment-specific (kinematics) | Cross-task, Cross-embodiment (semantic) |

*Source* Author's analysis based on Bjorck et al. (2025), Kawaharazuka et al. (2025), and Li et al. (2026)

This agentic framework can apply across different morphologies, from single-arm manipulators and quadrupedal robots to bipedal humanoids, wheeled warehouse robots, and autonomous drones. When a new capability is required, a specialized skill can be trained, verified, and added to the robot's repertoire without retraining the entire foundation-model stack from scratch. The agent can then compose that skill with existing perception, planning, memory, and control modules. This distinction matters: monolithic end-to-end policies can be powerful, but they are not always the easiest place to add a narrow edge-case capability. A skill-composition system can expose a newly learned capability as a callable tool or policy, making it available to the reasoning layer while preserving explicit interfaces for testing and safety review.

### 4.2.4 Failure Modes and Edge-Case Handling

Agentic humanoid architectures should be evaluated not only by their best-case demonstrations but by how they behave under ambiguous, adversarial, and physically marginal conditions. A useful failure-mode taxonomy spans at least five layers. First, perception failures occur when lighting, occlusion, reflective surfaces, motion blur, sensor dropout, or out-of-distribution objects corrupt the robot's world model. Second, semantic reasoning failures occur when an LLM or VLM produces a plausible but unsafe interpretation, such as treating a hot pan, a fragile glass, or a nearby child as irrelevant context. Recent embodied-planning evaluations show that high task feasibility does not automatically imply safety awareness: models may generate executable plans that still introduce physical or normative danger (Zhang et al., 2026). Third, planning failures occur when a linguistically valid plan is kinematically unreachable, collision-prone, under-specified, or impossible under current payload, workspace, battery, or tool constraints. Fourth, execution failures occur when a correct plan is degraded by actuator lag, backlash, slip, contact instability, network latency, or state-estimation drift. Fifth, orchestration failures occur when multiple agents, skills, or robots contend for the same physical resource without deterministic arbitration.

These edge cases imply a practical design rule: agentic capability must be surrounded by explicit monitors and fallback paths. Before execution, plans should pass through feasibility checks, safety constraints, resource locks, and task-and-motion validation. During execution, controllers should monitor contact forces, proprioceptive errors, tool state, human proximity, and progress against expected milestones. After a failure, the system should classify the failure and return structured feedback to the reasoning layer rather than simply retrying the same action. Safe fallback behaviors include stopping, backing away, releasing force, requesting human confirmation, switching to a verified primitive, or handing control to a deterministic supervisor. This is why current commercial architectures remain hybrid systems: learned models provide generalization, but edge-case handling still depends on explicit constraints, runtime monitors, and conservative recovery policies.

## 4.3 Foundation Models and Vision-Language-Action Architectures

Foundation models provide the broad cognitive substrate for modern humanoid robots. In robotics, this family includes large language models for instruction understanding and task reasoning, vision-language models for scene interpretation, world models for prediction, and VLA models that map perception and language into action-conditioned representations or motor policies. VLAs are therefore best understood as one important subset of the larger foundation-model stack, rather than the singular cognitive engine for every agentic robot. This section introduces the general foundation-model role first and then narrows to VLA architectures, which are especially important when perception, language, and control must be learned jointly.

### 4.3.1 Architecture and Principles

Vision-language-action (VLA) models are the technical core of modern agentic humanoid robots. They function as multimodal foundation models that unify three previously disparate capabilities into a single, end-to-end, differentiable trainable architecture: advanced visual perception, deep natural language understanding, and continuous motor action generation. Given an image (or a streaming video context window) of the robot's immediate environment combined with a text instruction, a VLA model directly produces a coordinated sequence of robot actions—such as specific joint positions in radians, end-effector spatial velocities, and gripper closure commands—that, when physically executed, accomplish the requested task.

The deep architecture of a state-of-the-art VLA model typically consists of three intertwined functional components. First, a robust visual encoder—typically a Vision Transformer (ViT) or a Contrastive Language-Image Pretraining (CLIP)-derived architecture—processes raw high-resolution image inputs from the robot's head and wrist cameras. This encoder extracts physical features and produces high-dimensional dense embeddings that simultaneously capture both semantic content (understanding what objects are present, such as recognizing a "Phillips head screwdriver") and spatial, geometric structure (precisely where the screwdriver lies in 3D Euclidean space). Second, a language encoder and reasoning backbone—essentially a pre-trained large language model (LLM)—processes the tokenized text instruction and contextualizes it against the incoming visual embedding. This cross-attention mechanism produces a rich latent representation of the holistic task. Third, an advanced action decoder translates this latent intent into continuous, physical control signals. Common and successful decoder architectures include diffusion policies, which model action distributions as iterative denoising processes over a multidimensional trajectory space; flow-matching models, which learn optimal, direct vector mappings from Gaussian noise to deterministic action; and autoregressive decoders that treat robot movements similarly to language, generating discrete "action tokens" sequentially.

This tripartite architecture—visual encoding, linguistic semantic reasoning, and physical action generation—unifies perception and execution in an end-to-end framework for embodied intelligence (Exxact Corporation, 2026; Singh, 2026). Autoregressive action tokenization remains useful for aligning robot actions with language-model training objectives, but discretizing continuous joint commands into finite bins can introduce quantization or "snapping" that is unacceptable for sub-millimeter tasks such as threading, precision insertion, or delicate contact-rich assembly. This limitation has accelerated interest in continuous action decoders, including diffusion and flow-matching policies that generate smooth trajectories directly in continuous action space rather than forcing every movement through discrete action tokens (Brohan et al., 2022; Gao et al., 2025). The training paradigm for VLA models is equally revolutionary. To bypass the scarcity of physical robot data, these models are typically initialized from pre-trained VLMs to leverage internet-scale semantic knowledge (learned from billions of human images and texts). Only then are they fine-tuned via behavioral cloning or reinforcement learning on curated datasets of robot trajectories paired with rich language annotations. The major insight driving this field is that a model that already understands language and vision at an abstract level can transfer that world knowledge to physical robotic tasks. By unifying vision, language, and physical action data at scale—domains which have traditionally been studied in isolation—VLA models aim to learn robust, general-purpose policies that generalize broadly across diverse tasks, objects, robotic embodiments, and unstructured environments.

Figure 4.2 illustrates the canonical vision-language-action (VLA) model architecture: visual and linguistic inputs are encoded and fused through a transformer backbone, then decoded into continuous low-level robot commands.

### 4.3.2 From Modular Pipelines to End-to-End Control

The true historical and technical significance of the VLA paradigm is best understood when sharply contrasted with the classical systems it replaces. As previously established, traditional robotic manipulation pipelines utilized separated, hand-coded modules. An object detector (like YOLO) would identify what was present; a rigid pose estimator would calculate its orientation matrix; a specialized grasp planner would compute inverse kinematics to approach it; and a trajectory planner would generate smooth polynomial splines for the arm motion. Each module was trained on its own hyper-specific dataset, rigorously evaluated on its own isolated metrics, and connected through brittle, hand-designed interfaces. If a robot were asked to "pick up the blue cup and place it in the sink," it would require every single module to execute in sequence; a minor calibration error in the camera leading to a 2-cm failure in pose estimation would cause the entire physical grasp to fail.

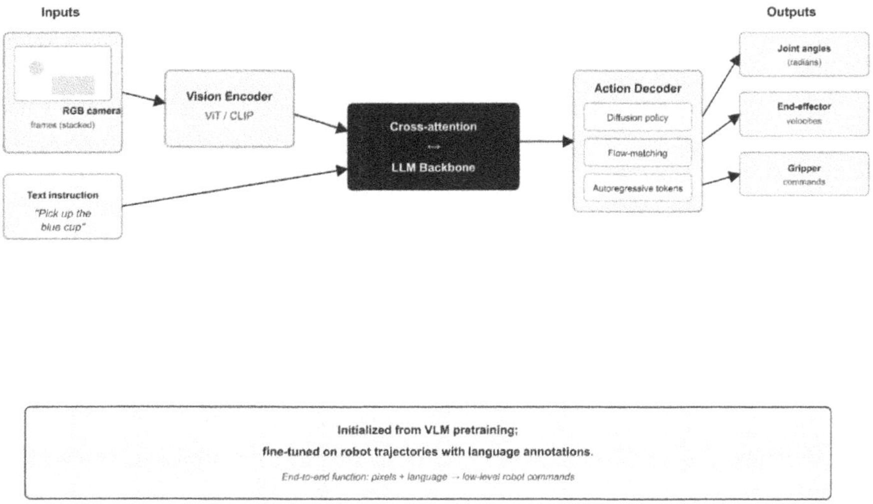


**Fig. 4.2** Vision-language-action model architecture

VLA models collapse this entire precarious pipeline into a single, learned mathematical function. The robot simply observes the scene through its RGB cameras, receives the high-level instruction in plain text, and the VLA directly outputs the low-level motor commands required to accomplish the task. There are no explicit bounded-box object detectors, no mathematically rigid inverse kinematic planners, and no hand-coded data interfaces between them. Everything, from visual feature extraction to motor torque application, is learned jointly from raw data.

This approach significantly differs from traditional robotics precisely because older systems split perception, planning, and control into discrete modules connected with hand-built rules, which almost always fail in messy, flexible human environments. VLAs build upon vision-language models by treating physical action as merely another output modality, effectively learning shared representations that support flexible, context-aware task execution. The practical benefits are substantial. End-to-end models can exploit subtle visual cues—like the specific shadow a cup casts to determine its depth—that explicit, rigid detectors might simply ignore. They generalize to novel object appearances, bizarre lighting conditions, and novel spatial configurations because they intrinsically learned from the full continuous distribution of robot-interaction data, not just the narrow, human-defined distributions that individual classical module designers artificially anticipated. Furthermore, they can handle vague instructions that contain heavy implicit cultural assumptions—such as "bring me something to drink," which requires deep reasoning about what objects physically constitute a safe, drinkable item—that rigid, rule-based systems cannot logically accommodate.

### 4.3.3 Key Design Decisions and Current State of Research

Cutting-edge research presented at leading conferences like ICLR 2026 illustrates how rapidly the VLA field is evolving. A 2026 survey of over 164 VLA model submissions to ICLR identifies converging trends toward discrete diffusion VLAs, larger reasoning backbones, and new physical benchmarks (Reuss, 2026). This research reveals the explosive, exponential growth in the field and highlights a rapidly widening resource gap between funded frontier industrial research labs and traditional academic university work.

Three specific, technical design decisions have proved especially consequential in modern VLA construction. First, the strategic choice of the neural backbone: larger VLM backbones consistently improve semantic generalization and zero-shot reasoning, but they impose inference latency that may be incompatible with safe, real-time physical control. Li et al. (2026), in Nature Machine Intelligence, rigorously examined over 8 VLM backbones, 4 distinct policy architectures, and over 600 distinct experimental trials, conclusively finding that backbone selection, precise VLA architecture formulation, and the use of cross-embodiment training data are the definitive key design choices that most significantly influence real-world performance.

Second, the mathematical representation of the action space itself is critical. Continuous action spaces require complex regression algorithms or heavy generative models, while tokenized, discrete actions enable fast autoregressive decoding (similar to ChatGPT outputting text) but may lose sub-millimeter precision in tasks requiring fine manipulation (like threading a needle). The dominant, most successful current approaches—specifically diffusion policies and continuous flow-matching models—treat continuous action generation as an iterative generative process. They produce smooth action trajectories by iterative refinement from a noise prior, which naturally captures the complex, multimodal distributions of multiple feasible physical behaviors for a given state (Kawaharazuka et al., 2025).

Third, the insatiable requirement for data: VLA model performance scales with both the quantity and the variance of physical robot demonstration data. VLA models, when trained on diverse multimodal datasets, can transfer knowledge across different physical tasks and environments, bringing the industry significantly closer to the holy grail of general-purpose robotic assistants. However, the immense cost, difficulty of collection, and overall quality of physical training data remain the primary bottlenecks for the entire global industry (Table 4.2).

## 4.4 Leading Humanoid Foundation Models and Industrial Systems

The broad foundation-model landscape has now been commercialized into several influential humanoid robotics systems. NVIDIA's GR00T N1, Google DeepMind's Gemini Robotics family, and Figure AI's Helix illustrate different ways to combine

**Table 4.2** Comparison of major VLA model architectures and their key characteristics

| Model | Developer | Year | Architecture | Key innovation | Embodiment focus |
|---|---|---|---|---|---|
| GR00T N1 | NVIDIA | 2025 | Dual-system (VLM + Diffusion Transformer) | Open, cross-embodiment; synthetic data flywheel | Humanoids (GR-1, NEO, Optimus) |
| Gemini Robotics | Google DeepMind | 2025 | VLA natively on Gemini 2.0 backbone | Action tokenized as a new core output modality; dexterity | Multi-embodiment (incl. Apollo) |
| Gemini Robotics 1.5 | Google DeepMind | 2025 | Dual-model (Embodied Reasoning + VLA) | Zero-shot cross-embodiment transfer; complex tool use | Multi-embodiment (incl. Apollo) |
| Helix | Figure AI | 2025 | Dual-system (Custom VLM + high-freq visuomotor) | High-frequency continuous whole-body humanoid control | Figure 01/ Figure 02 |
| $\pi$0 (pi-zero) | Physical Intelligence | 2024 | Flow-matching VLA | Cross-embodiment; sub-millimeter high-freq control | Multi-embodiment |
| SmolVLA | Hugging Face | 2025 | Compact VLA (450M parameters) | Fully open-source; democratized researcher access | General robotic arms |
| OpenVLA | Stanford/ Open X | 2024 | Open-source VLA (7B parameters) | Supports efficient LoRA fine-tuning for edge | General robotic arms |

*Source* Author's analysis based on Kawaharazuka et al. (2025), Li et al. (2026), and Exxact Corporation (2026)

general multimodal reasoning, VLA-style action models, task planners, and low-level control policies. The comparison below should therefore be read as a survey of leading foundation-model systems, not as evidence that one VLA architecture alone defines the field.

### 4.4.1 NVIDIA Isaac GR00T N1

NVIDIA's Isaac GR00T N1, released at GTC 2025, represents the global robotics field's first open, customizable foundation model designed and specifically for full-sized humanoid robots. Its underlying architecture implements the dual-system cognitive design described in Sect. 4.2.2. In GR00T's architecture, System 1 serves as a fast action model, while System 2 acts as a slower, high-parameter model for deliberative reasoning. Powered by a vision-language model, System 2 reasons about the unstructured environment and the instructions received to plan macro-actions. System 1, utilizing diffusion transformer technology, then translates these high-level macro-plans into precise, continuous, high-frequency robot joint movements (NVIDIA, 2025b).

The training pipeline designed for GR00T N1 is particularly notable for its reliance on heterogeneous data sources (Bjorck et al., 2025). GR00T N1 is trained with a complex mixture of millions of real-robot physical trajectories, hundreds of millions of human interaction videos, and billions of synthetically generated, physically accurate datasets. The integrated vision-language module interprets the chaotic environment through streaming vision and dynamic language instructions, while the specialized diffusion transformer module generates motor actions in real time. Both modules are coupled and trained end-to-end on NVIDIA H100 and B200 GPU clusters.

A major scientific contribution of GR00T N1 is its empirical demonstration that high-fidelity synthetic data can substitute for prohibitively expensive real-world human demonstrations. The model's baseline performance was rigorously evaluated using both simulated validation suites and physical real-world benchmarks, showing approximately a 40% performance improvement on NVIDIA's internal evaluation benchmarks when using a combination of synthetic and real data compared to using only real data (Bjorck et al., 2025; NVIDIA, 2025a). This finding has practical implications for the commercial industry: it means that the primary historical bottleneck of robot data collection—which requires delicate physical hardware, paid human teleoperators, and thousands of hours of extensive real-time logging—can now be largely bypassed through large-scale synthetic generation utilizing advanced physics simulators and generative world foundation models.

GR00T N1 was designed as an open foundation for the broader humanoid developer community. Early-access partners included Agility Robotics, Boston Dynamics, Mentee Robotics, and NEURA Robotics. Furthermore, NVIDIA announced "Newton," an open-source physics engine under development in partnership with Google DeepMind and Disney Research, purpose-built for developing generalist robots, alongside the comprehensive GR00T Blueprint pipeline for automatically generating synthetic data. The platform ecosystem has since expanded in 2026. GR00T-Dexterity provides reinforcement-learning-based sub-policies for fine-grained object manipulation; GR00T-Mobility supports bipedal navigation across varied terrains and embodiments; GR00T-Control addresses whole-body inverse-kinematic motion planning; and GR00T-Dreams generates synthetic trajectory data from a single uploaded image and text instruction using NVIDIA Cosmos world foundation models. This allows robots to acquire new tasks in unfamiliar environments without human teleoperation data (NVIDIA, 2025c).

### 4.4.2 Google DeepMind Gemini Robotics

Google DeepMind's Gemini Robotics family represents a different architectural philosophy from NVIDIA's approach. Rather than building a novel, robotics-specific architecture from scratch, DeepMind sought to extend their Gemini foundation model into the physical world by adding physical robotic actions as a new output modality—treating joint movements much like text or audio tokens. Gemini Robotics is a vision-language-action model built onto the Gemini backbone, extending the foundation model's existing multimodal capabilities by natively adding closed-loop robotic control (Google DeepMind, 2025a). This architectural choice means the model can generate motor actions and control robotic systems in physical environments alongside processing textual and audio inputs.

Alongside the primary VLA model release, Google DeepMind introduced Gemini Robotics-ER (Embodied Reasoning), a spatial-intelligence companion model that handles high-level cognitive and semantic demands of robotic operation. Gemini Robotics-ER enhances Gemini's inherent understanding of the physical world in ways necessary for safe robotics, focusing on 3D spatial reasoning, occlusion handling, and geometric understanding. It allows traditional roboticists to connect this "brain" with their existing classical low-level PID/MPC controllers. It supports 3D object detection, spatial pointing, computing corresponding points, and detecting partially occluded objects.

This two-model architecture reflects the aforementioned dual-system cognitive principle, but applied at the macro system-architecture level rather than within a single, monolithic model: Gemini Robotics-ER serves as the high-level cognitive layer (System 2), while Gemini Robotics acts as the continuous motor execution system (System 1). The embodied reasoning model orchestrates a robot's entire day of activities, analogous to a high-level cognitive layer, then systematically issues the VLA model dense natural language instructions and spatial coordinates for each discrete step. The VLA then utilizes its own refined vision and language understanding to perform the specific physical actions requested.

The anticipated September 2025 release of Gemini Robotics 1.5 brought a notable capability advance: zero-shot cross-embodiment transfer. Gemini Robotics 1.5 shows the ability to transfer skills across different robotic bodies—it can transfer motions learned on one robot to a different robot morphology without needing to retrain or specialize the model to the new body's specific kinematics. For example, insertion tasks demonstrated only on the dual-arm ALOHA 2 teleoperation robot during training also worked on the very first try on both the Apptronik Apollo humanoid and the bi-arm Franka, enabled by a Motion Transfer mechanism trained over heterogeneous robot data (Google DeepMind, 2025c, 2025d). Furthermore, to address severe latency constraints, in June 2025, Gemini Robotics On-Device was introduced as a compressed, lightweight localized variant. This model is optimized via quantization to run locally on embedded robotic devices, making it useful for latency-sensitive physical applications and ensuring safety and robustness in dangerous industrial environments suffering from intermittent, degraded, or zero internet connectivity. It can be fine-tuned on novel physical tasks with as few as 50–100 human demonstrations, indicating strong foundational spatial knowledge transfer (Google DeepMind, 2025b).

### 4.4.3 Figure AI Helix

Figure AI's proprietary Helix model, unveiled in February 2025, tackles a specific and technically demanding challenge: the real time, ultra-high-frequency, unified control of the full upper body of a bipedal humanoid robot. This entails simultaneously computing the torques and positions for two arms, two complex dexterous five-fingered hands, a twisting torso, an articulating head, and multiple individual finger joints with sub-millimeter precision. Helix uses a dual-system architecture, with two complementary systems trained end-to-end to communicate. System 2 is a VLM specialized in real-time scene understanding and spatial language comprehension, while System 1 is a low-latency visuomotor policy that translates the latent representations produced by System 2 into continuous robot actions at 200 Hz. The 200 Hz figure refers to Helix's visuomotor policy output rate; the underlying low-level joint controllers operate at 500–1000 Hz beneath the policy layer, consistent with the dual-system model described in Sect. 4.2.2. This decoupled architecture aims to combine semantic generalization with low-level reactive physical control. Helix was trained on approximately 500 h of teleoperation data paired with auto-generated language descriptions, comprising a System 2 reasoning model with about 7 billion parameters and a System 1 fast-control policy with about 80 million parameters running at 200 Hz over 35 upper-body degrees of freedom (Figure AI, 2025).

What makes Helix technically and historically significant is the mathematical scope of its control problem. Controlling a 70 kg bipedal humanoid's entire upper body—including complex, mechanically dense, dexterous five-fingered hands—at the high frequencies required for delicate dynamic manipulation is vastly more complex than controlling a single, static industrial robotic arm bolted firmly to a table. That a single, unified VLA architecture can accomplish this feat while still reliably generalizing to novel, unspoken instructions strongly suggests that the dual-system theoretical design has resolved at least the most immediate computational hardware barriers limiting the robotics field.

### 4.4.4 Comparing Architecture Philosophies

The three advanced systems just described reveal two contrasting architectural philosophies currently competing for adoption. GR00T N1 and Figure AI's Helix build heavy, robotics-specific neural architectures from the ground up. These are specifically and mathematically designed to meet the particular computational and physical data demands of dynamic humanoid control—particularly the strict requirement for high-frequency, continuous action generation and, in GR00T's case, cross-embodiment generalization. Conversely, Gemini Robotics takes the opposite approach: it extends a general-purpose foundation model directly into the physical domain. Google's bet is that the rich semantic world knowledge of a trillion-parameter frontier LLM will reliably transfer to physical robotics, overriding the need for hyper-specific local architectures.

As of early 2026, neither strategic approach has conclusively prevailed over the other. Heavily robotics-specific architectures (like Helix) tend to consistently show stronger, smoother performance on ultra-fine, delicate continuous motor tasks (like threading a needle or dynamically catching a falling object) and have clearer, more efficient paths to low-power on-device edge deployment. General-purpose foundation model extensions (like Gemini) tend to show stronger performance on complex, multi-step, ambiguous semantic reasoning tasks that require vast amounts of implicit cultural and physical world knowledge. However, the rapid convergence of these two distinct approaches—visible in Gemini Robotics 1.5's improved low-level dexterity and GR00T N1's expanded high-level logical reasoning capabilities—strongly suggests that the philosophical distinction between the two may eventually dissolve into a single, unified methodology within the next decade.

## 4.5 Hierarchical Task Planning and Long-Horizon Reasoning

While VLA models excel at direct, immediate physical execution, true autonomy requires sustained action over time. This section examines how robots bridge the gap between executing a single command and managing complex, multi-step operations over extended periods—a challenge known as the long-horizon problem.

### 4.5.1 The Long-Horizon Problem

A humanoid robot that is asked to "prepare the conference room on the 4th floor for a board meeting" faces a complex problem that no single VLA model call can address. The state space of this instruction is large. The task requires semantic reasoning about what a "board meeting" entails and what physical items a conference room needs (chairs arranged symmetrically, the projector powered on and focused, water bottles provided at each seat, notepads aligned). It must decompose that goal into many subtasks, execute each subtask in causally appropriate order, and adapt the plan when unexpected conditions arise—for instance, noticing a broken chair leg, an empty water pitcher that requires a detour to the kitchen, or people entering the room. This is the infamous long-horizon planning problem: tasks whose completion requires a sequence of continuous actions too long, and too dependent on changing variables, to be generated in a single pass by any current action policy.

Hierarchical architectures address this problem by decomposing the task representation across multiple communicating levels of abstraction. At the highest, most abstract level, a large language model interprets the abstract goal and produces a structured JSON-based decomposition into discrete subtasks. At intermediate computational levels, these subtasks are grounded in the specific robot's known physical kinematic capabilities and the current geometric environmental state. At the lowest, physical level, individual VLA inferences or high-frequency motor primitive policies execute each grounded action using closed-loop feedback.

### 4.5.2 LLM-Based Task Decomposition

Large language models have proved effective as high-level task planners for robots. Their training on large corpora of human-authored text endows them with implicit knowledge of causal task structures, logical dependencies, and human procedural sequences that would be impractical to encode manually as rule trees. The LLM is explicitly prompted via system instructions to act as a master task decomposer and resource allocator, mathematically mapping high-level fuzzy goals to specific, robot-capable subtasks based on their known APIs. To ensure the LLM generates physically executable, safe action plans, the available physical motion functions for each specific robot morphology are actively provided directly into the context prompt. This mathematical bounding prevents dangerous logic hallucinations, prevents the generation of physically invalid actions, and rigidly constrains the model's vast reasoning process to grounded, system-supported kinematic functions.

Borate et al. (2025) introduced CoMuRoS, an advanced hierarchical architecture that is inspired by traditional human corporate organizational structures. In this framework, the human user simply provides high-level, ambiguous goals in plain natural language to a "Manager LLM." This manager, possessing complete database knowledge of every single agent's physical capabilities, instantly allocates discrete tasks accordingly. Each "Team Member LLM" then autonomously decides how to best logically combine its own available internal physical skills for local execution. Crucially, they proactively and continuously report back up the hierarchy to the manager whenever significant physical events, errors, or anomalies occur. A centralized Task Manager LLM is responsible for high-level macro-planning, while each individual, localized robot securely maintains its own smaller, fast LLM for low-level local execution, real-time Python code generation, and rapid physical event classification.

This organizational metaphor—a high-level executive manager LLM coordinating specialized worker LLMs—has become a common design pattern because it mirrors how human organizations handle complex tasks: via centralized goal-setting paired with decentralized execution. It also provides mechanisms for autonomous error recovery. When a worker robot encounters a physical failure (e.g., dropping a heavy box), it immediately reports the physical state back to the manager LLM, which can instantly replan the entire operation and smoothly reassign a different, closer robot to help recover the dropped item without human intervention.

### 4.5.3 Task and Motion Planning Integration

A persistent challenge for pure LLM-based planning is the phenomenon of physical hallucination: LLMs can generate detailed plans that are syntactically valid and logically sound in text, but physically impossible to execute. A generated text plan that confidently commands "pick up the 100 kg steel box using only your left hand" or "reach the small object located on the top shelf 4 meters away" may be linguistically valid, but it is physically impossible given a standard 1.7-m humanoid's strict joint torque limits and kinematic constraints. This gap between linguistic plausibility and physical feasibility necessitates the deep, mathematical integration of advanced LLM planning with rigorous, classical Task and Motion Planning (TAMP) methodologies.

Kawabe and Takano (2026) propose an advanced hierarchical multi-robot planner that addresses this gap. They merge the semantic capabilities of LLMs at the decomposition and task allocation layers while employing rigorous, mathematically provable classical PDDL (Planning Domain Definition Language) planners for the actual physical execution scheduling. Their robust three-layer hierarchy features: a global planner layer that interprets the ambiguous user instruction; a type layer that intelligently distributes physical subtasks to specific robot typologies based on their known hardware skills and payload capacities; and a strict robot layer that automatically generates precise PDDL logic code for individual robot execution. This decomposition reduces the latency and context-window scalability bottlenecks of utilizing a single monolithic LLM planner. Equally important, the lower task-and-motion-planning layer should return structured failure explanations rather than only a binary failure signal: if a proposed subtask is infeasible because the target is unreachable, the arm is too short, a collision-free path cannot be found, or the grasp violates payload constraints, that reason should be passed back to the reasoning layer as semantic feedback. This semantic back-propagation prevents the LLM from repeatedly hallucinating kinematically impossible plans and allows it to repair the task decomposition using motion-planning evidence (Wang et al., 2024). This decomposition increases physical safety, and securely enables parallel execution across hundreds of heterogeneous agents simultaneously.

This hybrid approach—LLM semantic reasoning for macro-decomposition paired with classical TAMP planning for physical feasibility—is increasingly recognized as a reliable path to achieving robust, industrial-scale long-horizon manipulation.

Figure 4.3 summarizes the five-layer hierarchical planning stack that bridges abstract natural-language goals at the top with millisecond-scale motor torques at the bottom, with feedback and replanning flowing back upward through the layers (Table 4.3).

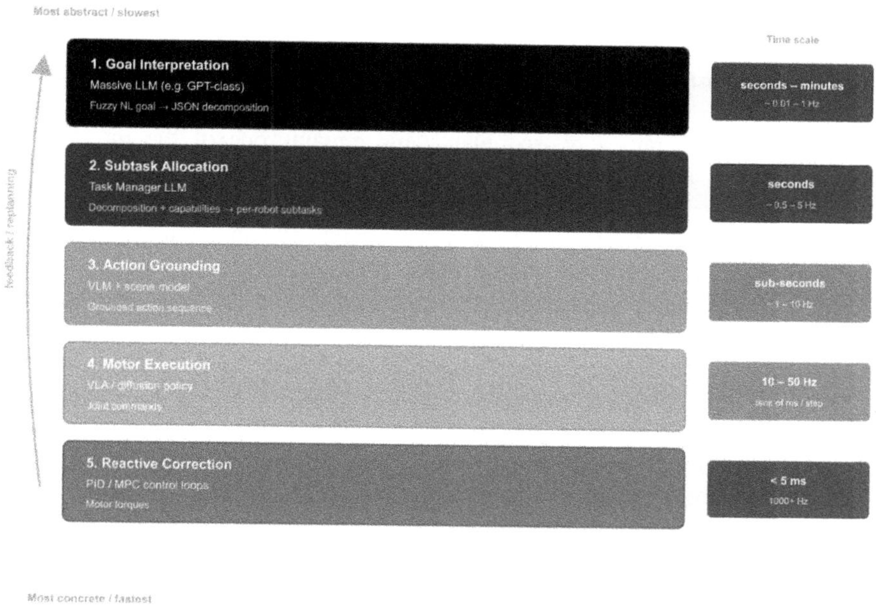


**Fig. 4.3** Hierarchical task planning stack

**Table 4.3** Hierarchy of abstraction in agentic humanoid robot task planning

| Planning level | Primary system | Input data | Output generated | Operating timescale |
|---|---|---|---|---|
| Goal interpretation | Large LLM (System 2) | Ambiguous natural language instruction | Structured JSON task decomposition | Seconds to Minutes |
| Subtask allocation | Specialized Task Manager LLM | Task decomposition tree, Fleet robot capabilities | Precise per-robot subtask assignments | Seconds |
| Action grounding | VLM + local geometric environment model | Subtask assignment, Current 3D visual scene | Grounded, sequential kinematic action sequence | Sub-seconds |
| Motor execution | VLA/fast diffusion policy | Grounded action embedding, real-time sensorimotor state | Precise joint angle commands/target velocities | Milliseconds (10–50 Hz) |
| Reactive correction | High-frequency Low-level classical PID/MPC controller | Instantaneous sensorimotor torque/current feedback | Absolute corrective electrical motor torques | <5 ms (1000+ Hz) |

*Source* Author's analysis based on Borate et al. (2025) and Kawabe and Takano (2026)

### 4.5.4 Feedback, Replanning, and Recovery

Achieving robust long-horizon execution requires not merely initial planning, but continuous replanning—the ability to detect when a subtask has failed and autonomously generate a revised plan. This remains one of the most practically important and historically under-studied capabilities in agentic robotics. Unexpected physical task failures, sudden environmental perturbations, or immediate user intent changes must trigger replanning, allowing multiple robots to assist teammates, resume interrupted tasks, or request human help when stuck. Recent factory studies report autonomous recovery from disruptive events, filtering of visual distractions, and coordinated transport with emergent human–robot cooperation.

The core autonomous recovery mechanism requires that the robot's perception system continuously monitors physical task progress and produces low-latency semantic feedback to the planning LLM. A robot that physically notices via force sensors that it has grasped a delicate object incorrectly, or feels it slipping, must instantly communicate this physical anomaly back to the overarching task planner. The planner must then generate a safe correction (e.g., "gently place the object down immediately and re-attempt the grasp from a 45-degree angle"). This closed-loop cycle between physical execution and semantic planning distinguishes agentic systems from legacy systems that execute rigid plans generated before operation begins.

## 4.6 Multi-Agent Coordination for Humanoid Robot Fleets

As agentic robots transition from laboratory experiments to industrial deployment, they rarely operate in isolation. This section explores the complex coordination required when multiple intelligent agents—whether they are distinct physical robots or independent software agents controlling different limbs of the same robot—must collaborate to achieve shared goals.

### 4.6.1 From Single Robots to Heterogeneous Fleets

The deployment scenarios envisioned for humanoid robots at scale—including large logistics warehouses, urban hospitals, automotive factory floors, and eldercare facilities—are not single-robot environments. They are multi-agent environments where many robots of different morphologies operate alongside human workers. They must share physical space, equipment, and interleaved tasks. Managing these environments requires software architectures for multi-agent coordination. These orchestration systems must allocate tasks among a heterogeneous fleet, detect and resolve physical or logical conflicts, communicate state and intent between agents, and maintain system efficiency while allowing each agent to operate semi-autonomously.

As physical AI matures into 2026, organizations are increasingly deploying heterogeneous fleets consisting of bipedal humanoids, autonomous wheeled vehicles, delivery drones, and stationary AI software agents. Crucially, these will frequently originate from multiple different corporate vendors, each utilizing proprietary, closed communication protocols (Chauhan, 2026). This creates interoperability challenges that can lead to physical incidents, downtime, system congestion, and operational inefficiency. Vendor-agnostic fleet-management and orchestration systems are therefore needed to address these interoperability issues.

### 4.6.2 Intra-Robot Multi-Agent Learning

A distinctive application of multi-agent reinforcement learning (MARL) operates at the level of a single robot's internal control. The Multi-Agent reinforcement learning for Single Humanoid (MASH) framework treats each limb of a humanoid—the two arms and two legs—as independent, self-interested agents operating within a cooperative MARL system (Liu Q et al., 2025). MASH reformulates humanoid locomotion as a cooperative-heterogeneous MARL problem. This enables more computationally efficient coordination learning than conventional, monolithic single-agent Reinforcement Learning by treating each limb as a largely independent agent with its own action space and local reward function. Reported results show improved bipedal gait execution, higher task performance, better training efficiency, improved robustness in dynamic environments, and zero-shot sim-to-real transfer of the learned policy to a physical humanoid.

The Slow-Fast Two-Agent (SoFTA) framework (LeCAR Lab, 2025) takes a similar approach to the difficult challenge of maintaining stable end-effector control during bipedal locomotion. SoFTA decouples the action spaces and value-function spaces of the heavy lower body and the dexterous upper body. This decoupling enables different execution frequencies and distinct, separate reward structures. The upper-body agent runs at 100 Hz, prioritizing end-effector precision, while the lower-body agent runs at 50 Hz, prioritizing stable bipedal locomotion. The reported result is a 2–5× reduction in end-effector acceleration relative to monolithic baselines, enabling tasks such as carrying liquids without spilling and capturing stable first-person video while walking.

These biologically inspired intra-robot multi-agent approaches reveal an important insight: the degrees-of-freedom problem in full humanoid control is large enough that centralized single-agent RL struggles with coordination. Distributing control responsibility across several semi-independent local agents, each optimizing a local objective while sharing sensory observations, mirrors the distributed control architecture of the human motor system.

### 4.6.3 Inter-Robot Coordination and Fleet Orchestration

Scaling up to the macro fleet level, multi-agent coordination for heterogeneous robot swarms must handle dynamic task allocation, collision avoidance, conflict resolution, and latent communication across many agents with different capabilities and action spaces. The dominant approach uses hierarchical architectures with a centralized planning layer and decentralized execution: a cloud-based task manager reasons about global state and distributes work; individual robot agents execute locally and report physical outcomes.

Figure 4.4 depicts the hub-and-spoke topology used by frameworks like COHERENT, in which a centralized manager LLM coordinates heterogeneous worker robots through a proposal-execution-feedback-adjustment (PEFA) loop.

The COHERENT framework (Liu, Y. et al., 2025) demonstrates a proposal-execution-feedback-adjustment (PEFA) mechanism explicitly designed for heterogeneous multi-robot systems that include flying quadrotors, quadruped robots, and stationary robotic arms. The framework uses structured dialog and debate between multiple LLM-enabled robots to improve reasoning and reduce hallucinations in interleaved tasks. The centralized task assigner logically decomposes complex goals into specific subtasks and assigns them directly to localized robot executors. Continuous self-reflection generated from low-level execution feedback enables bottom-up task correction and safe local adaptation.

Samsung's 2026 industrial AI strategy provides a real-world, large-scale example of this multi-agent fleet design. Samsung deploys fine-tuned AI agents for distinct factory functions: quality-control drone agents that detect defects,

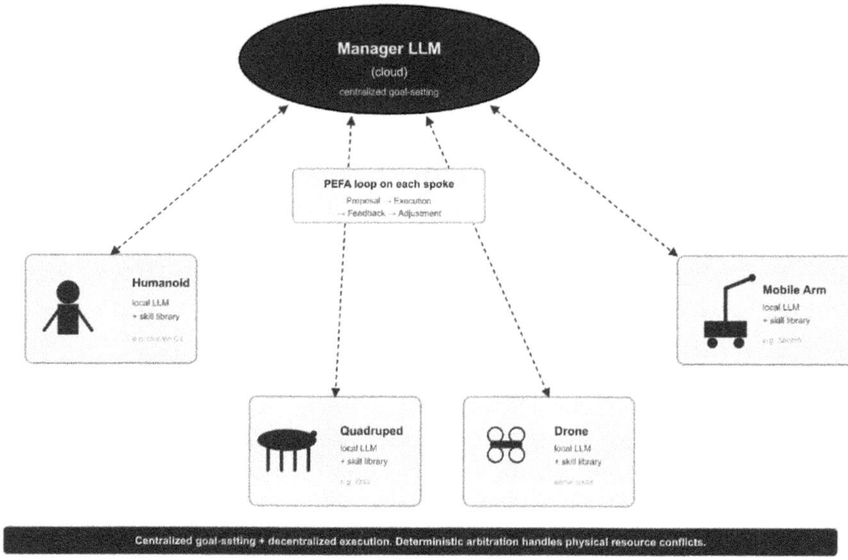


Fig. 4.4 Multi-agent robot fleet coordination (PEFA loop)

production-manager agents that optimize manufacturing workflows using reinforcement learning, and logistics agents that coordinate material handling via bipedal humanoids. The agent-based approach is intended to standardize manufacturing operations across global Samsung sites.

Coordination challenges in multi-robot humanoid environments extend beyond task allocation to physical interaction. When two humanoid robots must cooperate on a shared task—such as jointly lifting a heavy object or handing off a hazardous component—their motions must be synchronized to prevent failure. Recent research identifies what is termed the "isolation issue" in multi-humanoid interaction: methods that treat each humanoid as independent are ill-suited to dense multi-humanoid interactions because they neglect the coupled dynamics that emerge during shared contact. This isolation causes misaligned contacts, loss of coordinated bipedal behavior in closely coupled motions, and implausible kinematic executions due to morphological mismatches and shifting center-of-mass requirements. Unified physics models address this by computing joint constraint matrices across multiple robot bodies simultaneously.

## 4.7 Simulation-to-Reality Transfer

Simulation-to-reality transfer is not unique to agentic AI, and it should not be treated as a defining property of agents. It matters here because modern agentic humanoid systems depend on large-scale synthetic experience for training, evaluation, safety testing, and policy refinement before deployment. This section therefore treats sim-to-real as an enabling methodology for agentic architectures: the mechanism by which foundation-model policies, skill libraries, planners, and safety monitors are exposed to rare physical conditions that would be expensive or unsafe to collect directly on hardware.

### 4.7.1 The Reality Gap

Training humanoid robots in the physical real world is impractical at industrial scale. Real-world training is slow, expensive, hardware-intensive, and physically dangerous—a 70 kg, six-figure robot attempting to learn to walk through trial-and-error reinforcement learning is likely to damage its gearboxes, structural linkages, and surrounding environment. Simulation offers an alternative: physics simulators such as MuJoCo and NVIDIA Isaac Gym can run many parallel training environments at well above real-time speed. This generates the large volume of training examples that deep reinforcement learning requires, without risking physical hardware.

However, policies learned in simulation often fail when deployed to the real world. This phenomenon is universally known as the reality gap, caused by differences between simulated and real physics, combined with discrepancies in sensing and electromechanical actuation.

The prevailing paradigm conducts reinforcement-learning training in simulated environments and then transitions to real-world deployment—a process known as sim-to-real transfer. This paradigm cannot fully bridge the gap between simulation and reality. The gap persistently manifests in several forms: physical parameters such as dynamic surface friction, non-linear joint damping, elastic contact stiffness, actuator hysteresis, gearbox backlash, motor current-to-torque delay, and compliance differ between simulation and reality; visual appearances differ between rasterized computer graphics and raw real-world photon captures; optical sensor noise distributions differ; and crucial actuation dynamics—the complex, non-linear relationship between commanded electrical current and the resulting physical joint torque—differ. These actuator effects are difficult to eliminate with domain randomization alone. Current sim-to-real research therefore combines system identification, residual dynamics learning, and differentiable physics engines that can optimize parameters and policies through simulation gradients (Howell et al., 2022; Schwarke et al., 2024). Even small per-timestep dynamics errors compound across multi-second trajectories and can cause catastrophic balance failures.

### 4.7.2 Techniques for Bridging the Gap

Several techniques have proved effective for reducing the sim-to-real gap. Domain randomization is the most widely used technique: it trains control policies across a wide distribution of simulation parameters. During training, friction coefficients, object center-of-masses, rigid-body inertias, lighting conditions, and camera poses are randomized. The goal is for the real world to appear to the policy as another sample from its training distribution. This approach trades some peak performance in any single configuration for robustness across the broader distribution. System identification uses real-world telemetry data to calibrate simulation parameters such as motor torque constants and joint friction, making the simulation more accurate before training begins.

He et al. (2025) introduced ASAP (Aligning Simulation and Real-World Physics), a two-stage framework that addresses dynamics mismatch through residual correction. ASAP pre-trains motion-tracking policies in simulation using retargeted human motion-capture data. It then deploys these baseline policies on hardware and collects real-world telemetry data to train a "delta action model" that compensates for residual dynamics mismatch. This approach improves physical agility and whole-body bipedal coordination, reducing tracking error relative to system identification, domain randomization, and delta-dynamics-learning baselines.

Taking a different approach, He et al. (2026) propose an alternative paradigm: rather than perfecting the simulation, train directly on the physical robot through safety-bounded online learning. Their online learning strategy conducts fine-tuning directly on the physical robot. It uses simulated pre-training and instruction learning to improve initial safety and efficiency, and includes an autonomous resetting system that lets the robot reconfigure its joints and resume learning after a fall

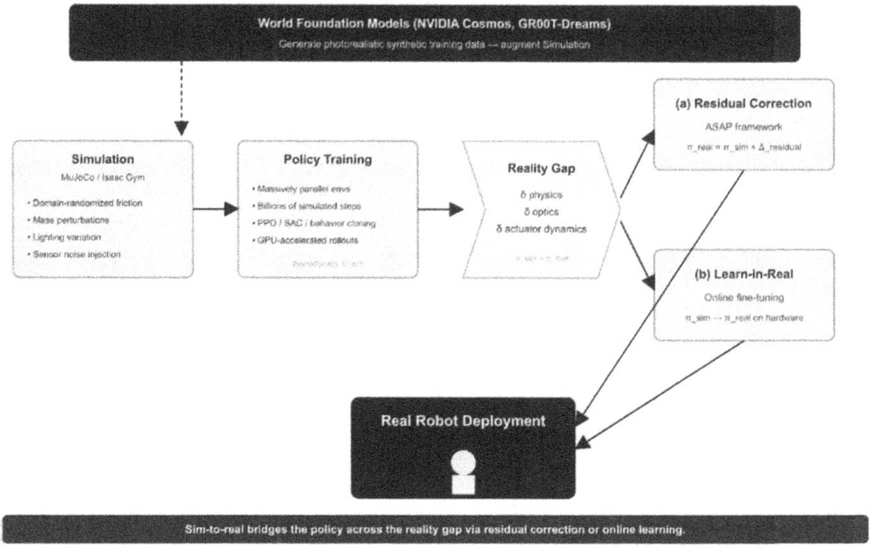


**Fig. 4.5** Sim-to-real transfer pipeline

without human intervention. This learn-in-real approach avoids the reality gap, at the cost of requiring rigorous safety management during all real-world training.

Figure 4.5 illustrates the sim-to-real transfer pipeline, from domain-randomized simulation through large-scale parallel policy training to real-robot deployment via residual correction or online learn-in-real fine-tuning.

### 4.7.3 The Role of World Foundation Models

A more recent approach to the sim-to-real problem uses world foundation models—generative models trained on large-scale video data—to produce photo-realistic, physically plausible synthetic training video. NVIDIA's Cosmos platform and the GR00T-Dreams blueprint exemplify this approach. They use world models to generate diverse visual environments and task scenarios from text or image prompts. This lets robots train visually on novel or hazardous situations not encountered in physical reality or hand-crafted simulation.

Tesla's Optimus humanoid robot similarly trains extensively in internal simulation infrastructure before transferring learned behaviors to physical hardware. This sim-to-real transfer, while limited by persistent optical and physical reality gaps, accelerates learning and reduces the cost of acquiring physical behaviors. The combination of deterministic physics simulation for motor-skill training with generative world foundation models for visual diversity is emerging as a standard data pipeline for next-generation humanoid systems (Table 4.4).

**Table 4.4** Sim-to-real transfer techniques for humanoid robots

| Technique | Mathematical mechanism | Key strengths | Severe limitations |
|---|---|---|---|
| Domain randomization | Heavily randomize physics parameters (friction, mass) during pure simulation training | Broad robustness; zero physical real-world data required | Significantly reduces peak physical performance; struggles with large unexpected distributional shifts |
| System identification | Use vast real-world telemetry data to calibrate simulation physics equations | Produces physically accurate simulation engines | Strictly requires expensive real hardware testing; naturally risks overfitting to specific lab conditions |
| Residual correction (ASAP) | Train a delta action model from real-world telemetry data post-deployment | Preserves the original trained policy; fixes dynamics mismatch precisely | Strictly requires safe deployment on real hardware; expensive data collection cost |
| Learn-in-real | Train directly on physical hardware utilizing strict, mathematical safety bounds | No reality gap; models are accurate to the specific robot hardware | Slow; requires extensive safety engineering; causes severe physical hardware wear and tear |
| World foundation models | Synthesize photorealistic, physically accurate synthetic environments at scale | Effectively unlimited data visual diversity; zero physical robotic hardware needed | Mind-blowing computational cost (thousands of GPUs); physical contact dynamics remain only approximate |

*Source* Author's analysis based on He et al. (2025, 2026) and NVIDIA (2025c)

## 4.8 World Models and Predictive Cognition

As models scale in size and capability, agentic systems are transitioning from merely reacting to their environments to predicting them. In this chapter, the term world model refers specifically to a predictive internal model that estimates how physical states may evolve under candidate actions, often in a latent representation used for planning or policy evaluation. This usage is narrower than other contemporary meanings of world model, such as 3D spatial scene representations, generative visual reconstructions, or broader world foundation models for physical AI. Those representations can support robot cognition, but the focus here is on predictive transition models and learned simulators that help robots evaluate possible futures before acting (Huang, 2026b; NVIDIA, 2026a; Unitree Robotics, 2025).

### 4.8.1 From Reactive Systems to Predictive Agents

The various architectures thoroughly examined in the preceding sections all operate, to varying degrees, in a reactive mode: they perceive the current physical state, generate an action, observe the outcome, and repeat. A more sophisticated form of robot cognition involves world models—internal representations that allow the robot to simulate the consequences of proposed actions before executing them. A robot equipped with a high-fidelity world model can ask itself, "if I push this glass object 5 cm to the left with 10 N of force, will its center of gravity cause it to fall off the table?" without physically pushing it, reasoning instead about the imagined outcome.

High-fidelity world models have been a long-standing aspiration in robotics and AI, but until recently they were too inaccurate to be practically useful. Recent generative-modeling advances that produced vision-language models have enabled world models of sufficient fidelity to inform robotic decision-making. Systems including NVIDIA's DreamZero framework and Physical Intelligence's MEM and MemER systems represent early commercial deployments of world-model-informed robot control (NVIDIA, 2026a; Physical Intelligence, 2026).

### 4.8.2 Architecture of Robotic World Models

At its core, a robotic world model is a function that takes the current state and a proposed action as inputs and predicts a future state. To be effective, it must model rigid-object interactions, frictional contact dynamics, material properties, and lighting changes with enough fidelity to guide real-world planning.

Current world models for robotics typically use video-prediction architectures based on Joint Embedding Predictive Architectures (JEPAs) or continuous diffusion mechanisms. They represent world state as a high-dimensional image or video latent embedding and learn to predict how future frames will change under specified motor commands. An end-to-end embodied model coupled with a humanoid robot is one plausible architecture for general-purpose embodied intelligence, though it is not yet established whether this approach will scale to open-world domestic and industrial deployment. Such systems aim to interpret ambiguous natural-language commands, learn new physical tasks from few demonstrations, and operate in unstructured human environments. High-fidelity world models are central to this vision because they provide the substrate for mental simulation, allowing robots to plan many steps ahead without committing to potentially unsafe actions.

### 4.8.3 Memory Systems and Long-Term Contextual Adaptation

Memory systems are closely related to world models, but they are architecturally distinct. A world model predicts possible future states; memory stores and retrieves past experience, task context, user preferences, and semantic knowledge that can condition planning. In agentic humanoid robots, working memory, episodic memory, and semantic memory typically sit alongside the world model rather than inside it. Physical Intelligence's MemER framework addresses this complementary problem by scaling episodic memory for physical robot control through experience retrieval (Physical Intelligence, 2026). This allows robots to recall relevant prior interactions when facing new physical situations and use those retrieved episodes to guide current decisions.

The practical implications are significant. A home assistant robot may need to remember user preferences, task history, room layouts, object locations, or recurring household constraints. Those memories do not by themselves constitute a world model; instead, they provide context that a planner, VLA policy, or predictive model can use when choosing actions. Treating memory as its own architectural layer also makes the safety problem clearer: designers must decide what should be retained, when it should be retrieved, how it should be updated, and how privacy-sensitive human interaction histories should be protected.

## 4.9 The Open-Source Ecosystem: China, the United States, and the Geopolitics of Agentic Robot Architectures

The development of humanoid robotics does not happen in a vacuum. It is intertwined with global economics, national security, and sovereign strategic investments. This final section explores the geopolitical dimensions of the humanoid revolution, highlighting the contrasting approaches taken by the world's two largest technological superpowers.

### 4.9.1 A Strategic Divergence at the Infrastructure Layer

The development of agentic AI architectures for humanoid robots is not occurring in a geopolitically neutral space. As of early 2026, the global humanoid robot market is defined by a strategic geopolitical divergence: the People's Republic of China leads in open-source hardware manufacturing and high-volume physical deployment, while the United States leads in open-source software frameworks and frontier AI. This division is not merely commercial—it reflects different theories about where architectural and economic leverage lies within the agentic robotics stack.

Approximately 90% of humanoid robots sold in 2025 were manufactured by Chinese companies, and six of the highest-selling vendors in the sector were based in China; China accounted for over 80% of global humanoid robot installations and roughly half of all industrial robot installations (Bloomberg, 2026). China's deployment velocity is generating a self-reinforcing advantage that extends beyond commercial market share: whichever country physically deploys robots faster will inevitably mathematically collect more physical telemetry data, which in turn directly unlocks better, safer AI policy deployment. Major cities across China—including Beijing, Wuhan, and Shanghai—are opening state-funded training sites for acclimating humanoid robots to varied physical settings and collecting standardized data. China recorded over 610 robotics investment deals in the first nine months of 2025.

Global humanoid robot shipments in 2025 reached approximately 18,000 units, a year-over-year expansion of roughly 508% (IDC, 2026; Rest of World, 2026). The IDC headline total includes a long tail of small Chinese and international vendors not enumerated individually in Table 4.5, which lists only the leading open-source platforms; the difference between the table's named vendors (~13,500 units) and the IDC global figure is attributable to this unenumerated long tail of smaller producers. The top humanoid robot makers by 2025 physical shipments were led by China's AgiBot and Unitree, closely followed by UBTech, Leju Robotics, Engine AI, and Fourier Intelligence, underscoring Beijing's early dominance in the physical sector. This lead is state-backed: China's national AI strategy seeks to combine its strengths in AI software integration and robotics hardware manufacturing to achieve a durable form of technological leadership. The CCP Central Committee has recommended incorporating embodied AI as a primary new driver of economic growth in China's forthcoming 15th Five-Year Plan, signaling that Chinese leadership intends to make a long-term strategic commitment to embodied AI (Buchanan & Chou, 2025).

### 4.9.2 China's Full-Stack Open-Source Hardware Strategy

China's approach to agentic robot architecture development is anchored in what practitioners call full-stack open-sourcing: releasing not only software code but also hardware designs, mechanical blueprints, engineering bills of materials, and simulation environments. This strategy is intended to lower barriers to entry, build a global ecosystem of secondary developers, and establish Chinese physical platforms as a de facto standard.

Tiangong (Beijing Humanoid Robot Innovation Center). The Tiangong series, developed by the state-backed Beijing Humanoid Robot Innovation Center (also known as X-Humanoid), is the most prominent example of China's open-source hardware strategy applied to full-size humanoid platforms. Tiangong is positioned as the first electric-driven full-size humanoid capable of dynamic running, sustaining a bipedal running speed of 6 km/h with a reported top speed of 12 km/

**Table 4.5** Key Chinese open-source humanoid robot platforms (2025–2026)

| Platform | Organization | Open-Source scope | Key capability | 2025 global units |
|---|---|---|---|---|
| Tiangong 3.0 | Beijing Humanoid Robot Innovation Center | Hardware blueprints, Pelican-VL VLM, RoboMIND dataset | 6 km/h sustained, 12 km/h peak; 43 DoF; advanced perceptual terrain navigation | 100+ secondary dev units |
| Roboto Origin | RoboParty | Full stack: hardware CAD, strict code, supplier lists, detailed SOP | 3 m/s running; advanced AMP gait algorithm; perfect full sim-to-real pipeline | Prototype (research grade) |
| G1+ UnifoLM | Unitree Robotics | UnifoLM-WMA-0 (world model), UnifoLM-VLA-0 (VLA) | Complex manipulation; flawless zero-shot cross-embodiment generalization | ~5,500 total humanoids |
| A2 Series | AgiBot | Genie Sim 3.0; AgiBot World dataset; open hardware modules | 49+ DoF; advanced WorkGPT multimodal AI; 96% semantic accuracy | 5,168 (Omdia, #1 globally) |

*Source* Author's analysis based on Kalil (2025), Malayil (2026), Qiao (2025), and AgiBot (2026)

h. The open-source release includes three distinct versions of the physical platform: Tiangong 1.0 Lite focuses on bipedal walking and running; Tiangong 1.1 Pro adds upper and lower limb coordination and visual interaction; and Tiangong 1.2 Max is described as an integrated embodied intelligent robot with closed-loop perception–action capabilities (Kalil, 2025). The open-source scope covers structural design documents, software API interfaces, and URDF kinematic models, and has attracted over 100 partner organizations conducting secondary development across diverse application scenarios. In February 2026, the center unveiled Tiangong 3.0, a 169 cm, 62 kg machine with 43 degrees of freedom designed for industrial and general-purpose use. The Tiangong 3.0 release focuses on accessibility, open-sourcing several core components including the design blueprints, the Pelican-VL vision-language model, and the RoboMIND dataset. It supports protocols such as ROS 2, MQTT, and TCP/IP to allow industrial partners to perform secondary development without rebuilding the underlying architecture.

RoboParty Roboto Origin. At the other end of the corporate maturity spectrum, RoboParty—a Xiaomi-backed Beijing startup founded by 21-year-old Yi Huang—represents the grassroots dimension of China's full-stack open-source movement. RoboParty open-sourced its flagship bipedal humanoid, Roboto Origin, after completing R&D in 120 days. Promoted as a full-stack open-source bipedal humanoid prototype, it has earned over 1,000 GitHub stars and nearly 100 kit pre-orders within months (Huang, 2026a). Roboto Origin is a 1.25 m, 34 kg prototype that runs at 3 m/s using RoboParty's self-developed AMP anthropomorphic gait algorithm. The open-source release covers hardware structural drawings, engineering bill of materials, domestic Chinese supplier lists, low-level control code, SMPL-X human-model adaptation, and a sim-to-real pipeline. The company claims its open-source approach could reduce global humanoid development costs by up to 80% (Malayil, 2026).

Unitree Robotics. Unitree emerged as one of the highest-volume humanoid robot vendors of 2025, shipping over 5,500 humanoids—though analyst rankings differ, with Omdia placing AgiBot first by shipments and other reports placing Unitree first (Deng et al., 2026). Open-sourcing its world-model-action architecture is central to building a broader global developer ecosystem. The company open-sourced two AI components in late 2025: UnifoLM-WMA-0, a world-model-action architecture spanning multiple embodiments, and UnifoLM-VLA-0, a vision-language-action model designed for general-purpose humanoid manipulation (Unitree Robotics, 2025). UnifoLM-WMA-0 is Unitree's open-source world-model–action architecture for general-purpose robot learning. It features a core world model that captures interactions between robots and environments, providing both a simulation engine for synthetic data generation and a policy-enhancement head for improving decision-making (Qiao, 2025). Unitree's G1 humanoid, which integrates UnifoLM natively, targets developers including those using Apple Vision Pro headsets for teleoperation and data collection.

AgiBot. Shanghai-based AgiBot led global humanoid robot producers by shipment volume in 2025, with 5,168 units delivered according to Omdia, capturing approximately 39% of total global market share (AgiBot, 2026). The AgiBot A2 features 49+ degrees of freedom and the WorkGPT multimodal AI system, which is reported to reach 96% multimodal semantic accuracy and a 99% face wake-up rate in service deployments. AgiBot's open-source strategy extends beyond software to include modular hardware designs. The company's Genie Sim 3.0 simulation environment and AgiBot World dataset provide the training infrastructure underpinning its robots' agentic capabilities. AgiBot made its U.S. debut at CES 2026, where it received multiple Best of CES awards, and is now commercially available in six countries.

### 4.9.3 The United States: Open-Source Software and Intelligence Infrastructure

Where China's open-source strategy operates primarily at the physical hardware layer, the United States' open-source contributions to agentic humanoid robotics operate primarily at the software and intelligence layers. The U.S. approach bets that the most durable form of architectural and economic leverage lies not in the commoditized physical platform, but in the cognitive operating system that runs on top of any platform—including subsidized Chinese ones.

OpenMind OM1. The clearest embodiment of this strategy is OpenMind's OM1, which the company launched in beta in September 2025 (with a $20M seed round announced in August 2025), describing it as an open-source operating system designed for agentic robots. OpenMind describes OM1 as a hardware-agnostic open-source operating system that supports fleet autonomy and developer collaboration. It works across quadrupeds, humanoids, wheeled robots, and drones, with plug-and-play support for LLMs including OpenAI, Gemini, DeepSeek, and xAI models. It includes preconfigured agents tuned for platforms including the Unitree G1 humanoid, Go2 quadruped, TurtleBot, and UBTech mini humanoid. The company, founded by Stanford professor Jan Liphardt and backed by a $20 million seed round led by Pantera Capital (Pantera Capital, 2025; Robot Report, 2025), frames its mission as building an Android-equivalent platform for robotics— a universal software layer that allows different robot embodiments to run the same intelligence stack (Openrobotics.org, 2025).

OpenMind launched a hardware-agnostic "App Store for robots" in January 2026 as a distribution layer for robotic intelligence, with launch partners including UBTECH Robotics. This positions the platform to allow humanoid and quadruped robots to gain new capabilities through over-the-air software updates rather than hardware overhauls. The OM1 architecture uses lightweight Python configuration files to define models and task logic, letting developers package behaviors—such as autonomous movement, social interaction, and environmental reasoning—into portable applications that run across heterogeneous hardware.

The strategic insight is clear: if the agentic intelligence layer is standardized and open-sourced at the U.S. software level, hardware vendors become commoditized. A subsidized Chinese humanoid running American OM1 software is, from the developer's perspective, the same platform as an American one. This makes the U.S. software layer potentially more durable than any hardware-bound architecture, because it can ride the volume advantage of subsidized Chinese manufacturing rather than competing against it.

Google DeepMind and Open Datasets. Google DeepMind has also open-sourced foundational robotics research and large-scale physical demonstration datasets that serve as training infrastructure for many agentic robot policies worldwide. The Open X-Embodiment dataset and RT-X initiative provide cross-embodiment robot demonstration data that U.S. and Chinese firms use to train generalist robot policies. Gemini Robotics On-Device (introduced in June 2025), while not open-source, is designed for local deployment on third-party robot hardware and supports fine-tuning with as few as 50–100 demonstrations, making it accessible to independent robot developers.

Stanford ALOHA and Academic Infrastructure. Stanford's ALOHA bimanual teleoperation system, while initially an academic research platform, has become a reference design for dexterous-manipulation data collection used by humanoid robot developers globally. Its simple, open hardware design has influenced manipulation data-collection standards at U.S. and Chinese research institutions.

### 4.9.4 Structural Strengths, Limitations, and Convergence Risks

The two-ecosystem global structure has important implications for how agentic humanoid architectures will evolve. China's hardware-volume advantage generates deployment telemetry that trains better VLA models—creating a virtuous cycle where more robots deployed yields better policies, which attract more buyers, who deploy more robots. Despite these manufacturing advantages, analysts note that China lags in frontier AI software and in next-generation logic-chip development (Schmidt & Xu, 2026). U.S. and allied companies including Nvidia, TSMC, Palantir, and Qualcomm dominate these compute areas, while Chinese humanoid robot research remains largely focused on physical hardware actuators and application scenarios rather than the maturity and reliability of control software.

The U.S. software advantage, conversely, depends on adoption at scale—an operating system that runs only on a small number of expensive American robots has limited real-world training data and limited global commercial value. The open-source strategy of platforms like OM1 is therefore a deliberate attempt to achieve global scale by running on subsidized Chinese hardware, trading platform control for telemetry data access and ecosystem breadth.

China's "bot playbook" is reminiscent of how it built its global lead in the EV industry: early state financial support enabled many small entrants, created subsidized domestic demand, boosted production volume, and cultivated manufacturing experience. This produced intense domestic competition and global economies of scale (Park, 2026). Industry analysts note structural risks in the Chinese approach: over 150 humanoid robot startups competing in the same market as of late 2025 suggests excessive fragmentation. Critical observers note that many Chinese humanoid robots, while visually impressive at trade shows, cannot yet reliably perform skilled manipulation tasks autonomously in unstructured human environments without teleoperation (Table 4.6).

Figure 4.6 contrasts China's hardware-first and the United States' software-first open-source strategies, and shows how the two ecosystems are converging at the seams.

The convergence risk is real: the two geopolitical strategies are increasingly interdependent. OpenMind's OM1 runs on Unitree hardware. Unitree's UnifoLM-VLA-0 is fine-tuned on American Open X-Embodiment data. Gemini Robotics On-Device is deployed on both Chinese and American robots. The architectural paradigms discussed throughout this chapter—VLA models, dual-system cognition, hierarchical planning, world models—do not belong to either country alone; they represent a shared global technical commons that both ecosystems build upon. The geopolitical divergence defining the robotics industry is therefore not primarily a divergence of technical architecture, but of ecosystem strategy: who controls the platform, who owns the telemetry data, and who sets the de facto global standards for what it means for a physical humanoid robot to be intelligent.

**Table 4.6** Comparative landscape of China versus U.S. open-source humanoid robot ecosystem (Early 2026)

| Strategic dimension | People's Republic of China | United States of America |
|---|---|---|
| Primary open-source layer | Physical hardware blueprints, telemetry datasets, localized embodied AI models | Operating systems, universal software frameworks, frontier AI backbone models |
| Leading open platforms | Tiangong (X-Humanoid), UnifoLM (Unitree), AgiBot World | OM1 (OpenMind), Open X-Embodiment (Google DeepMind), ALOHA (Stanford) |
| 2025 global market share | ~80–90% of global installations/physical units sold | <10% of physical hardware market (led by Tesla Optimus) |
| Deployment volume (2025) | ~12,000+ units physically deployed (Unitree + AgiBot alone: ~10,600) | ~1,000–1,500 total units estimated (Tesla, Figure, Agility) |
| Core Strategic thesis | Hardware volume scale → telemetry data flywheel → better AI | Universal Software layer → embodiment-agnostic intelligence → rides hardware volume |
| Key vulnerability | Frontier AI software maturity gap; severe high-end chip dependency; intense market fragmentation | Low physical deployment volume limits training data; ecosystem dependent on hardware partners |
| Government support | Heavy Direct: Made in China 2025, 15th Five-Year Plan, aggressive MIIT targets | Largely Indirect: DARPA funding, NSF grants, strict export controls on AI chips |

*Source* Author's analysis based on Lo (2026), Schmidt and Xu (2026), Buchanan and Chou (2025), and AgiBot (2026)

## 4.10 OpenClaw: The Rise of Multi-Agent Orchestration

The geopolitical and technical divergence between hardware-centric and software-centric architectures has recently been disrupted by the viral emergence of OpenClaw (formerly known as Clawdbot and Moltbot). Released in late 2025 and achieving developer adoption by early 2026, OpenClaw is positioned as a next-generation agentic interface that functions as a decentralized, model-agnostic orchestration layer for physical hardware (CNBC, 2026; Steinberger, 2025). While systems like GR00T and Gemini focus on the end-to-end "brain" of the robot, OpenClaw provides an orchestration layer that coordinates multiple specialized models and agents to achieve high-level goals.

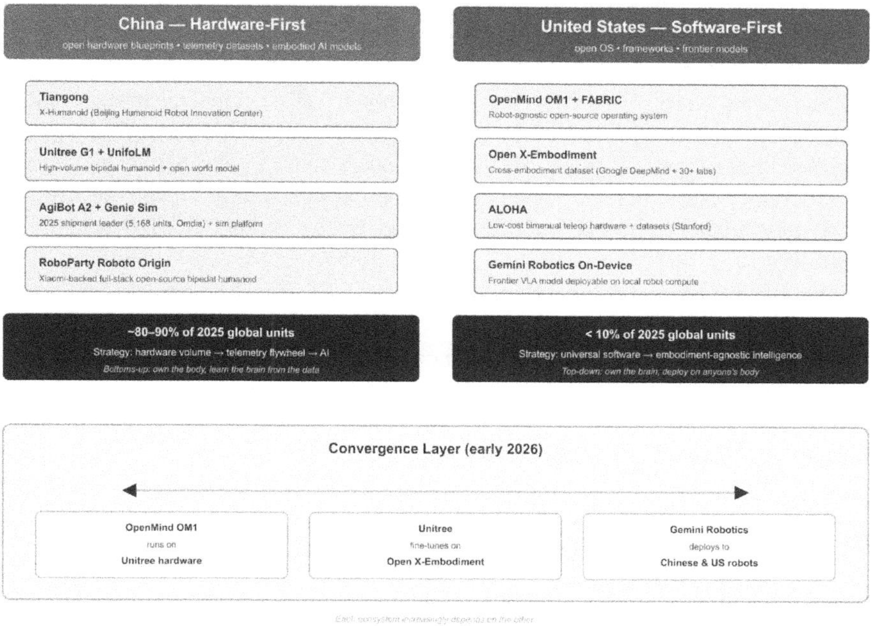


**Fig. 4.6** China versus U.S. open-source humanoid ecosystem

### 4.10.1 Architecture: Multi-Agent, Multi-Model, and Multi-Modal

Unlike monolithic VLA models, OpenClaw is built on a modular Skill system (Humanoid.guide, 2026). It operates as a local-first agentic runtime that allows developers to swap between leading models—such as Anthropic's Claude, OpenAI's o-series reasoning models, or local DeepSeek instances—depending on the task's complexity or privacy requirements. Its core architectural innovation is the Agent Communication Protocol (ACP), which enables a primary "Coordinator Agent" to spawn and manage specialized sub-agents. In a humanoid context, this allows the robot to delegate high-level visual reasoning to one agent while a separate, faster agent manages real-time environmental monitoring or specific tool-use scripts.

OpenClaw's multi-modality is cascaded: by integrating directly with messaging platforms such as Telegram and Discord (Cardenas, 2026), it serves as a bidirectional gateway: it receives high-level natural language instructions and returns real-time RGB-D visual snapshots from the robot's cameras for human verification or remote inspection. This architecture treats the humanoid not just as a machine, but as an active participant in a digital-physical multi-agent workflow.

### 4.10.2 From Desktop to Dexterity: The Unitree G1 Case Study

In March 2026, the open-source community integrated OpenClaw with the Unitree G1 humanoid via the RosClaw bridge (Goertzel, 2026). This integration allows the G1—a capable $16,000 bipedal platform—to be controlled through an agentic interface rather than traditional SDKs. Commands sent as plain text (e.g., "inspect the laboratory and report any open containers") are parsed by OpenClaw, decomposed into spatial sub-goals, and transmitted to the robot's ROS2-based motor controllers.

This hybrid approach effectively implements a System 3 layer:

System 1 (Reactive): Low-level balance and gait (Unitree proprietary).
System 2 (Reasoning): The VLA/VLM backbone (e.g., GR00T).
System 3 (Orchestration): OpenClaw, which manages long-term memory, external API calls, and fleet-wide coordination.

### 4.10.3 An Android-Equivalent for Robotics and Community Governance

At GTC 2026, NVIDIA CEO Jensen Huang positioned OpenClaw as the foundational open-source layer of the agentic computing era, comparing its growth trajectory to early Linux adoption and noting its role in lowering the barrier to complex robot control (Fierce Network, 2026; NVIDIA, 2026b). However, decentralized orchestration introduces safety and liveness risks that do not appear in ordinary software agents. Two agents may both generate locally reasonable actions while competing for the same physical resource: one arm, one gripper, one base trajectory, a shared workspace, or a limited power/thermal budget. In robotics, this is an agentic deadlock or resource-conflict problem rather than a mere scheduling inconvenience, because conflicting commands can damage hardware or injure nearby people. Practical systems therefore need a deterministic arbitration layer above the AI logic and below the execution controllers. This layer should own physical resource locks, priority rules, emergency-stop authority, command preemption, and conflict resolution, analogous to the dispatcher and task-allocation logic used in multi-robot frameworks such as Open-RMF (Open Robotics, 2026). Because OpenClaw agents have broad system access and the ability to modify their own behavior via skill, memory, and other agentic harness technologies, the future of humanoid intelligence may depend less on unrestricted agent autonomy than on community-driven agent ecosystems bounded by deterministic safety arbitration.

## 4.11 Conclusion

The agentic AI architectures examined throughout this chapter define the cognitive infrastructure driving the humanoid robot revolution of 2026. Foundation-model systems—including VLMs, VLAs, embodied-reasoning models, hierarchical planners, memory systems, and low-level control policies—are dissolving some of the old engineering boundaries between perception, language understanding, planning, and continuous motor control. This shift enables physical robots to respond more flexibly to natural-language instructions, novel scenes, and changing task conditions, while still depending on classical control and safety mechanisms for reliable execution.

The widespread implementation of the biologically inspired dual-system cognitive architecture has helped reconcile the high-frequency requirements of safe physical control (System 1) with the higher-latency demands of semantic reasoning (System 2). This synthesis produces robotic systems that are simultaneously fast and cognitively flexible. Hierarchical task-planning frameworks have extended robotic capability toward long-horizon goals that exceed the temporal scope of any single foundation-model inference.

At the same time, multi-agent coordination frameworks—both within a single robot's subsystems and across fleets of heterogeneous robots—are beginning to address the organizational and safety problems of large-scale deployment. Advances in high-fidelity simulation, synthetic data generation, predictive world models, and sim-to-real transfer can expand the range of conditions under which robot policies are trained and tested, but they do not remove the need for real-world validation. Finally, the contrasting open-source ecosystem strategies championed by China and the United States are shaping which architectural choices become widely adopted, not through technical merit alone but through data flywheels, developer communities, deployment volume, and platform lock-in effects.

What unites these advances is a shift in the engineering epistemology of robot design. The older paradigm assumed that reliable robot behavior should be specified through explicit models, rules, interfaces, and controllers wherever possible. The emerging paradigm does not eliminate that discipline; instead, it combines explicit safety and control structures with learned foundation models trained on large-scale multimodal and physical data. Current commercial systems remain hybrids: they use classical control, simulation, verification, and safety monitors alongside learned policies, VLMs, VLAs, memory systems, and planning models. The direction of progress is therefore not a simple replacement of engineered control with scale, but a layered architecture in which learned generalization is bounded by testing, constraints, and fallback mechanisms.

The coming years will test these architectures under conditions that controlled academic research cannot fully anticipate: unstructured domestic environments, time-varying industrial manufacturing conditions, the physical and legal liability complexity of operating heavy machines alongside humans, and the geopolitical pressures of a global technology competition that extends from AI chip supply chains to open-source software licensing. The architectural foundations detailed in this chapter will remain the substrate on which those tests are conducted. Their success or failure will determine whether humanoid robots fulfill the societal potential that 2025 and 2026 have begun to demonstrate.

## Comprehension Questions

1. Explain the practical limitations of the traditional sense-plan-act robotic pipeline and describe how modern agentic foundation-model architectures, including VLA components, planners, memory, feedback, and classical controllers, address those limitations. Use concrete physical examples to illustrate differences in generalization capability.
2. The dual-system cognitive architecture borrows explicitly from human cognitive science. What is the deep theoretical and biological basis for this specific design, and what specific computational hardware constraints (e.g., inference latency, GPU memory) make it mathematically necessary for safe physical humanoid robots rather than an aesthetic software design choice?
3. Compare and contrast the specific architectural philosophies of NVIDIA's proprietary GR00T N1 and Google DeepMind's Gemini Robotics. What are the relative strengths and limitations of each approach, and under what specific deployment conditions (edge vs. cloud) might each be preferred?
4. VLA models often combine foundation vision-language backbones with action heads or visuomotor policies trained on multimodal robot datasets. What advantages can this integration provide compared with traditional modular pipelines, and what new failure modes, such as physically unsafe hallucinated actions or poor out-of-distribution generalization, must designers manage?
5. Section 4.5 introduces the challenge of translating linguistically plausible semantic plans into physically feasible, safe robot kinematic behavior. Design a hybrid planning architecture that mathematically uses both LLM reasoning and strict classical PDDL motion planning to address this challenge, clearly explaining the computational role of each component.
6. The advanced MASH and SoFTA frameworks apply multi-agent reinforcement learning directly to mathematically control a single physical humanoid robot. Exactly why would multi-agent mathematical methods be useful for controlling merely one robot, and what coordination challenges does this distributed approach introduce compared to centralized single-agent RL?
7. Deeply analyze the five advanced sim-to-real mathematical transfer techniques described thoroughly in Table 4.4. For a humanoid robot intended for safe deployment in unstructured domestic home environments featuring visual diversity and unpredictable physical contact dynamics, which approach or combination of mathematical approaches would you strongly recommend, and mathematically why?
8. Predictive and generative world models may help robots "think before acting" by simulating possible physical outcomes before execution. What are the current mathematical and computational limitations of world models for safe physical

robot control, and what research progress would be required to make world-model-based planning reliable enough for safety-critical industrial applications?

9. Samsung's industrial AI strategy deploys specialized intelligent agents for drone quality control, production-manager agents for manufacturing-workflow optimization, and complex logistics coordination. Deeply evaluate this specialized multi-agent architectural design against a single, general-purpose humanoid approach for the same physical factory environment. What economic and computational factors should determine which specific architecture is chosen?
10. The chapter argues that safe behavior in agentic humanoid robots cannot be treated either as hand-specified control logic or as an emergent property of scale. What risks does this hybrid paradigm introduce for human safety, legal accountability, and regulatory compliance, and how should organizations deploying agentic humanoid fleets bound those risks through verification, monitoring, fallback control, and governance?

## References


AgiBot. (2026). *AGIBOT makes its U.S. debut with more than 5,100 robots shipped*. The Robot Report.

Bloomberg. (2026, January 8). *Chinese firms dominated global humanoid robot shipments in 2025*. Bloomberg.

Bjorck, J., Castañeda, F., Cherniadev, N., Da, X., Ding, R., Fan, L., Fang, Y., Fox, D., Hu, F., Huang, S., Jang, J., Jiang, Z., Kautz, J., Kundalia, K., Lao, L., Li, Z., Lin, Z., Lin, K., Liu, G., Luo, E., Magne, L., Mandlekar, A., Narayan, A., Nasiriany, S., Reed, S., Tan, Y. L., Wang, G., Wang, Z., Wang, J., Wang, Q., Xiang, J., Xie, Y., Xu, Y., Xu, Z., Ye, S., Yu, Z., Zhang, A., Zhang, H., Zhao, Y., Zheng, R., & Zhu, Y. (2025). *GR00T N1: An open foundation model for generalist humanoid robots*. arXiv preprint arXiv:2503.14734

Borate, S., Narayanan, V., & Panangaden, A. (2025). *LLM-based generalizable hierarchical task planning and execution for heterogeneous robot teams with event-driven replanning*. arXiv preprint.

Brohan, A., Brown, N., Carbajal, J., et al. (2022). *RT-1: Robotics transformer for real-world control at scale*. arXiv preprint arXiv:2212.06817.

Buchanan, B., & Chou, E. (2025). *Embodied AI: China's big bet on smart robots*. Carnegie Endowment for International Peace.

Cardenas, I. (2026). *RosClaw: A ROS 2 bridge connecting OpenClaw agents to robotic software stacks*. GitHub. https://github.com/PlaiPin/rosclaw

Chauhan, A. (2026). *Physical AI in 2026: How embodied intelligence is redefining industrial operations, healthcare, and smart cities*. TechAhead.

CNBC. (2026, February 2). *From Clawdbot to Moltbot to OpenClaw: Meet the AI agent generating buzz and fear globally*. CNBC.

Deloitte. (2025). *Physical AI and humanoid robots*. Deloitte Insights, Tech Trends 2026.

Deng, I., Feng, C., & Chen, W. (2026, January). *China's Unitree ships more than 5,500 humanoid robots in 2025, surpassing US peers*. South China Morning Post.

Exxact Corporation. (2026). *Vision language action (VLA) models powering robotics*. Exxact Blog.

Fierce Network. (2026, March). *NVIDIA GTC: OpenClaw is the new Linux, says Jensen Huang*. Fierce Network.

Figure AI. (2025, February). *Helix: A vision-language-action model for generalist humanoid control*. Figure AI Blog. https://www.figure.ai/news/helix

Gao, D., Zhao, B., Lee, A., Chuang, I., et al. (2025). *VITA: Vision-to-action flow matching policy.* arXiv preprint arXiv:2507.13231
Goertzel, B. (2026, March). *When AI gets physical hands: A review of OpenClaw on the Unitree G1 and other robots.* Medium (evoailabs).
Google DeepMind. (2025a). *Introducing Gemini Robotics and Gemini Robotics-ER.* Google DeepMind Blog.
Google DeepMind. (2025b). *Gemini Robotics On-Device brings AI to local robotic devices.* Google DeepMind Blog.
Google DeepMind. (2025c). *Gemini Robotics 1.5 brings AI agents into the physical world.* Google DeepMind Blog.
Google DeepMind. (2025d). *Gemini Robotics 1.5: Pushing the frontier of generalist robots with advanced embodied reasoning, thinking, and motion transfer.* arXiv preprint arXiv:2510.03342
He, T., Xie, Y., Zhu, L., Ye, L., Li, Q., & Peng, Y. (2026). From sim-to-real to learn-in-real: Real-world online learning for humanoid robots. In T. Matsuno et al. (Eds.), *Intelligent robotics and applications.* ICIRA 2025, Lecture Notes in Computer Science, vol. 16075. Springer.
He, T., Zhang, Z., Luo, Z., Ma, W., Zhang, M., Fu, Z., & Leng, Z. (2025). *ASAP: Aligning simulation and real-world physics for learning agile humanoid whole-body skills.* arXiv preprint.
Howell, T. A., Le Cleac'h, S., Brüdigam, J., Chen, Q., Sun, J., Kolter, J. Z., Schwager, M., & Manchester, Z. (2022). *Dojo: A differentiable physics engine for robotics.* arXiv preprint arXiv: 2203.00806
Humanoid.guide. (2026). *OpenClaw connects AI agents to Unitree G1 humanoid robot.* https://humanoid.guide/openclaw-connects-ai-agents-to-unitree-g1-humanoid-robot/
Huang, Y. (2026a). *Xiaomi-backed RoboParty launches ORIGIN: World's first-tier full-stack open-source bipedal humanoid robot.* GlobeNewswire/Manila Times.
Huang, K. (2026b). *World models, architectures, and the next phase of AI. Agentic AI.* https://kenhuangus.substack.com/p/world-models-architectures-and-the
Malayil, J. (2026). *World's first full-stack humanoid robot open-sourced for developers.* Interesting Engineering.
IDC. (2026). *Worldwide humanoid robot market shipments, 2025.* IDC Market Analysis.
International Federation of Robotics. (2026). *Top 5 global robotics trends 2026.* IFR Press Release. Retrieved from https://ifr.org/ifr-press-releases/news/top-5-global-robotics-trends-2026
Kalil, M. (2025). *Tiangong: World's first electric humanoid robot goes open source.* Mike Kalil Blog.
Kawabe, T., & Takano, R. (2026). *Hierarchical LLM-based multi-agent framework with prompt optimization for multi-robot task planning.* arXiv preprint.
Kawaharazuka, K., Oh, J., Yamada, J., Posner, I., & Zhu, Y. (2025). Vision-language-action models for robotics: A review towards real-world applications. *IEEE Access, 13*, 162467–162504.
LeCAR Lab. (2025). *Hold my beer: Learning gentle humanoid locomotion and end-effector stabilization control* [SoFTA framework]. arXiv preprint arXiv:2505.24198.
Li, X., Li, P., Qian, L., Liu, M., Wang, D., Liu, J., Kang, B., Ma, X., Wang, X., Guo, D., Kong, T., Zhang, H., & Liu, H. (2026). What matters in building vision-language-action models for generalist robots. *Nature Machine Intelligence, 8*, 158–172.
Liu, Y., Chen, X., Zhou, M., Wang, T., & Li, J. (2025). *Hierarchical language models for semantic navigation and manipulation in an aerial-ground robotic system.* Advanced Intelligent Systems.
Liu, Q., Zhang, X., Tan, M., Ma, S., Ding, J., & Li, Y. (2025). *MASH: Cooperative-heterogeneous multi-agent reinforcement learning for single humanoid robot locomotion.* arXiv preprint arXiv: 2508.10423
Lo, K. (2026). *China is winning the humanoid robot race while Tesla's Optimus lags.* Rest of World.
NVIDIA. (2025a). *NVIDIA announces Isaac GR00T N1—The world's first open humanoid robot foundation model.* NVIDIA Newsroom.
NVIDIA. (2025b). *Accelerate generalist humanoid robot development with NVIDIA Isaac GR00T N1.* NVIDIA Technical Blog.
NVIDIA. (2025c). *NVIDIA computing platforms for humanoid robots.* NVIDIA Use Cases.
NVIDIA. (2026a). *NVIDIA Cosmos: World foundation models for physical AI.* NVIDIA.

NVIDIA. (2026b). *NVIDIA GTC 2026 keynote: Live updates on what's next in AI*. NVIDIA Blog.
Open Robotics. (2026). *Programming multiple robots with ROS 2: Tasks in RMF. Open-RMF documentation*. https://osrf.github.io/ros2multirobotbook/task.html
Openrobotics.org. (2025). *OpenMind launches OM1 Beta open-source, robot-agnostic operating system*. The Robot Report.
Pantera Capital. (2025). *Investing in OpenMind*. Pantera Capital Blog.
Park, K. (2026, February 28). *Why China's humanoid robot industry is winning the early market*. TechCrunch.
Physical Intelligence. (2026). *MemER: Scaling episodic memory for embodied reasoning*. Physical Intelligence Research.
Qiao, X. (2025). *China's Unitree open-sources world model to advance robotics ecosystem*. Yicai Global.
Reuss, M. (2026). *State of vision-language-action research at ICLR 2026*. Personal blog. https://mbreuss.github.io/blog_post_iclr_26_vla.html
Robot Report. (2025, August 4). *OpenMind raises $20M to connect intelligent machines*. The Robot Report.
Schmidt, E., & Xu, S. (2026). *China could dominate the physical AI future*. Time. https://time.com/7382151/china-dominates-the-physical-ai-race/
Schwarke, C., Klemm, V., Bagajo, J., Sleiman, J.-P., Georgiev, I., Tordesillas, J., & Hutter, M. (2024). *Learning deployable locomotion control via differentiable simulation*. arXiv preprint arXiv:2404.02887
Singh, R. (2026). *Vision-language-action (VLA) models: The AI brain behind the next generation of robots & physical AI*. Stackademic (Medium).
Steinberger, P. (2025). *OpenClaw: An open-source agentic AI runtime (formerly Clawdbot, Moltbot)*. GitHub. https://github.com/openclaw/openclaw
TechBuzz AI. (2026). *Samsung bets on agentic AI to transform global factories by 2030*. TechBuzz AI.
Unitree Robotics. (2025). *UnifoLM-WMA-0: A world-model-action framework*. GitHub.https://github.com/unitreerobotics/unifolm-world-model-action
Wang, S., Han, M., Jiao, Z., Zhang, Z., Wu, Y. N., Zhu, S.-C., & Liu, H. (2024). *LLM3: Large language model-based task and motion planning with motion failure reasoning*. arXiv preprint arXiv:2403.11552
Zhang, T., Qu, K., Li, Z., Wu, J., Hutter, M., Li, M., & Shi, F. (2026). *Using large language models for embodied planning introduces systematic safety risks*. arXiv preprint arXiv:2604.18463

**Grace Huang** is a seasoned product and technology professional with extensive experience at organizations including PIMCO (a global investment management firm managing over $2 trillion in assets), IBM, SoftBank, and DataRobot. Throughout her career, she has successfully launched multiple products and managed large-scale projects, leveraging her expertise in product management, market analysis, strategic planning, and cross-functional team leadership. Her work brings a finance and product strategy perspective to physical AI, particularly the operational, risk, and adoption considerations involved in bringing advanced robotics into enterprise environments. Always at the forefront of emerging technology, she has explored innovative tools including the implementation of Agentic AI powered solutions within the product management process—streamlining communication, improving decision-making, enhancing customer satisfaction, and driving business growth. In addition to her professional accomplishments, Grace holds a degree from Babson College, where she built a strong foundation in business management and entrepreneurship. She continues to lead at the intersection of product strategy and advanced technology, bringing a rare combination of financial rigor and deep technical acumen to complex, high-stakes initiatives.

Her work brings a finance and product strategy perspective to physical AI, particularly the operational, risk, and adoption considerations involved in bringing advanced robotics into enterprise environments. Always at the forefront of emerging technology, she has explored innovative tools including the implementation of Agentic AI powered solutions within the product management process—streamlining communication, improving decision-making, enhancing customer satisfaction, and driving business growth.

In addition to her professional accomplishments, Grace holds a degree from Babson College, where she built a strong foundation in business management and entrepreneurship. She continues to lead at the intersection of product strategy and advanced technology, bringing a rare combination of financial rigor and deep technical acumen to complex, high-stakes initiatives.

**Ken Huang** is a globally recognized expert in AI and Web3 security, a prolific author, and a leading figure in shaping industry standards. He is the CEO and Chief AI Officer (CAIO) of DistributedApps.ai, specializing in Generative AI training and consulting. Ken is deeply involved in driving the development of secure AI systems. He serves as a Research Fellow and Co-Chair of the AI Safety Working Groups at the Cloud Security Alliance (CSA), leading Agentic AI initiatives. He is also a Co-Chair of the AI STR Working Group at the World Digital Technology Academy under the UN Framework, and a core contributor to the OWASP GenAI project, focusing on Agentic AI security. His expertise extends to his contributions to OWASP's Top 10 Risks for LLM Applications and his active participation in the NIST Generative AI Public Working Group. He is also a member of Open AI forum.

**Tongzhou Mu** is a researcher at the forefront of world modeling and robot simulation. Currently serving as a Research Scientist at Rhoda AI, his work focuses on advancing video world models for robot manipulation. Tongzhou completed his doctoral studies at UC San Diego under Prof. Hao Su. Throughout his career, Tongzhou has contributed to high-impact research teams at NVIDIA, 1X, and Microsoft Research. He is perhaps best known for his work on ManiSkill, a benchmark he originally developed that has since evolved into a popular open-source framework for the robot learning field.

**Manish Bhatt** Based in Silicon Valley, Manish Bhatt is a technical leader dedicated to making autonomous systems both safe and secure. He currently serves as the Application Security Tech Lead for AI & Crypto at Amazon Leo. With over 15 years in the field, Manish specializes in the critical space where complex hardware meets intelligent software.

He has been at the forefront of AI safety as a founding engineer of Purple Llama and the lead behind the CyberSecEval series, which helped set the standard for evaluating risks in frontier models. His research is widely recognized, covering everything from kernel rootkits to the safety of Large Language Models, and has earned over 20,000 citations.

As a founding member of the OWASP AI Security Verification Standard (AISVS), Manish is now focused on building the Artificial Intelligence Vulnerability Scoring System (AIVSS). His goal is to better understand risk in Agentic AI, ensuring that as AI technology moves into humanoid robotics, these systems remain resilient and trustworthy for everyone.

# Integration and Control Systems

# 5

Ken Huang, Grace Huang, Chuer Pan, and Manish Bhatt

## 5.1 Introduction

The preceding chapters established humanoid robots' physical capabilities (Chapter 2: actuators, structures, power systems), perceptual capabilities (Chapter 3: vision, LiDAR, tactile sensing, sensor fusion), and cognitive capabilities (Chapter 4: AI foundation models, reinforcement learning, planning, memory). However, these sophisticated components in isolation do not create functioning robots capable of autonomous operation.

Integration—combining perception, cognition, and action into coherent systems executing coordinated behaviour while meeting hard real-time constraints and strict stability and safety requirements—represents the critical challenge that separates research prototypes from commercially deployable robots.

Integration spans multiple dimensions. Temporal integration coordinates processes operating at vastly different timescales: high-level planning at 1–10 Hz (hundreds of milliseconds per decision), mid-level motion planning at 10–100 Hz (tens of milliseconds per update), and low-level control at 100–1000 + Hz (milliseconds or sub-millisecond per cycle). Spatial integration coordinates control

K. Huang (✉)
DistributedApps.Ai, Fairfax, VA, USA
e-mail: ken@distributedapps.ai

G. Huang
PIMCO, Austin, TX, USA

C. Pan
Stanford University, Stanford, CA, USA
e-mail: chuerpan@stanford.edu

M. Bhatt
Amazon Kuiper Security, OWASP, Seattle, WA, USA



K. Huang et al. (eds.), *Humanoid Robots and Physical AI*, Advances in Data Analytics, AI, and Smart Systems, https://doi.org/10.1007/978-3-032-33669-9_5

across roughly 30–56 body degrees of freedom, ensuring whole-body motion maintains balance while executing tasks; dexterous hands can add more than 40 additional degrees of freedom in two-handed humanoid systems. Modular integration connects perception, planning, and control modules with well-defined interfaces enabling independent development while ensuring coherent system behaviour. Safety integration ensures that all levels of the control hierarchy respect safety constraints preventing harmful behaviours.

The challenge magnitude should not be understated. Contemporary humanoid robots run dozens to hundreds of software processes concurrently operating at different priorities and frequencies: multiple perception pipelines processing camera streams, LiDAR point clouds, and tactile sensor arrays; large AI foundation models providing language understanding, task planning, and learned behavioural control, including end-to-end manipulation policies; multiple real-time balance, force, and grasp-closure controllers operating in tight feedback loops; comprehensive safety monitoring systems checking for constraint violations, hardware failures, and anomalous behaviours; and human interface components enabling natural interaction and remote monitoring. These numerous concurrent processes exchange gigabytes of data per second through shared memory and message passing, must meet strict microsecond to millisecond timing deadlines for safety-critical functions, and cannot fail or violate timing constraints even under worst-case loading conditions. Achieving this level of reliable real-time integration requires sophisticated layered software architectures, real-time operating systems with deterministic scheduling, and rigorous engineering practices adapted from aerospace and automotive safety-critical system development.

As of 2026, the field exhibits diverse integration approaches reflecting different company backgrounds, priorities, and strategic philosophies. Boston Dynamics leverages decades of accumulated robotics expertise through proprietary real-time control frameworks achieving 1000 + Hz control loop rates, but its deeper differentiation lies in hardware design, in-house actuator expertise, and tightly integrated control rather than in dynamic showcase behaviours alone, which are increasingly common across the industry. Tesla applies automotive software engineering discipline including vertically integrated development processes, extensive continuous integration and automated testing, and over-the-air update capabilities; these practices were an early perceived advantage for Optimus, though comparable software infrastructure is increasingly standard among serious humanoid programs. Startups and research institutions continue to use open-source frameworks, particularly ROS 2, after ROS 1 Noetic reached end of life on May 31, 2025 (Open Robotics Discourse, 2025), but production-oriented systems often combine ROS 2 with proprietary or newer middleware optimised for deterministic real-time control and learned-policy dataflow.

This chapter examines integration across multiple perspectives. Section 5.2 analyses real-time operating systems providing deterministic execution. Section 5.3 explores control hierarchies coordinating multiple timescales. Section 5.4 addresses low-level control algorithms. Section 5.5 examines whole-body control frameworks. Section 5.6 investigates simulation and digital twins.

Section 5.7 covers sim-to-real transfer challenges. Section 5.8 addresses software development practices. Section 5.9 examines computational infrastructure. We conclude (Sect. 5.10) synthesising insights about integration challenges and evolution trajectories.

## 5.2 Real-Time Operating Systems and Middleware

Real-time operating systems (RTOS) provide the software foundation enabling deterministic, predictable execution of time-critical control code. Unlike general-purpose operating systems optimising for average-case throughput, RTOS prioritises worst-case latency guarantees essential for safety–critical robotics applications.

### 5.2.1 RTOS Requirements and Characteristics

Humanoid robot control imposes stringent timing requirements. Balance control must run at 100–1000 Hz with bounded latency—delays exceeding a few milliseconds risk falls. Manipulation force control requires similar rates for stability. These hard real-time constraints demand RTOS capabilities.

Determinism: The same inputs must produce outputs at predictable times with bounded jitter (timing variation). General OS may delay time-critical tasks when higher-priority system operations pre-empt them. RTOS guarantees that highest-priority tasks execute within specified deadlines.

Priority-based scheduling: Tasks assigned priorities execute in priority order, with highest-priority tasks always pre-empting lower-priority ones. This ensures critical safety loops never wait for less critical operations like logging or communication.

Bounded interrupt latency: External events (sensor readings, timer ticks) trigger interrupts requiring rapid response. RTOS bounds maximum interrupt latency—guaranteeing responses within microseconds—where general OS may take milliseconds due to interrupt masking or processing other interrupts.

Minimal jitter: Control loops require consistent timing. Running at "1000 Hz" means executing every 1 ms—not 0.9 ms then 1.3 ms. RTOS minimises jitter through precise timers and priority enforcement.

Hardware-induced jitter mitigation: Deterministic timing also depends on hardware and firmware configuration, not only the RTOS scheduler. Practical robot controllers isolate real-time threads on dedicated CPU cores, pin EtherCAT or motor-control interrupts away from noncritical workloads, disable or limit CPU frequency scaling and deep sleep states, pre-allocate memory to avoid page faults, and measure worst-case latency under stress using tools such as cyclictest. These practices reduce jitter caused by network interrupts, DMA activity, USB/GPU drivers, cache contention, and firmware-level events that can otherwise break millisecond control deadlines even on a real-time kernel (Intel, 2026; Linux Kernel Documentation, 2026).

The importance of real-time operating systems (RTOS) in robotics is reflected in strong market growth. The RTOS market was valued at approximately $5.97 billion in 2024 and is projected to grow from $6.41 billion in 2025 to $12.21 billion by 2034, representing a CAGR of about 7.41% (Market Research Future, 2025). This expansion is driven largely by applications in robotics and autonomous systems, where RTOS platforms enable deterministic control of robot motion, real-time sensor processing, and low-latency communication required for high-precision operation in industrial automation, medical robotics, and autonomous vehicles.

### 5.2.2 ROS 2 and Emerging Middleware Strategies

Robot Operating System (ROS) provides not an operating system but middleware—software layers facilitating communication between components, providing common libraries, and standardising interfaces. ROS 2, the second-generation architecture, remains one of the most important open-source frameworks for robotics research, prototyping, and integration as of 2026, but it is increasingly part of hybrid stacks rather than the sole production middleware. The industry trend is towards hardened middleware: ROS-compatible interfaces for development speed, alternative transport layers such as DDS and Zenoh, and production runtimes that emphasise deterministic execution, fault tolerance, safety certification, and fleet deployability (Apex.AI, 2025; Eclipse Foundation, 2023; ROS 2 Documentation, 2026).

Support for ROS 1 officially ended on May 31, 2025 with the conclusion of long-term support for ROS 1 Noetic. This milestone accelerated migration to ROS 2, while also pushing industrial teams to evaluate which parts of their stacks should use open middleware and which require custom deterministic implementations. Fleet-scale deployments also expose DDS discovery limits: default ROS 2 discovery relies on multicast advertisements across the subnet, and packet growth can degrade WiFi performance when many robots or nodes start simultaneously. Production systems therefore increasingly restrict discovery scope, use discovery servers, or route traffic through Zenoh-style bridges and rmw_zenoh to reduce broadcast traffic while preserving ROS-compatible development workflows.

Figure 5.1 illustrates a three-layer control architecture commonly implemented in ROS 2 systems. In the figure and discussion, motion and force control describe controlled variables, while balance describes a control objective; the reactive layer may combine all three to maintain stability while executing contact-rich manipulation. Recent manipulation surveys similarly describe lower-level controllers that maintain stable grasps or execute manipulation primitives using tactile and force feedback, while higher-level learned policies select or adapt those behaviours (Weinberg et al., 2024).

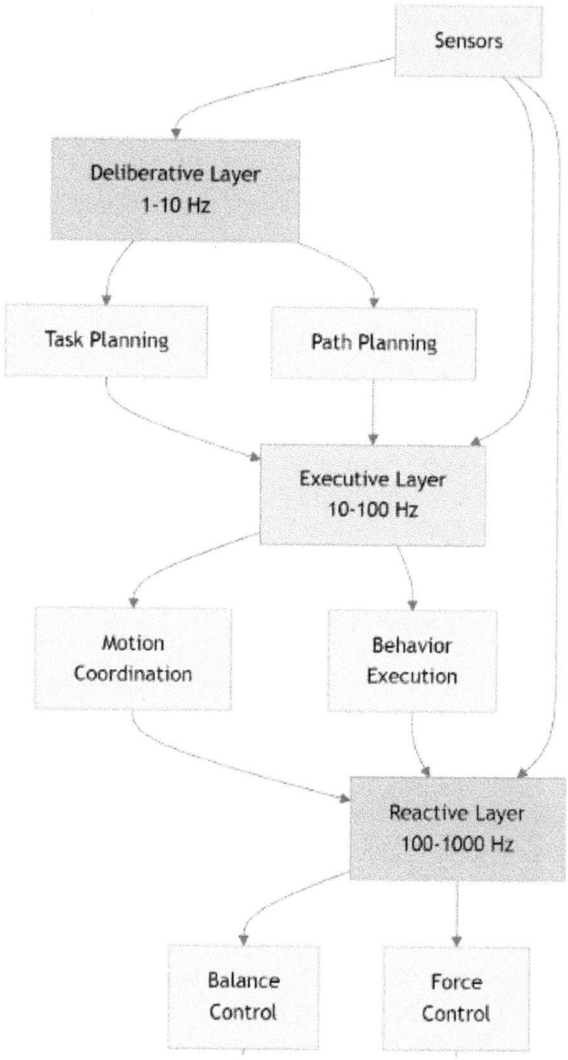


**Fig. 5.1** Three-layer robot control architecture

This architecture separates concerns across timescales: deliberative planning reasons about task decomposition and long-term strategies at slow rates; executive coordination manages ongoing task execution at medium rates; reactive control ensures safety and stability at high rates.

ROS 2 introduces several major improvements compared with ROS 1, including better real-time performance, support for distributed system architectures, and stronger security and scalability features. It also provides native compatibility with major operating systems such as Linux, Windows, and macOS (Robotnik Automation, 2025). One of the most significant architectural changes is the transition from the TCP/UDP communication approach used in ROS 1 to DDS (Data Distribution

Service) as the default middleware. DDS enables configurable Quality of Service (QoS) policies that support more reliable real-time communication, improved scalability, better performance, and enhanced security in robotic systems.

DDS supports more predictable communication timing for real-time control—publishers and subscribers negotiate QoS contracts specifying latency and reliability behaviour. This is critical when control commands must reach actuators within milliseconds or sensor data must propagate to controllers without delays. At the same time, ROS 2 now supports multiple middleware implementations, including DDS-based implementations and Zenoh through the ROS middleware API, reflecting a broader move towards selectable communication backends for different latency, bandwidth, topology, and deployment constraints.

However, challenges persist: A key concern among developers is that a substantial portion of the ROS 1 package ecosystem has yet to be ported to ROS 2, and this dependency on legacy packages continues to discourage many from making the transition (Portugal et al., 2024). This ecosystem maturity gap means some functionality available in ROS 1 remains unavailable in ROS 2, forcing developers to reimplement or wait for community migrations (Portugal et al., 2024).

### 5.2.3 Proprietary vs. Open-Source Approaches

Companies face strategic choices between open-source frameworks (ROS 2, generic RTOS like FreeRTOS, VxWorks) and proprietary custom systems. Each approach offers distinct advantages creating different competitive positions.

Boston Dynamics employs proprietary real-time control framework achieving 1 kHz + control rates, optimised specifically for Atlas' demanding dynamic capabilities including backflips, parkour, and rapid whole-body manoeuvres. This custom framework reflects decades of robotics expertise accumulated since the company's 1992 founding, creating deep technical moats competitors cannot easily replicate. The proprietary approach enables optimisation impossible with general frameworks—custom scheduling algorithms tailored to specific control loop structures, hand-optimised communication protocols minimising latency, and tight integration with custom hardware.

Tesla likely employs custom Linux-based RTOS with FSD heritage, leveraging extensive automotive real-time systems expertise from Full Self-Driving development. The automotive pedigree brings proven approaches for safety–critical real-time software developed and validated across millions of vehicles. Tesla's vertical integration extends to software—controlling entire stack from operating system through application layer enables optimisation and rapid iteration impossible when depending on external frameworks.

Proprietary systems provide several strategic advantages beyond pure technical performance. Intellectual property protection prevents competitors from easily copying control strategies or integration approaches that may represent competitive differentiators. Freedom from external dependencies means development priorities are controlled internally rather than subject to open-source project governance and release schedules. Deep optimisation for specific applications enables extracting maximum performance from hardware. And proprietary systems can employ novel approaches without constraining to community standards or compatibility requirements.

However, proprietary development imposes substantial costs and risks. Development requires large engineering teams with expertise spanning operating systems, middleware, control algorithms, and application software—potentially dozens of engineers over multiple years representing millions in development costs. The burden of maintaining and evolving the entire software stack rests on the company rather than being shared across community. Integration of third-party components (sensors, actuators, tools) may be complicated without standard interfaces. And hiring engineers familiar with proprietary systems is harder than hiring for widely-known open-source frameworks.

Open-source approaches, particularly ROS 2, remain widely used among startups and companies prioritising rapid development. ROS 2 provides mature, well-tested frameworks reducing development time from years to months by leveraging community work. The ecosystem offers vast libraries: perception packages for standard sensors, motion planning libraries, manipulation primitives, navigation stacks, and simulation integration. Driver support for hundreds of sensors and actuators eliminates custom driver development. Development tools including visualisation (RViz), debugging, and recording/playback facilitate development and testing.

The ROS 2 ecosystem also facilitates hiring and collaboration. Engineers with ROS experience are abundant—universities teach ROS, online tutorials are extensive, and community support through forums and documentation is comprehensive. Code sharing across projects and companies accelerates progress—solutions developed by one group benefit entire community. Standards compliance ensures components work together—sensors, actuators, planners from different vendors interoperate through standard ROS 2 interfaces.

However, ROS 2 trade-offs include potential performance limitations versus custom-optimised systems for demanding applications, dependency on community development priorities which may not align with specific company needs, security concerns from running open-source code with public-known vulnerabilities (though also benefits from community security review), and compatibility constraints where maintaining ROS 2 compliance may limit architectural choices.

The industry increasingly employs hybrid approaches recognising that binary choice between fully proprietary and fully open-source represents a false dichotomy. Companies use ROS 2 for non-time-critical subsystems (perception preprocessing, user interfaces, data logging, remote monitoring) where community packages provide immediate value, while implementing custom or safety-engineered real-time control layers for safety–critical loops (balance control, force control, emergency stops) where performance, fault containment, certification, and reliability requirements exceed what a default ROS 2 stack can guarantee. This hybrid strategy captures open-source ecosystem benefits while hardening the middleware and runtime layers that matter most for production deployment.

The strategic calculus depends on company resources, technical requirements, and competitive positioning. Well-funded companies targeting maximum performance may justify proprietary investments. Startups with limited resources often employ ROS 2 to accelerate time-to-market. Companies in the middle ground often start with ROS 2 then gradually replace components with custom implementations as resources permit and requirements demand.

As the field matures, we may see convergence towards standard interfaces with proprietary implementations—similar to how automotive industry has standard ECU interfaces but proprietary control algorithms, or smartphone industry has standard apps ecosystems but proprietary operating systems. Robotics could develop standard middleware and application interfaces enabling ecosystem development while allowing proprietary optimisation of underlying implementations.

## 5.3 Control Hierarchies and Architectures

Humanoid robot control must simultaneously handle multiple objectives across different timescales. Hierarchical architectures decompose this complexity into manageable layers, each operating at appropriate rates and abstraction levels.

### 5.3.1 Three-Layer Architecture Details

The three-layer architecture shown in Fig. 5.1 warrants detailed examination:

Deliberative Layer (1–10 Hz): This top layer implements high-level cognition—task understanding from natural language, task decomposition into subgoals, long-horizon planning over seconds to minutes, and learning from past experiences. Processing rates are slow because decisions involve complex reasoning potentially requiring large AI model inference. Example components include language understanding models parsing verbal instructions, task planners decomposing complex objectives, and strategic decision-making selecting which tasks to attempt given current state.

Executive Layer (10–100 Hz): The middle layer manages ongoing task execution—monitoring progress towards subgoals, coordinating multiple behaviours (locomotion while manipulating), selecting appropriate low-level controllers, and handling exceptions or failures. This layer bridges abstract plans from above with concrete control below. Example components include behaviour trees managing state transitions, motion planners generating collision-free trajectories, and grasp planners selecting contact points.

Reactive Layer (100–1000 + Hz): The bottom layer ensures immediate safety and stability—maintaining balance through joint torque control, regulating manipulation forces through compliance control, making fine grasp and contact adjustments, avoiding collisions through emergency stops, and tracking motion plans through precise trajectory following. This layer operates at rates matching or exceeding sensor sampling to provide fast feedback loops. Example components include balance controllers adjusting centre of mass, impedance controllers regulating contact forces, grasp-closure and in-hand adjustment controllers responding to motor current, force-torque, or tactile feedback, and safety monitors triggering emergency responses. As learned vision-language-action (VLA) policies move higher in the stack, the reactive layer also needs a safety governor that treats deliberative commands as proposals rather than trusted facts: reachability maps, control barrier functions, torque/contact limits, and collision checks can filter or minimally modify commands that would drive the robot outside physically safe states (Borquez et al., 2024).

### 5.3.2 Information Flow and Coupling

The layers communicate bidirectionally. Top-down flow carries commands: deliberative layer specifies tasks, executive layer decomposes into behaviours, reactive layer executes control. Bottom-up flow carries status: reactive layer reports execution progress and safety violations, executive layer reports subgoal completion or failures, deliberative layer updates plans based on outcomes.

Tight coupling between layers risks instability—reactive control changes could disrupt executive plans, executive adaptations could conflict with deliberative objectives. Loose coupling risks incoherence—layers operating independently without coordination. Achieving appropriate coupling requires careful interface design: hierarchical layers specify objectives and constraints without micromanaging lower layers, lower layers provide sufficient status information for higher layers to assess progress without overwhelming with details, and safety constraints propagate top-down as hard boundaries while bottom-up feedback informs but doesn't control higher layers.

### 5.3.3 Behaviour Trees and State Machines

Executive layer implementation has traditionally employed behaviour trees—hierarchical graphical structures representing task logic through composable nodes that can be developed, tested, and refined modularly. Behaviour trees remain useful for explicit task orchestration, safety gating, and interpretable state transitions. However, manipulation systems increasingly use end-to-end learned policies that map images, tactile signals, force-torque readings, and robot state directly to actions, reducing the amount of manually specified behaviour logic required for contact-rich tasks.

Behaviour tree structures consist of several node types serving distinct functions, while newer learned-policy systems often replace some of this manually authored structure with behaviour learned from demonstrations. This shift moves engineering effort from writing explicit task logic towards collecting, curating, labelling, replaying, and scaling robot data:

Action nodes represent atomic behaviours that the robot can execute directly: navigate to a specified location, grasp a detected object with a particular strategy, place a held object at a target position, wait for a specified duration, or check a sensor reading against a threshold. These leaf nodes connect directly to robot capabilities. In end-to-end manipulation, an imitation-learned policy can instead learn portions of this action selection implicitly from observation-action demonstrations, including robot teleoperation data, third-person video, egocentric human video, and portable end-effector-centric collection systems. The Universal Manipulation Interface (UMI), for example, uses a handheld gripper and a latency-aware policy interface to collect in-the-wild manipulation demonstrations without requiring an in-the-wild robot, then transfers those demonstrations into deployable robot policies through embodiment-specific interfaces such as end-effector inverse kinematics (Chi et al., 2024). This is why data has become a central bottleneck: policies such as RT-X, UMI-style manipulation systems, and generalist VLA models improve when trained on diverse demonstrations, but collecting high-quality real-world robot-deployable trajectories remains slower and more expensive than scaling text or image data (Black et al., 2024; Chi et al., 2024; Open X-Embodiment Collaboration et al., 2023).

Data infrastructure therefore becomes part of the control stack. Tooling such as LeRobot emphasises recording teleoperated trajectories, visualizing datasets, training policies from demonstrations, and replaying episodes on real hardware; these workflows make imitation learning operational rather than purely algorithmic (Hugging Face, 2026). For humanoid companies, the defensible asset increasingly becomes the closed loop of robot deployment, data collection, policy improvement, and validation, not just the model architecture itself.

Composite nodes combine multiple child behaviours through different logic:

- Sequence nodes execute children in order (do A, then B, then C), succeeding only if all children succeed and failing if any child fails
- Selector nodes try children in order until one succeeds (try A, if fails try B, if fails try C), succeeding when any child succeeds
- Parallel nodes execute multiple children simultaneously (walk while monitoring for obstacles), defining success/failure conditions based on children outcomes

Decorator nodes modify child behaviour execution:

- Repeat decorators execute child multiple times (repeat 3 times) or until condition changes (repeat while battery sufficient)
- Inverter decorators reverse child success/failure status
- Timeout decorators limit child execution time, failing if child exceeds allocated time
- Precondition decorators check conditions before attempting child execution

This compositional structure enables modular task design where behaviours at different abstraction levels develop and test independently then compose into complex task trees. A "prepare workspace" task might sequence: check workspace state, clear existing items (selector trying different clearing strategies), fetch required tools (parallel fetching of multiple items), and arrange in working configuration. Each component subtree can be developed separately and reused across tasks.

Advantages of behaviour trees include: modularity where behaviours are developed and tested independently then composed, avoiding monolithic code; visual clarity allowing stakeholders including non-programmers to understand and validate task logic through graphical tree representations; runtime flexibility where trees can be modified dynamically based on environmental conditions or generated procedurally from high-level specifications; and debuggability where execution state (which nodes active, which succeeded/failed) is explicitly visible facilitating troubleshooting.

Limitations create challenges for some applications: behaviour trees can become very large and complex for sophisticated multi-step tasks with many conditionals and error-handling branches, potentially creating tree structures with hundreds of nodes difficult to comprehend; they naturally represent discrete decisions (try this behaviour, if fails try alternative) but struggle with continuous optimisation problems better suited to numerical methods; reactivity to environmental changes may be limited since tree evaluation proceeds top-down at finite rates rather than reacting instantaneously to events; and memory and computational overhead from tree traversal and state management can be substantial for very large trees.

Finite state machines (FSMs) provide alternative executive layer implementations through explicitly representing system states and transitions. States correspond to distinct robot operating modes: idle waiting for task assignment, navigating towards goal location, manipulating object, recovering from error, charging battery, or shut down for maintenance. Transitions between states trigger based on events or conditions: task assigned transitions idle → navigating, destination reached transitions navigating → manipulating, object grasped enables manipulation progress, error detected transitions any state → error recovery, battery critically low forces transition to charging.

FSMs excel for systems naturally described as discrete modes with clear transitions. Robot operating modes (startup, operational, error, shutdown) map naturally to states. Safety-critical mode management benefits from FSM explicit state representation enabling formal verification of state reachability and transition safety. However, FSMs struggle with continuous behaviours that don't decompose into discrete states, parallel task execution requiring complex state products, and hierarchical decomposition where state machines at different levels must coordinate.

Modern robotic systems increasingly combine these approaches synergistically rather than choosing one exclusively: behaviour trees implement task-level logic describing task decomposition and sequencing, finite state machines manage robot operating modes and safety-critical transitions, and continuous optimisation handles motion planning and low-level control. The integration leverages each approach's strengths while avoiding limitations—discrete task logic in trees, mode management in state machines, continuous planning in optimisation.

## 5.4 Low-Level Control Algorithms

The reactive layer implements low-level control algorithms translating desired behaviours (maintain balance, follow trajectory, apply specific force) into motor commands (joint torques, velocities). Multiple algorithmic approaches address different control objectives and requirements.

### 5.4.1 PID Control and Classical Approaches

Proportional-Integral-Derivative (PID) control remains fundamental despite its age. PID controllers drive errors towards zero through proportional response to current error, integral response accumulating past errors (eliminating steady-state offsets), and derivative response to error rate of change (damping oscillations and improving transient response).

For robotics, PID controllers regulate joint positions and velocities—comparing desired versus actual joint angles, computing control signals from error terms, and commanding actuator torques or currents. Proper PID tuning determines performance: too aggressive gains create oscillations or instability, too conservative gains yield slow response and large tracking errors.

However, PID limitations become apparent for complex multi-degree-of-freedom systems. PID treats joints independently, ignoring dynamic coupling between joints. For humanoid robots where moving one joint creates forces affecting others through body dynamics, independent PID control can create suboptimal or unstable behaviour. More sophisticated model-based approaches address these limitations.

### 5.4.2 Model Predictive Control (MPC)

Model Predictive Control optimises control inputs over a finite future horizon, explicitly accounting for system dynamics, constraints, and objectives. Figure 5.2 illustrates the MPC loop.

Recent research demonstrates MPC effectiveness for robotics: A novel Spatial Operator Algebra Nonlinear Model Predictive Control (SOA-NMPC) strategy was developed to overcome the heavy computational demands traditionally associated with nonlinear MPC (NMPC), with its performance validated through experiments on a five-degree-of-freedom robot manipulator (Yaren & Kizir, 2025). The spatial operator approach efficiently computes robot dynamics enabling real-time nonlinear MPC previously considered too computationally expensive.

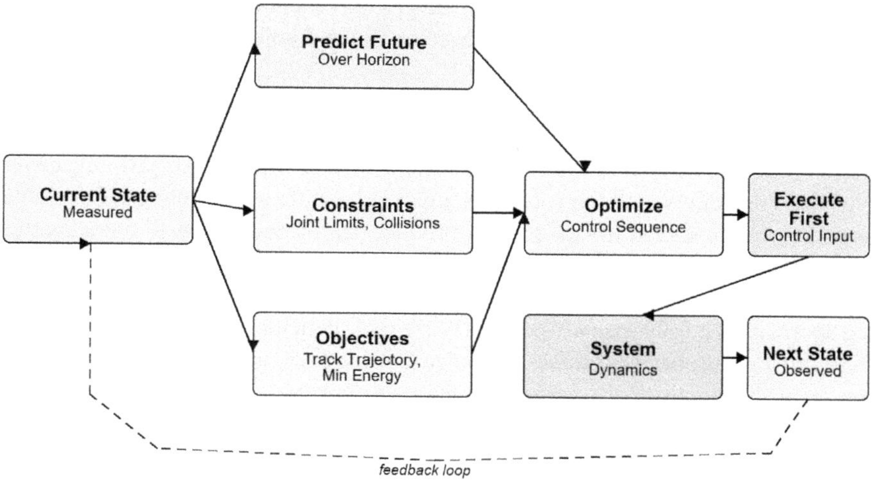


**Fig. 5.2** Model predictive control receding horizon loop

Work on fast sampling-based MPC through rollout differentiation showed promising results across both simulated environments and physical platforms, including a four-legged robot navigating irregular terrain and an aerial drone performing high-agility manoeuvres using only onboard processing power (Belvedere et al., 2026). This proves MPC can run on embedded computers in real time for dynamic behaviours, not just quasi-static applications.

The applications span domains: MPC has become a preferred control strategy in robotics precisely because modern robotic systems must operate in unpredictable environments while satisfying strict safety margins and balancing competing objectives—tasks that conventional PID controllers are ill-equipped to handle. It is widely deployed across mobile robots, manipulators, drones, and autonomous vehicles where constraint satisfaction is essential (Schwenzer et al., 2021).

Recent work addresses safety: A combined safe and disturbance-compensated MPC framework pairs a disturbance observer with standard MPC to actively estimate and counteract real-time uncertainties, while embedding a control barrier function as an online constraint to guarantee that the system remains within safe operating bounds (Cao et al., 2025). This ensures MPC maintains safety even under disturbances and modelling errors.

### 5.4.3 Whole-Body Optimisation-Based Control

For humanoid robots with 30–56 coupled degrees of freedom, whole-body control frameworks treat the entire robot as unified system rather than controlling joints independently. Traditional approaches controlling each joint with separate PID loops ignore dynamic coupling—moving an arm shifts the body's centre of mass affecting balance, leg motions create ground reaction forces affecting whole-body dynamics, and contact forces between robot and objects create complex force distributions across the body.

Whole-body control addresses these couplings through optimisation formulations. Figure 5.3 depicts a Whole-Body Controller based on Quadratic Programming (QP). Sensor data is fused into a state estimate, which feeds task objectives, physical constraints, and the robot's kinematic/dynamic model into a QP formulation. The QP is solved in real time to produce joint torques sent to the actuators, with a feedback loop closing back to the sensors. Practical whole-body controller (WBC) implementations must also handle kinematic singularities: when an arm or leg approaches a fully extended or otherwise rank-deficient configuration, the Jacobian can lose usable directions and naive inverse kinematics or QP formulations can request unrealistically large velocities or torques. Singularity-robust damping, task reweighting, joint-limit margins, and torque saturation keep the solver well conditioned while preserving the highest-priority balance and safety tasks (Chiaverini, 1997).

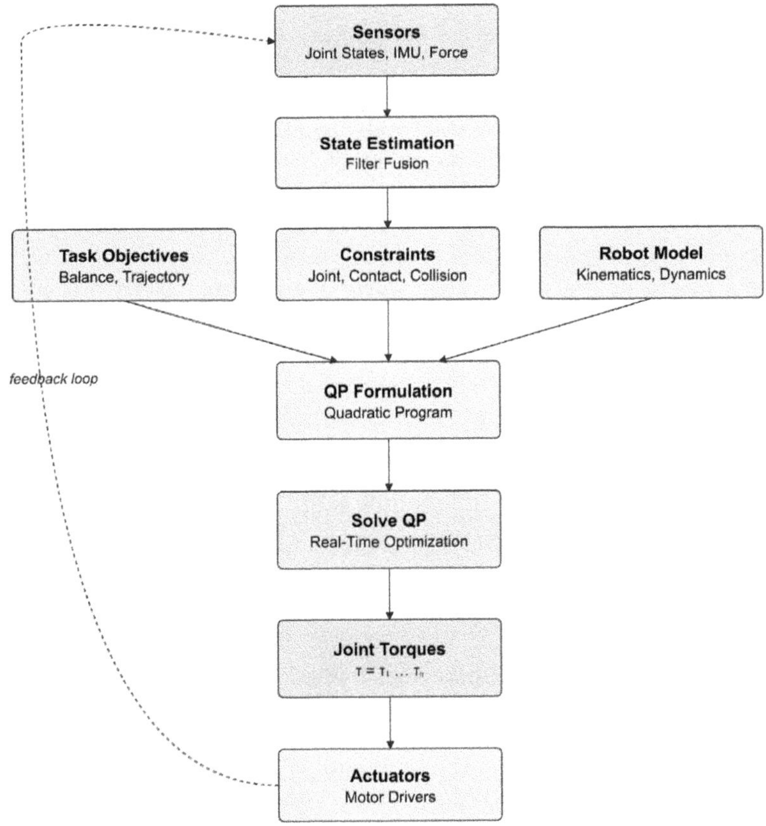


**Fig. 5.3** Whole-body control optimisation framework

The following is another way to describe this framework:

Objective: Minimise weighted sum of multiple costs—tracking error between desired and actual motion, joint velocities (for smoothness), joint accelerations (for compliance), and energy consumption (for efficiency).

Subject to constraints:

- Joint position limits (mechanical stops preventing damage)
- Joint velocity limits (actuator speed limitations)
- Joint torque limits (actuator force limitations)
- Contact force constraints (friction cone constraints preventing foot slippage)
- Zero-moment point constraints (ensuring dynamic balance)
- Collision avoidance (minimum distances to obstacles and self-collision avoidance)
- Task-specific constraints (maintaining grasp forces, following path waypoints)

The resulting optimisation problem—minimise multi-objective cost subject to potentially dozens of inequality and equality constraints over 30–56 control variables—solves in real-time through quadratic programming. Modern QP solvers leverage problem structure (sparsity patterns in constraint matrices, warm-starting from previous solutions) achieving sub-millisecond solve times for humanoid-scale problems.

The framework provides several capabilities impossible with decoupled control:

Task prioritisation: When multiple objectives conflict (maintain balance while reaching for distant object), the optimisation automatically resolves trade-offs based on specified priorities. Critical tasks (balance) receive hard constraints; optional tasks (reaching full distance) may be partially violated if necessary.

Null-space projection: For redundant robots (more DOF than strictly needed for tasks), whole-body control exploits redundancy for secondary objectives. While primary task (hand position/orientation requires 6 DOF, provided by 7-DOF arm) is satisfied, remaining DOF optimise secondary criteria (collision avoidance, joint limit avoidance, energy minimization) through null-space projections ensuring secondary objectives don't conflict with primary task.

Contact handling: When robot contacts environment (foot–ground contact during walking, hand-object contact during manipulation, body-obstacle contact during confined navigation), whole-body control explicitly models contact forces and constraints. This enables smooth transitions between contact states and stable control of contact forces.

Model-based coordination: By incorporating full robot dynamics models, whole-body control accounts for Coriolis forces, centrifugal effects, and gravity compensation across all joints. This achieves coordinated motion where joint torques properly account for dynamic coupling.

Achieving efficient whole-body control in humanoid robots remains a major challenge due to their complex dynamics, inherent underactuation, and the wide variety of tasks they must perform. More generally, underactuation means a system has fewer independent actuators than degrees of freedom, so it cannot be commanded to follow arbitrary trajectories in configuration space. For humanoids, this often appears in a contact-centric form: the robot cannot generate arbitrary body forces without environmental contact. The difficulty increases further when robots must integrate locomotion and manipulation simultaneously, such as walking while carrying objects, interacting with the environment during navigation (e.g. opening doors or moving obstacles), or coordinating bimanual manipulation while maintaining balance.

Recent advances employ learning-based approaches augmenting optimisation: Behavioural foundation models represent an emerging paradigm in which large-scale pre-training enables robots to acquire reusable motion primitives and general behavioural knowledge, allowing them to tackle new tasks with little or no additional training (Yuan et al., 2025). For locomotion, these policies are typically trained offline in simulation using reinforcement learning; reliable online real-world learning remains limited because exploration is sample-inefficient, slow, and costly for hardware. Learned models can discover coordination strategies that hand-designed optimisation formulations miss, while optimisation-based controllers still provide high-frequency constraint handling for safety-critical reactive layers such as balance, walking stabilisation, and whole-body control (Yuan et al., 2025).

Boston Dynamics' Atlas exemplifies state-of-the-art whole-body control: One of the most notable developments of 2025 was Atlas's collaboration with Toyota Research Institute to deploy Large Behaviour Models, enabling the humanoid robot to unify locomotion and manipulation under a single AI-driven control architecture rather than relying on separate, independently designed motion systems (Boston Dynamics, 2025). In the released demonstrations, Atlas executed continuous task sequences combining locomotion, balance, and object handling, including responding to mid-task perturbations introduced by researchers, illustrating capabilities that are difficult to encode through hand-authored control alone.

The integration of learned and optimisation-based whole-body control represents a promising direction. Optimisation provides stability and constraint satisfaction guarantees essential for safety, and can often operate at higher frequencies than large learned models. Learning provides adaptability and generalisation beyond hand-designed approaches. Combining them—learned models proposing behaviours subject to optimisation-based verification and refinement—potentially achieves both safety and capability.

## 5.5 Simulation and Digital Twins

Simulation environments enable developing, testing, and training robot systems without requiring physical hardware, but their value depends heavily on sim-to-real fidelity and accurate system identification. Digital twins—virtual representations of physical robots and environments—facilitate design iteration, policy learning, and validation before deployment. Before sim-to-real transfer and sysID are fully mature, simulation is often most reliable for comparing relative design choices—such as hardware designs, control approaches, or data collection strategies—rather than predicting absolute real-world performance.

### 5.5.1 Physics Simulation Engines

Multiple simulation platforms serve robotics development with different trade-offs. Current stacks often combine high-fidelity tools, lightweight simulators, and custom physics rather than relying on a single engine:

NVIDIA Isaac Sim features a GPU-accelerated physics engine capable of rendering physically based environments alongside multi-sensor RTX visuals at scale, while supporting synthetic data generation, software- and hardware-in-the-loop validation, robot learning through Isaac Lab, ROS integration, and digital twin workflows (NVIDIA Developer, 2025). Isaac Sim is valuable for testing, rendering, and data generation, but production-oriented stacks may use lighter or custom simulators for execution-critical workflows.

Isaac Lab, the successor framework, brings together high-performance parallel physics simulation, photorealistic rendering, and a flexible modular design that simplifies the creation of training environments and the development of robot control policies (Mittal et al., 2025). The photorealistic rendering enables training vision systems on simulated imagery approaching real-world visual complexity.

MuJoCo is widely used for robotics research and development because it combines fast contact-rich physics, model-based optimisation support, system identification workflows, and open-source extensibility (Google DeepMind, 2022; MuJoCo, 2026). Gazebo serves as another widely used open-source alternative in robotics education and research, offering reliable physics simulation without licensing fees. However, its architecture prioritises physics calculations over high-fidelity graphics. Because of this limitation, Gazebo can sometimes overestimate prediction accuracy compared to real-world outcomes. In contrast, platforms like NVIDIA Isaac Sim provide environments that match reality much more closely, making them the preferred choice for robotics research where advanced computer vision and photorealistic rendering are essential.

A 2025 study explored a complementary workflow pairing Gazebo and Isaac Sim, assigning Gazebo the role of lightweight prototyping and algorithm validation while reserving Isaac Sim for demanding, large-scale scenarios that require visual fidelity and robust simulation performance (Sun et al., 2025). This hybrid strategy optimises development speed versus simulation fidelity.

### 5.5.2 Digital Twins in Manufacturing and Development

Digital twins extend beyond physics simulation to comprehensive virtual representations synchronised with physical counterparts. For learning-based manipulation, digital-twin workflows increasingly include scene reconstruction, rendering, and novel-view synthesis using neural radiance fields or Gaussian splatting, allowing real-world captures to support data generation, policy training, visual servoing, and motion-planning queries from reconstructed scenes. For example, Pan et al. (2025) use Gaussian splatting with trajectory optimisation to generate visually realistic novel-view image sequences and physically feasible action trajectories from real-world eye-in-hand demonstrations, improving data augmentation for visuomotor manipulation policies. A notable example is Siemens Digital Twin Composer, set for release on the Siemens Xcelerator Marketplace in mid-2026, which integrates Siemens' own digital twin technology with NVIDIA Omniverse simulation capabilities and real-world engineering data into a unified platform (Walter, 2026).

Manufacturing and medical robotics applications highlight the value of simulation and digital twins. Using NVIDIA Isaac for Healthcare together with Omniverse libraries, Johnson & Johnson MedTech can build high-fidelity digital twins that simulate how surgical robotic systems will operate in real operating room environments (Johnson & Johnson, 2025). Testing these systems in virtual operating rooms before physical deployment helps reduce risk and speeds up development by allowing extensive validation in simulation.

The broader digital twin market is expanding rapidly across robotics and industrial systems. Independent industry analyses estimate the global digital twin market at roughly \$20–35 billion in 2025 and project compound annual growth rates (CAGRs) between approximately 30% and 45% over the following decade, with estimates varying by scope and methodology; the narrower AI-enhanced digital-twin subset is estimated at approximately \$2.95 billion in 2025 growing to \$3.74 billion at a 26.7% CAGR (Grand View Research, 2025). Integrating digital twins with edge AI is expected to reduce dependence on centralised cloud computing while enabling millisecond-level autonomy, which is essential for robotics, autonomous systems, and real-time adaptive control (Salamone, 2025).

At the same time, new research frameworks are emerging to structure digital twin development. For example, a four-layer Robot Digital Twin (RDT) architecture shows how Industry 4.0 technologies can enhance robotic digital twin capabilities in manufacturing systems (Qin et al., 2026). Similarly, model-based approaches using co-simulation-driven digital twins allow robotic systems to be validated through interconnected simulation environments before deployment (Gil et al., 2026).

### 5.5.3 Parallel Simulation for Training at Scale

Training robot policies through reinforcement learning requires millions of environmental interactions—fundamentally infeasible with physical robots limited to real-time operation but achievable through massively parallel simulation leveraging GPU acceleration. NVIDIA Isaac Gym pioneered massively parallel GPU-accelerated simulation, running 2,000–10,000 + independent robot instances simultaneously on single workstation GPUs; NVIDIA has since deprecated Isaac Gym in favour of Isaac Lab as the supported successor framework (see Sect. 5.5.1).

Scale alone is insufficient, however: locomotion and contact-rich manipulation policies need stable contact modelling, appropriate physics sub-stepping, solver-iteration tuning, and non-smooth multibody dynamics or equivalent contact formulations so simulated feet and grippers do not learn artifacts such as sliding, missed impacts, or unrealistically smooth friction that fail on hardware (NVIDIA Developer, 2024).

The computational efficiency transformation warrants detailed analysis. Consider training a locomotion policy requiring 10 million environmental interactions to converge to good performance. With a single physical robot:

- At 2 steps per second, 10 M steps requires 5 million seconds = 1,389 h = 58 days of continuous operation
- Accounting for battery charging, maintenance, and 8-h work days, calendar time extends to 6–12 months
- One robot available, no parallelisation possible
- Risk of hardware damage during failed attempts requiring repair downtime
- Total cost includes robot amortisation, electricity, facility costs, and human supervision

With massively parallel simulation training 10,000 robots simultaneously:

- Simulated robots can run faster than real time (often 2–10 × depending on complexity)
- 10 M total interactions = 1,000 interactions per robot
- At 60 Hz simulation rate running 5 × faster than real-time (300 Hz effective), 1,000 interactions takes ~ 3.3 s per robot
- All 10,000 complete training in ~ 3.3 s of actual wall-clock time
- Accounting for training infrastructure (backpropagating gradients, updating policies), total training time is hours not months
- No hardware damage risk—failed attempts are costless
- Cost is primarily GPU compute (hundreds to thousands of dollars) versus tens of thousands for physical robot operation

This efficiency gap—hours versus months, thousands versus tens of thousands of dollars—explains why reinforcement learning (RL) became viable for robotics only after GPU-accelerated parallel simulation. Earlier CPU-based simulation limited to tens or hundreds of parallel instances couldn't achieve the necessary scale.

However, parallelisation imposes constraints and trade-offs:

Simulation fidelity reduction: Simulating 10,000 robots simultaneously requires simplified physics versus single-robot high-fidelity simulation. Contact models may use faster approximate algorithms rather than detailed finite-element analysis. Visual rendering may employ simplified graphics versus photorealistic ray-tracing. Sensor simulation may use simplified models. These simplifications widen the reality gap, potentially degrading transfer quality despite the scale benefits.

Memory limitations: Each simulated robot requires memory for state, observations, and computation. GPUs have finite memory (16–80 GB for high-end cards in 2026). This bounds maximum robot count and environment complexity. Complex environments with detailed visual geometry or large numbers of interactive objects may reduce the maximum parallel count from 10,000 to hundreds or thousands.

Load balancing: Distributing simulation across multiple GPUs requires careful load balancing ensuring all GPUs remain utilised. Unbalanced loads leave some GPUs idle while others are overloaded, reducing efficiency. When scaling to very large simulations (100,000 + robots across dozens of GPUs), load balancing becomes a sophisticated distributed systems problem.

Synchronisation overhead: Parallel robot instances must occasionally synchronise for policy updates, data collection, or coordinator communication. This synchronisation creates bottlenecks that scale poorly—synchronising 10 robots is trivial, synchronising 100,000 robots requires careful engineering.

Despite these challenges, massively parallel simulation has become standard practice for RL-based robot learning, particularly for locomotion and full-body control. Many contemporary RL systems for robotics employ this approach, while manipulation systems still rely heavily on physical data because the sim-to-real gap remains difficult to close. The policy learning demonstrated in Chapter 4—such as box climbing (Shanghai Tech Robotics Lab, 2025), sit-stand transitions, and locomotion on varied terrain (Radosavovic et al., 2024)—all relied on GPU-accelerated parallel simulation. Without this infrastructure, RL for robotics would remain a research curiosity rather than a commercial technology.

The trajectory towards even more massive parallelization continues. Distributed simulation across GPU clusters enables 100,000 + parallel robots. Cloud-based simulation infrastructure allows companies without local GPU clusters to access massive compute. And simulation fidelity improvements through better physics engines, learned simulation corrections, and hybrid approaches (high-fidelity simulation for policy evaluation, simplified simulation for bulk training) address fidelity-scale trade-offs.

The implications extend beyond RL training to system validation and testing. Parallel simulation enables testing policies across thousands of scenarios simultaneously—varied environments, different object configurations, randomised perturbations—providing statistical confidence about success rates and failure modes that testing with a single physical robot cannot achieve. Safety-critical validation increasingly employs such massive parallel testing before physical deployment.

## 5.6 Sim-to-Real Transfer

Bridging simulation and reality—ensuring behaviours learned or tested in simulation work on physical robots—represents a critical challenge. The reality gap arises from inevitable approximations in physics simulation, sensor simulation, hardware parameters, contact dynamics, and environmental complexity. This gap remains especially large for manipulation, where contact-rich interactions, object variation, tactile sensing, and perception errors make transfer harder than for many locomotion and full-body control tasks. Figure 5.4 illustrates the sim-to-real pipeline.

Recent research demonstrates increasingly effective transfer techniques, but sim-to-real transfer remains an active research problem rather than a solved deployment pathway. A January 2025 study outlined a pipeline for deploying reinforcement learning policies by training in Isaac Sim, validating in Gazebo, and executing inference on physical robots via ROS 2, with results showing that policies learned entirely in simulation could be transferred directly to real hardware without any additional training (Salimpour et al., 2025). Zero-shot transfer—deploying directly from simulation to physical hardware without any real-world fine-tuning—represents the ideal scenario that research is progressively approaching, though perfect zero-shot transfer across all behaviours remains elusive.

The methodology combines complementary simulation strengths: Isaac Sim for high-fidelity physics and photorealistic rendering during policy training, Gazebo

**Fig. 5.4** Simulation-to-reality transfer pipeline

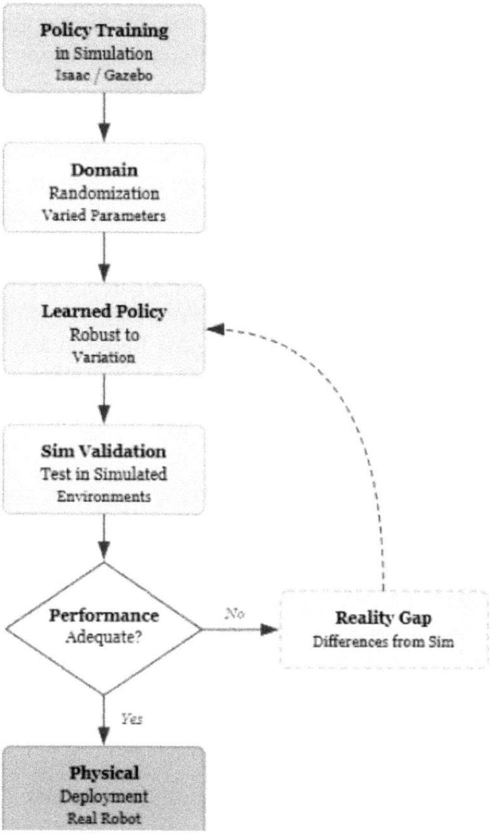

for lightweight intermediate testing and ROS 2 integration validation, and direct deployment to physical robots with ROS 2 providing consistent interfaces across simulation and reality. This multi-stage pipeline catches sim-specific issues before physical testing while leveraging each tool's strengths.

Domain randomisation, introduced in Chapter 4's reinforcement learning discussion, remains the dominant technique enabling robust transfer. By exposing policies to wide variation during simulation training—randomised physics parameters (friction coefficients varying 10x, masses varying 2x, actuator delays and gains varying 20–30%), visual appearance randomisation (lighting, textures, colours), sensor characteristic variation (noise levels, failure rates, calibration errors), and environmental diversity (object arrangements, obstacle placements, terrain properties)—policies develop robustness to simulation-reality discrepancies. When the physical world's characteristics fall within the randomised training distribution envelope, transfer succeeds because the policy has already encountered similar variation during training.

However, practical limitations prevent perfect randomisation. Some parameters cannot be randomised without destroying training (e.g. randomising gravity breaks physics fundamentals policies must learn). The randomisation range must be wide enough to encompass reality but not so wide that it includes impossible configurations diluting training signal. Finding optimal randomisation strategies requires domain expertise and often empirical tuning based on transfer results.

Despite sophisticated techniques, a residual reality gap persists—most systems deployed in practice combine sim-to-real transfer with subsequent real-world fine-tuning rather than relying on simulation alone (Chapter 4). This real-world refinement phase typically demands anywhere from hundreds to thousands of additional robot episodes in physical environments, yielding roughly 5–20% gains in performance over pure simulation transfer (Chapter 4). This suggests simulation accounts for the vast majority of final performance, with real-world training contributing the remainder by correcting systematic simulator errors, accommodating the minor physical differences between individual robot units introduced during manufacturing, and adapting to environmental phenomena that simulation does not capture—such as air currents, vibrations, and thermal effects.

## 5.7 Hardware-in-the-Loop Testing

Hardware-in-the-loop (HIL) testing combines physical hardware components with simulated environments, enabling validation of control systems using real sensors and actuators without requiring complete physical robots or environments.

### 5.7.1 HIL Methodology and Applications

HIL connects physical robot components (actuators, sensors, computing hardware) to simulated physics and environment models. Control code runs on actual robot computers, reads from real sensors (measuring simulated environment), and commands real actuators (whose effects feed back to simulation). This hybrid approach tests control algorithms with physical hardware dynamics—delays, noise, nonlinearities—while maintaining simulation flexibility and safety.

Hardware-in-the-loop (HIL) simulation is increasingly used in robotics to connect real hardware components with simulated environments during development and testing. In some systems, robots use their own sensors and actuators to collect real-world data, which is then used to construct more accurate physical simulations reflecting the robot's morphology and environmental characteristics rather than relying on idealized models.

HIL systems are widely applied in both research and industry. Universities and research institutions use HIL test benches to study advanced control methods in robotics, mechatronics, and emerging technologies, while industrial manufacturers employ HIL simulation to optimise machinery performance and production processes before deploying new manufacturing lines (GM Insights, 2025; OPAL-RT, 2025).

Market growth reflects this increasing adoption. Research Nester (2025) estimates the global hardware-in-the-loop testing market at approximately $819.8 million in 2025, projected to reach $2.32 billion by 2035 at an 11.1% compound annual growth rate (CAGR). (Note: market sizing methodologies differ across analyst firms, and alternative estimates—for example a $948 million 2024 valuation projecting 9.7% CAGR through 2034—exist; figures should be read as order-of-magnitude estimates rather than reconciled forecasts.)

### 5.7.2 Recent HIL Developments for Robotics

Industry developments during 2025 significantly expanded HIL capabilities and accessibility for robotics applications:

Recent product developments illustrate how hardware-in-the-loop (HIL) testing is becoming a mature industrial practice. In May 2025, OPAL-RT introduced the OP1400-BM Power-HIL system, designed as cost-efficient real-time hardware for testing power electronics and control systems in both academic and industrial environments. Power-HIL setups are particularly relevant for robotics because they allow engineers to test electrical subsystems—such as motor controllers, power distribution architectures, and battery management systems—while simulated mechanical loads represent the robot's physical dynamics.

These systems enable testing scenarios that would be risky or costly with fully assembled robots. Engineers can simulate actuator overloads, evaluate emergency shutdown responses during fault conditions, conduct thermal stress testing of power electronics, and optimise efficiency across wide operating ranges. Because the electrical hardware interacts with simulated mechanical loads, potential issues can be identified earlier in development before full system integration, when corrections are far less expensive.

Commercial software tools are also emerging to support HIL workflows. In March 2025, Emerson introduced the NI LabVIEW + Suite for HIL, a software platform designed to support the development and testing of embedded systems. The platform integrates test orchestration, real-time simulation, hardware data acquisition, and automated analysis, providing a standardised environment that significantly reduces the time required to deploy HIL testing infrastructure.

Infrastructure for large-scale validation is expanding as well. In August 2025, Tata Elxsi and Suzuki Motor Corporation opened the SUZUKI–TATA ELXSI Cloud HIL Center in Trivandrum, India—Suzuki's first cloud-enabled full-vehicle HIL validation facility (GM Insights, 2025). Although focused on automotive systems, the model highlights a broader trend towards cloud-accessible HIL infrastructure, which could allow robotics companies to remotely access shared testing hardware and perform distributed validation across global engineering teams.

For humanoid robots specifically, HIL provides several critical validation capabilities:

Balance controller validation: Physical inertial measurement units (IMUs) and actuators connect to simulated robot dynamics and physics. The controller reads real IMU data (with authentic noise, drift, and latency characteristics), computes torque commands using actual control hardware and software, and commands real actuators (experiencing actual motor dynamics, compliance, and delays). The simulated physics engine computes resulting body motion, feeding back to IMU. This tests the complete control loop with physical hardware dynamics while maintaining simulation safety and flexibility—the simulated robot can "fall" harmlessly, be subjected to extreme disturbances, or operate in dangerous configurations that would damage physical hardware.

Manipulation controller testing: Physical force-torque sensors, tactile sensors, end-effectors, and actuators connect to simulated objects and contact physics. The controller attempts grasps, insertions, and force-controlled assembly using real sensors and motors. Simulated contact physics determines object responses. This reveals sensor noise issues, force control instabilities, actuator limitations, and end-effector mechanical or morphological design problems that pure simulation (using idealized sensor models) would miss, while avoiding physical hardware damage from controller errors. HIL is therefore useful not only for validating controllers but also for comparing gripper, hand, fingertip, and tool geometries before committing to integrated humanoid hardware.

Sensor integration validation: Physical cameras, LiDAR, or other sensors capture data from simulated scenes rendered in real time. Perception algorithms process physical sensor data (with authentic noise, artefacts, and timing characteristics) while environments can be varied systematically (lighting conditions, object arrangements, clutter levels) impossible to control in physical testing. This validates that perception algorithms work with physical sensor characteristics across wide condition variations.

Communication and timing validation: Physical computing hardware and network infrastructure run actual software communicating with simulated components. This tests communication protocols, timing, bandwidth requirements, and failure modes using real network with authentic latencies, packet loss, and bandwidth limitations. Critical for distributed robotic systems where robots coordinate over networks.

HIL testing thus occupies middle ground between pure simulation (fast, safe, flexible but potentially unrealistic) and pure physical testing (realistic but slow, expensive, risky). By combining physical hardware experiencing real dynamics with simulated environments providing safety and flexibility, HIL enables validation scope between these extremes—more realistic than pure simulation, safer and faster than pure physical testing.

The market growth trajectory—$819.8 M (2025) to projected $2.32B (2035)—signals increasing industry recognition of HIL value as robotics systems become more complex, safety requirements intensify, and development cycles must accelerate while maintaining quality. Companies deploying safety–critical robotics (healthcare, collaborative manufacturing, eldercare) will likely make HIL testing standard practice, following automotive and aerospace precedents where HIL is mandatory.

## 5.8 Software Development and Testing

Developing reliable robot software requires rigorous engineering practices adapted from aerospace and automotive industries where software failures can cause injuries or deaths—similar stakes to robotics.

### 5.8.1 Software Architecture Patterns

Modern robot software employs established architectural patterns ensuring maintainability, testability, and safety:

Layered architecture: Separating perception, planning, control, and hardware interface layers (as shown in Fig. 5.1) enables independent module development and testing. Well-defined layer interfaces allow upgrading perception without modifying control code, or vice versa.

Component-based architecture: Decomposing functionality into reusable components (grasp planner, path planner, vision processor) that can be composed in different configurations. ROS 2's node-based architecture exemplifies this pattern.

Safety monitor pattern: Dedicated safety processes monitor robot state and can override normal control when safety violations are detected (unexpected forces, excessive speeds, human proximity). These monitors run at high priority ensuring rapid response.

State machine pattern: Explicit robot modes (startup, operational, error, shutdown) with defined transitions and behaviours in each mode. This ensures systematic handling of different operational phases.

Figure 5.5 illustrates the software development workflow integrating these patterns.

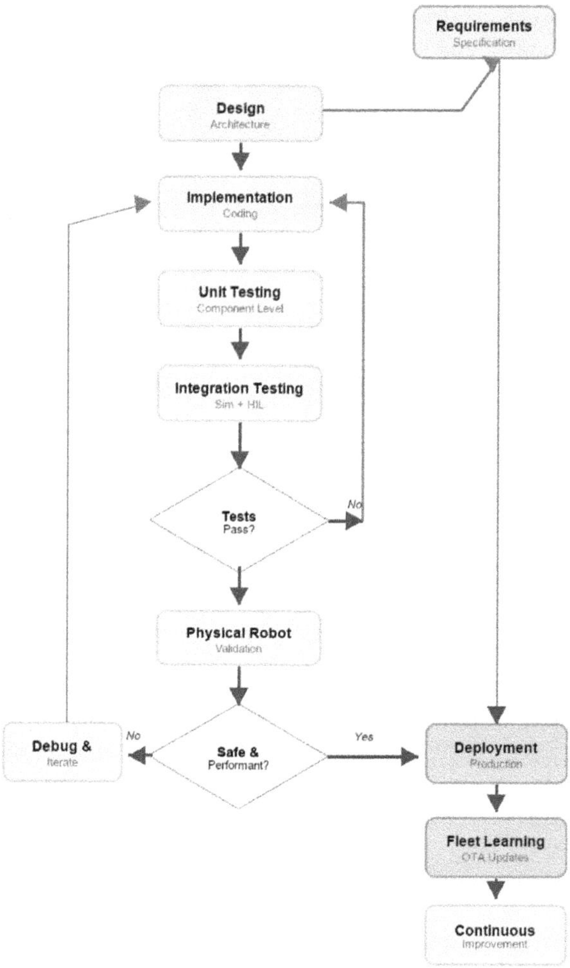


**Fig. 5.5** Robot software development and testing workflow

## 5.8.2 Continuous Integration and Automated Testing

Tesla Optimus leverages fleet learning and OTA (over-the-air) update pipelines, representing automotive software engineering practices applied to robotics at scale. This approach—enabled by Tesla's experience deploying software to millions of vehicles—is a strong example of robotics borrowing from mature automotive software infrastructure. However, the ultimate test for humanoid deployment is not software infrastructure alone, but manufacturing throughput and reliable operation at customer sites, where the industry is still early.

Continuous Integration (CI) automatically builds and tests software on every code change submitted by engineers. Modern CI pipelines execute within minutes of code commits: code compiles across all target platforms (×86 development machines, ARM embedded computers, custom accelerators), unit tests execute verifying individual component correctness, static analysis scans for common bugs and security vulnerabilities, and integration tests verify modules work together. Failures block merging changes into main codebase, ensuring main branch always builds and passes tests.

For robotics, CI extends to simulation-based testing and staged hardware validation: automated test scripts spawn hundreds or thousands of simulated scenarios (varied environments, object arrangements, task sequences), deploy candidate software to simulated robots, execute behaviours, and verify outcomes match specifications—success rates exceed thresholds, safety constraints are respected, execution times meet requirements, and no crashes or undefined behaviours occur. Similar practices increasingly extend to component robustness, endurance testing, and integrated hardware tests before subsystems are assembled into a humanoid product.

This simulation-based testing provides capabilities impossible with physical robots:

Regression testing ensures new features don't break previous capabilities. When engineers add new manipulation skills, regression tests verify existing locomotion, perception, and other behaviours still work correctly. With physical robots, comprehensive regression testing would require hours or days per software version—impractical for rapid iteration. Simulation enables testing thousands of scenarios in minutes.

Corner case testing exposes edge conditions and rare failures. Tests systematically vary environmental parameters (friction coefficients, object masses, sensor noise levels, lighting conditions) across ranges spanning normal operation to extremes. Scenarios unlikely to occur during typical operation but critical for safety (sudden obstacles appearing, sensor failures, actuator faults) are tested programmatically. Physical testing struggles to create these corner cases reproducibly; simulation generates them systematically.

Performance benchmarking tracks quantitative metrics across software versions. Automated benchmarks measure task success rates, completion times, energy consumption, and other metrics, plotting trends over time. This enables objective comparison—does new version improve or degrade performance?—and prevents gradual degradation from accumulating unnoticed changes.

Stress testing subjects systems to extreme loads revealing breaking points. Simulations run thousands of parallel robot instances executing tasks continuously for simulated days or weeks of operation, condensed to hours of real time. This exposes memory leaks, resource exhaustion, numerical instabilities, and other issues only appearing after extended operation—problems that wouldn't surface in short physical tests.

However, simulation testing doesn't eliminate physical validation needs. The reality gap means passing all sim tests provides confidence but not certainty about real-world performance. Physical robot testing remains essential but expensive and time-consuming—each test requires robot time, supervision, safety precautions, and potential repairs if tests fail damaging hardware. As a result, companies increasingly design real-world hardware tests and resettable test fixtures that can run with minimal human intervention.

The industry balances simulation and physical testing strategically: extensive automated sim testing (thousands of tests, millions of scenarios) catches most issues at low cost; targeted physical testing validates critical safety behaviours, sim-to-real transfer quality, and integration with real hardware; progressive rollout deploys updates to small robot subsets initially (1–10 robots), monitoring closely for failures, before fleet-wide updates to hundreds or thousands; and comprehensive logging and remote monitoring enables detecting issues in deployed robots, rolling back problematic updates, and collecting failure data informing future testing.

Tesla's fleet learning approach exemplifies this: deployed Optimus robots (initially in Tesla factories, eventually external customers) collect operational data, software updates deploy over-the-air after validation through sim and limited physical testing, fleet performance monitoring detects issues rapidly, problematic updates rollback automatically if failure rates exceed thresholds, and successful updates propagate fleet-wide, improving all robots simultaneously. This creates continuous improvement feedback loop impossible with traditional release cycles requiring physical media distribution.

### 5.8.3 Version Control and Safety Certification

Version control for robotics presents unique challenges beyond traditional software. In pure software, version control tracks code changes and any version runs on standard computing hardware. Robotics requires hardware-software co-design where software versions pair with specific hardware configurations—particular actuator firmware versions, sensor calibrations, structural configurations, or even mechanical revisions.

Configuration management becomes critical: tracking which software versions are compatible with which hardware configurations, preventing deployment of software to incompatible hardware, documenting dependencies between components, and maintaining compatibility matrices across versions. Figure 03—the next-generation Figure AI humanoid introduced in Chapter 2, with redesigned forearms among other revisions—illustrates this hardware-software coupling. When Figure AI updates Fig. 03 hardware based on BMW deployment lessons (forearm redesign per Chapter 2), software must account for new kinematics, updated actuator characteristics, and changed sensor configurations. Deploying Fig. 02 software to Fig. 03 hardware could cause failures or damage.

Modern robotics companies employ sophisticated configuration management systems: unique identifiers for each hardware and software version, automated compatibility checking preventing invalid combinations, comprehensive documentation linking hardware and software changes, and testing matrices verifying all supported hardware-software combinations. This infrastructure overhead—often managed by dedicated DevOps teams—represents significant engineering investment but is essential for reliable deployment.

Safety certification presents even greater challenges. Aerospace and automotive industries have mature safety certification processes developed over decades: DO-178C provides software development standards for aircraft systems requiring demonstration of development rigour, comprehensive testing, traceability from requirements through implementation and testing, and verification of safety–critical functionality. ISO 26262 similarly defines automotive software safety requirements from concept through decommissioning.

Robotics lacks a single mature humanoid-specific software certification pathway as of 2026. Existing standards like ISO 10218 (robot safety) and ISO 13482 (personal care robots) provide general safety requirements but limited detailed software certification guidance for humanoid autonomy. Companies deploying humanoid robots therefore assemble safety processes from existing robotics, electrical, mechanical, automotive, and customer-specific standards; large industrial customers with aerospace or automotive experience may effectively impose their own safety expectations as deployment requirements.

This regulatory gap creates risks and uncertainties. Companies uncertain whether their safety practices will satisfy future regulations may over-invest in conservative processes. Those under-investing risk future liability when regulations tighten. Inconsistent safety rigour across companies could undermine public trust if high-profile failures occur.

As robotics deployment scales and regulations mature—likely driven by safety incidents, public pressure, and legislative action—formal certification will probably become mandatory for commercial robots, particularly those interacting with untrained humans. This will require companies to document software development processes comprehensively (requirements traceability, design documentation, code reviews, testing records), maintain detailed test results demonstrating safety validation across operating envelope, potentially employ formal verification for safety-critical components proving mathematical correctness, implement ongoing safety monitoring in deployed systems detecting degradation or emerging hazards, and submit to third-party audits validating safety practices.

Companies establishing rigorous software development and safety practices now—even before mandatory—position themselves advantageously: reduced retrofit costs when regulations arrive, enhanced public trust from demonstrated safety commitment, reduced liability exposure from robust practices, and competitive differentiation in safety–critical markets (healthcare, eldercare) where customers demand proof of safety.

## 5.9 Computational Infrastructure

The AI, perception, and control systems examined in previous sections require substantial computation. Allocating computational resources, managing thermal dissipation, and ensuring real-time performance within power budgets creates architectural challenges.

### 5.9.1 Heterogeneous Computing Architectures

Contemporary humanoid robots employ heterogeneous computing—multiple specialised processors handling different workloads:

CPUs run general-purpose code including task planning, system management, and non-performance-critical logic. Modern embedded CPUs (ARM, × 86) provide multi-core parallelism and reasonable single-thread performance at moderate power (5–20 watts).

GPUs accelerate vision processing and AI model inference, providing 100–1000 × speedup versus CPUs for parallel workloads. Embedded GPUs like NVIDIA Jetson provide 20–200 TOPS (trillion operations per second) at 10–30 watts enabling real-time deep learning.

Neural Processing Units (NPUs) specialise further for neural network inference, achieving better efficiency than GPUs through dedicated matrix multiplication hardware and reduced precision (8-bit, 4-bit) versus full floating point.

FPGAs (field-programmable gate arrays) enable custom hardware acceleration for specialised algorithms, offering lower latency, deterministic execution, and higher efficiency than programmable processors at the cost of significant development complexity and longer development cycles. They are especially useful during prototyping because they provide reprogrammable flexibility before model architectures stabilise enough to justify fixed-function ASIC development.

Microcontrollers handle low-level real-time control running safety–critical loops with guaranteed timing. These simple processors (typically running at 100–500 MHz) execute PID control loops or simple state-space control laws at 1–10 kHz rates with microsecond-level determinism impossible on complex processors running full operating systems with unpredictable interrupt handling and task scheduling.

The fundamental challenge is intelligently partitioning workloads across this heterogeneous computing architecture: determining which processes run on which processor types, designing communication mechanisms between processors with different capabilities and constraints, implementing fault tolerance strategies handling processor failures gracefully without total system failure, and carefully managing total power consumption to stay within battery budgets while meeting all computational and timing requirements. Suboptimal partitioning leaves some processors idle while others are overloaded, wastes power on unnecessary communication, or violates timing constraints creating unstable or unsafe operation.

### 5.9.2 Power Budget Allocation and Thermal Management

Humanoid robots face stringent power budgets where total power consumption must stay within battery capacity enabling multi-hour operation. As detailed in Chapter 2, batteries store 2–4 kWh with robots consuming 500-2000W during active operation. Computation represents a significant fraction of this budget, but legged locomotion and actuators usually dominate energy use; walking can substantially reduce runtime compared with wheeled mobile robots, which generally achieve longer battery life because locomotion is mechanically more efficient.

Breaking down computational power consumption by subsystem reveals allocation challenges:

Perception processing (10-40W): Vision processing through transformer networks requires substantial GPU computation. Multi-camera systems processing 8 cameras at 30 FPS through vision models consume 20-40W. LiDAR point cloud processing adds 5-15W. Tactile sensor processing is relatively low (1-2W) due to lower data rates. Sensor fusion and Simultaneous Localization and Mapping (SLAM) add 5–10W.

AI inference (20-100W): Foundation models for task understanding and planning represent the largest computational load. A 300 M parameter VLA model running at 10 Hz might consume 20-30W. Larger models (1B + parameters) or higher frequencies push consumption to 50-100W. Companies must trade off model capability versus power consumption—larger more capable models drain batteries faster, limiting runtime.

Control computation (5-20W): Low-level controllers (PID, MPC, whole-body optimisation) running at 100–1000 Hz require modest power on modern processors. Multi-core CPUs running all control loops consume 10-20W total. Optimisation solvers for whole-body control add 5-10W.

Communication and overhead (5-10W): Wireless communication (WiFi, cellular), data logging, user interface, system management, and overhead processes consume 5-10W baseline power even during idle periods.

Summing these components: 40-170W total computational power from 2–4 kWh batteries. At 40W (conservative perception, small AI models, efficient control), a 2.4 kWh battery provides 60 h of computation-only operation—far exceeding actuator runtime limits. At 170W (aggressive multi-camera perception, large AI models, complex control), the same battery provides only 14 h computational budget. Average energy calculations can also hide peak-power failures: when many actuators accelerate simultaneously, transient current draw can sag the DC bus below compute or motor-drive thresholds, causing brownout resets even when the battery has sufficient remaining energy. Humanoid power architectures therefore need peak-current budgeting, local hold-up capacitance, battery-management limits, and in some designs supercapacitor buffers that shave short bursts while the battery supplies lower-frequency energy demand (Wang et al., 2014).

However, actuators dominate power consumption during movement: walking consumes 500-1000W, manipulation adds 200-500W, and dynamic behaviours peak at 2–3 kW. The computational budget must fit within total power envelope while leaving majority for actuators. If computation consumed 200W, only 300-800W remains for movement at 1000W total budget—severely constraining operation.

The allocation optimisation becomes multi-dimensional: reducing vision frame rates from 60 to 30 Hz cuts vision processing power nearly in half (processing half as many frames) while degrading perception only modestly for many tasks; running smaller AI models (300 M versus 1B + parameters) may cut inference power from 100 to 30W but sacrifices reasoning capability; adaptive computing throttles non-critical processes during high-demand periods—during intense manipulation or rapid locomotion, perception frame rates reduce, AI planning defers to reactive control, and logging suspends, preserving power for actuators; and duty cycle management recognises that peak power matters less than average—brief high-power periods are acceptable if followed by low-power recovery.

Companies optimise these trade-offs differently based on priorities. Tesla, emphasising AI capability, may allocate more budget to large models accepting shorter runtime. Agility Robotics, emphasising runtime for all-day warehouse operation, may limit AI models to preserve battery for longer shifts. The optimal allocation depends on application requirements—healthcare prioritising safety might emphasise perception and control redundancy over runtime, manufacturing prioritising productivity might optimise for speed accepting more frequent battery swaps.

Thermal management for computing hardware, discussed in Chapter 2 for actuators, creates related challenges, but compute is not the only bottleneck. Actuator thermals are often a primary operational constraint, and tactile sensors can add additional heat and packaging challenges. GPUs running intensive AI inference at 20-100W generate equivalent heat requiring dissipation in confined robot bodies. Unlike desktop computers with large heatsinks and fans, mobile robots face volume and weight constraints limiting cooling capability.

Multiple cooling approaches combine in contemporary designs:

Passive cooling conducts heat from processors through thermal interfaces to robot structural components (aluminium frames, body panels) acting as heatsinks. Heat then dissipates through convection and radiation to environment. This adds no power consumption, noise, or complexity but provides limited cooling capacity—perhaps 10-20W per component depending on thermal design.

Active air cooling employs fans directing airflow over heatsinks on processors. Carefully designed airflow paths maximise cooling efficiency. Small high-speed fans can dissipate 30-60W + from compact processors but consume 1-5W themselves and generate noise (often 40–60 dB) potentially unacceptable for service robots in quiet environments.

Liquid cooling, rare in humanoid robots as of 2026 but emerging, circulates coolant through cold plates on processors, transferring heat to radiators with forced air cooling or to structural elements. This enables dissipating 100W + from compact processors but adds system complexity (pumps, lines, radiators, coolant), weight, and potential leak failure modes.

Thermal throttling serves as last resort: when processor temperatures exceed safe limits (typically 80–100 °C), firmware automatically reduces operating frequency, lowering power and heat generation. This prevents thermal damage but degrades performance—AI inference slows, potentially violating timing constraints. Robust thermal design prevents throttling under normal operation, but throttling provides safety margin for worst-case conditions.

The power and thermal co-design becomes critical system-level optimisation. Allocating 100W to AI inference requires cooling infrastructure dissipating 100W heat plus potentially 5-10W fan power. Total battery drain is 110W with thermal overhead. If thermal design can't dissipate 100W reliably (ambient temperatures too high, inadequate heatsink area, insufficient airflow), processor throttles and AI performance degrades unpredictably.

Companies with thermal management expertise—Tesla from automotive battery packs and powertrains, Boston Dynamics from hydraulic systems and high-power actuators—have advantages in computational thermal design. Startups may underestimate thermal challenges, discovering during extended operation that processors overheat and throttle, degrading deployed robot performance versus laboratory testing in air-conditioned environments.

## 5.10 Conclusion

Integration and control systems—including real-time operating systems, hierarchical control architectures, whole-body optimisation, simulation environments, edge computing, data pipelines, and hardened middleware—enable humanoid robots to function as coherent autonomous systems suitable for commercial deployment. Across the industry, architectural convergence is emerging: three-layer control hierarchies are widely adopted, ROS 2 remains important for research and integration while production stacks increasingly use hybrid, hardened, or proprietary middleware, optimisation-based whole-body control remains essential for high-frequency safety loops, and end-to-end imitation-trained policies are becoming central for manipulation. Simulation platforms such as Isaac Sim, Gazebo, and MuJoCo support large-scale AI training, validation, and sim-to-real development, but the benefits depend on careful system identification, physical validation, disciplined deployment practices, and access to high-quality robot data.

Despite this convergence, companies differentiate through implementation strategies. Boston Dynamics emphasises highly optimised proprietary real-time frameworks for maximum performance, Tesla applies vertically integrated automotive-style software development and fleet learning, while many startups rely on open-source ecosystems to reduce cost and accelerate deployment. Together with growing investment in RTOS, simulation, hardware-in-the-loop testing, and edge computing, this infrastructure forms the technological foundation for the emerging robotics industry.

However, significant challenges remain. Increasing AI model size strains real-time constraints, sim-to-real transfer remains imperfect—especially for manipulation—software complexity grows rapidly with robot capability, and mobile robots face strict power, actuator, and thermal limits. Data is now a major bottleneck: imitation-based robot policies need diverse, high-quality demonstrations across objects, environments, embodiments, and failure modes, yet collecting and validating such data is expensive and hardware-constrained. Safety certification also lags technological progress. As humanoid robots advance, improvements in simulation fidelity, edge AI efficiency, hardware–software co-design, data infrastructure, and safety verification will be critical.

Ultimately, integration quality determines real-world robot performance. Even advanced hardware and AI systems fail without robust system architecture, reliable software, and carefully coordinated control. Companies that excel at integrating complex subsystems—rather than simply developing the most advanced individual components—are likely to lead successful commercial deployment of humanoid robotics.

**Comprehension Questions**

1. Three-Layer Architecture: Referring to Fig. 5.1, explain why humanoid robots separate control into deliberative (1–10 Hz), executive (10–100 Hz), and reactive (100–1000 + Hz) layers. Why can't a single control loop operating at one frequency handle all tasks? What happens if reactive control runs too slowly?
2. ROS 2 Migration and Hybrid Middleware: Explain why ROS 1 Noetic reaching end of life on May 31, 2025 accelerated migration towards ROS 2. What key improvements does ROS 2 provide through DDS (Data Distribution Service) middleware with QoS profiles? Why might production humanoid systems still combine ROS 2 with proprietary or newer deterministic middleware?
3. MPC Advantages: Describe Model Predictive Control and why it is increasingly used in robotics for dynamic environments, tight safety requirements, and multi-objective trade-offs that PID cannot handle. Using Fig. 5.2, explain the receding horizon approach. What computational challenges does real-time nonlinear MPC face?
4. Whole-Body Control: Study Fig. 5.3 and explain how whole-body optimisation coordinates roughly 30–56 body DoF through quadratic programming, with dexterous hands adding additional DoF in some systems. Why is achieving efficient whole-body control a fundamental challenge due to complex dynamics, underactuation, and diverse task requirements? How do behavioural foundation models address this, and why do optimisation-based controllers remain important for high-frequency safety–critical control?
5. Simulation Fidelity: Compare Isaac Sim, Gazebo, and MuJoCo as simulation tools with different trade-offs in rendering fidelity, physics speed, contact modelling, and integration workflows. Why does photorealistic rendering matter

for vision-based policies? When might lightweight or custom simulation be preferable despite lower visual fidelity?

6. Sim-to-Real Transfer: Analyse Fig. 5.4's transfer pipeline. How can accurate system identification, domain randomisation, and policy validation support zero-shot transfer from simulation to real robots? Why does sim-to-real transfer remain harder for manipulation than locomotion, and why do deployed systems still rely on physical data and fine-tuning?
7. HIL Testing Value: Explain hardware-in-the-loop testing and why the HIL market is projected to grow from $819.8 M (2025) to $2.32B (2035). What specific validation does HIL provide that pure simulation or pure physical testing cannot? Design an HIL setup for testing a humanoid robot balance controller.
8. Real-Time Constraints: A humanoid robot must simultaneously run: balance control (1000 Hz, $<1$ ms latency), force control (500 Hz, $<2$ ms), vision (30 Hz, $<33$ ms), and planning (10 Hz, $<100$ ms). Using Fig. 5.1's architecture, assign these to appropriate layers and explain how the system ensures each meets its timing requirements even under worst-case loads.
9. Digital Twin Applications: Describe how Johnson & Johnson MedTech creates high-fidelity digital twins using NVIDIA Omniverse for surgical robotics systems. Why is coupling digital twins with edge AI for millisecond-level autonomy critical for robotics? What specific advantages do digital twins provide for surgical robot development?
10. Software Development Workflow: Study Fig. 5.5 and describe the complete development pipeline from requirements through deployment. Why are fleet learning and OTA updates valuable for companies like Tesla? What safety risks do OTA updates create, how should they be mitigated, and why will customer-site reliability and manufacturing throughput be the decisive tests for humanoid deployment?

## References

Apex.AI. (2025). *Apex.AI and QNX team up to fast-track deterministic AI for autonomous and robotic systems*. https://www.apex.ai/press-release/apex.ai-and-qnx-team-up-to-fast-track-deterministic-ai-for-autonomous-and-robotic-systems

Belvedere, T., Ziegltrum, M., Turrisi, G., & Modugno, V. (2026). Feedback-MPPI: Fast sampling-based MPC via rollout differentiation. *IEEE Robotics and Automation Letters, 11*(1), 1–8. https://arxiv.org/abs/2506.14855

Black, K., Brown, N., Driess, D., Esmail, A., Equi, M., Finn, C., Fusai, N., Groom, L., Hausman, K., Ichter, B., et al. (2024). *$\pi$0: A vision-language-action flow model for general robot control*. arXiv Preprint. https://arxiv.org/abs/2410.24164

Borquez, J., Chakraborty, K., Wang, H., & Bansal, S. (2024). On safety and liveness filtering using Hamilton-Jacobi reachability analysis. *IEEE Transactions on Robotics*. https://doi.org/10.1109/TRO.2024.3454470

Boston Dynamics. (2025). *Large behavior models and Atlas find new footing*. Boston Dynamics Blog. https://bostondynamics.com/blog/large-behavior-models-atlas-find-new-footing/

Cao, S., Wang, F., Li, X., Yao, D., & Xie, M. (2025). Safety-constrained disturbance-compensated model predictive control for flexible-joint robots. *Applied Sciences, 15*(24), 13238. https://www.mdpi.com/2076-3417/15/24/13238

Chi, C., Xu, Z., Pan, C., Cousineau, E., Burchfiel, B., Feng, S., Tedrake, R., & Song, S. (2024). Universal manipulation interface: In-the-wild robot teaching without in-the-wild robots. *Proceedings of Robotics: Science and Systems.* https://doi.org/10.15607/RSS.2024.XX.045

Chiaverini, S. (1997). Singularity-robust task-priority redundancy resolution for real-time kinematic control of robot manipulators. *IEEE Transactions on Robotics and Automation, 13*(3), 398–410. https://doi.org/10.1109/70.585902

Eclipse Foundation. (2023). *Eclipse Zenoh selected as the alternate ROS 2 middleware.* https://newsroom.eclipse.org/eclipse-newsletter/2023/october/eclipse-zenoh-selected-alternate-ros-2-middleware

Gil, S., Badyal, A., Miyazawa, A., Larsen, P. G., & Cavalcanti, A. (2026). A model-based approach for co-simulation-driven digital twins in robotics. *Robotics and Autonomous Systems, 196*, 105240. https://www.sciencedirect.com/science/article/pii/S0921889025003379

GM Insights. (2025). *Hardware-in-the-loop (HIL) testing market size.* Global Market Insights. https://www.gminsights.com/industry-analysis/hardware-in-the-loop-testing-market

Google DeepMind. (2022). *Open-sourcing MuJoCo.* https://deepmind.google/blog/open-sourcing-mujoco

Grand View Research. (2025). *Digital twin market size, share & trends analysis report. Grand View Research Industry Report.* https://www.grandviewresearch.com/industry-analysis/digital-twin-market

Hugging Face. (2026). *LeRobot documentation.* https://huggingface.co/docs/lerobot/index

Intel. (2026). *Linux OS runtime optimisations.* Edge Controls for Industrial documentation. https://eci.intel.com/docs/3.3/development/performance/optimisations.html

Johnson & Johnson. (2025). *Johnson & Johnson to advance robotics development with NVIDIA Isaac for healthcare.* J&J Media Center. https://www.jnj.com/media-center/press-releases/johnson-johnson-to-advance-robotics-development-with-nvidia-isaac-for-healthcare

Linux Kernel Documentation. (2026). *CPU isolation.* https://www.kernel.org/doc/html/next/admin-guide/cpu-isolation.html

Market Research Future. (2025). *Real-time operating system market research report—forecast to 2034.* https://www.marketresearchfuture.com/reports/real-time-operating-system-market-34731

Mittal, M., Roth, P., Tigue, J., et al. (2025). *Isaac Lab: A GPU-accelerated simulation framework for multi-modal robot learning.* arXiv Preprint. https://arxiv.org/abs/2511.04831

MuJoCo. (2026). *MuJoCo: Advanced physics simulation.* https://mujoco.org/

NVIDIA Developer. (2024). *Supercharge robotics workflows with AI and simulation using NVIDIA Isaac Sim 4.0 and NVIDIA Isaac Lab.* https://developer.nvidia.com/blog/supercharge-robotics-workflows-with-ai-and-simulation-using-nvidia-isaac-sim-4-0-and-nvidia-isaac-lab/

NVIDIA Developer. (2025). *Isaac Sim: Robotics simulation and synthetic data generation.* NVIDIA Developer Platform. https://developer.nvidia.com/isaac/sim

OPAL-RT. (2025). *A guide to hardware-in-the-loop testing in 2025.* OPAL-RT Blog. https://www.opal-rt.com/blog/a-guide-to-hardware-in-the-loop-testing-in-2025/

Open X-Embodiment Collaboration, Padalkar, A., Pooley, A., Jain, A., Bewley, A., Herzog, A., Irpan, A., et al. (2023). *Open X-Embodiment: Robotic learning datasets and RT-X models.* https://robotics-transformer-x.github.io/

Open Robotics Discourse. (2025). *ROS Noetic end-of-life: May 31, 2025.* Open Robotics Discourse. https://discourse.ros.org/t/ros-noetic-end-of-life-may-31-2025/43160

Pan, C., Liang, L., Bauer, D., Cousineau, E., Burchfiel, B., Feng, S., & Song, S. (2025). One demo is worth a thousand trajectories: Action-view augmentation for visuomotor policies. *Conference on Robot Learning (CoRL).* https://chuerpan.com/1001-demos.github.io/

Portugal, D., Rocha, R. P., & Castilho, J. P. (2024). Inquiring the robot operating system community on the state of adoption of the ROS 2 robotics middleware. *International Journal of*

*Intelligent Robotics and Applications, 9*(2), 454–479. https://doi.org/10.1007/s41315-024-00393-4

Qin, Q., Liu, Z., Zhong, R., Wang, X. V., Wang, L., Wiktorsson, M., & Wang, W. (2026). Robot digital twin systems in manufacturing: Technologies, applications, trends and challenges. *Robotics and Computer-Integrated Manufacturing, 97*, 103103. https://www.sciencedirect.com/science/article/pii/S0736584525001577

Radosavovic, I., Xiao, T., Zhang, B., et al. (2024). Real-world humanoid locomotion with reinforcement learning. *Science Robotics, 9*(88). https://doi.org/10.1126/scirobotics.adi9579

Robotnik Automation. (2025). *ROS 2 robot operating system: Overview and key features of the robotics software*. Robotics Tomorrow News. https://www.roboticstomorrow.com/news/2025/05/20/ros-2-robot-operating-system-overview-and-key-features-of-the-robotics-software/24795/

Research Nester. (2025). *Hardware-in-the-loop market size and share, growth trends 2035*. https://www.researchnester.com/reports/hardware-in-the-loop-market/3831

ROS 2 Documentation. (2026). *ROS 2 middleware implementations*. https://docs.ros.org/en/rolling/Concepts/Advanced/About-Middleware-Implementations.html

Salamone, S. (2025). *Digital twins in 2026: From digital replicas to intelligent, AI-driven systems*. RTInsights. https://www.rtinsights.com/digital-twins-in-2026-from-digital-replicas-to-intelligent-ai-driven-systems/

Salimpour, S., Queralta, J. P., et al. (2025). *Sim-to-real transfer for mobile robots with reinforcement learning: From NVIDIA Isaac Sim to Gazebo and real ROS 2 robots*. arXiv Preprint. https://arxiv.org/abs/2501.02902

Schwenzer, M., Ay, M., Bergs, T., & Abel, D. (2021). Review on model predictive control: An engineering perspective. *The International Journal of Advanced Manufacturing Technology, 117*(5), 1327–1349. https://doi.org/10.1007/s00170-021-07682-3

Shanghai Tech Robotics Lab. (2025). *Learning to leap: Humanoid robot box climbing with reinforcement learning*. https://robotics.shanghaitech.edu.cn/sites/default/data/2025-04/moma2024project-h1-manipulation.pdf

Sun, C., Wu, H., Yang, Z., Li, X., Jin, H., & Tian, S. (2025). Robotic simulation systems and intelligent offline teaching for urban rail transit maintenance. *Electronics, 14*(12), 2431.

Walter, N. (2026). *Siemens digital twin software targets faster factory upgrades*. Interesting Engineering. https://interestingengineering.com/ai-robotics/siemens-digital-twin-composer-pepsico

Wang, G., Dai, Z., Guan, Y., Dong, P., & Wu, L. (2014). Power management of hybrid power systems with Li-Fe batteries and supercapacitors for mobile robots. *Advances in Mechanical Engineering, 6*, Article 270537. https://doi.org/10.1155/2014/270537

Weinberg, A. I., Shirizly, A., Azulay, O., & Sintov, A. (2024). Survey of learning-based approaches for robotic in-hand manipulation. *Frontiers in Robotics and AI, 11*, 1455431. https://doi.org/10.3389/frobt.2024.1455431

Yaren, T., & Kizir, S. (2025). Real-time nonlinear model predictive control of a robotic arm using spatial operator algebra theory. *Journal of Field Robotics, 42*(1). https://doi.org/10.1002/rob.22514

Yuan, M., Yu, T., Ge, W., Yao, X., Li, D., Wang, H., Chen, J., Li, B., Zhang, W., Zeng, W., Chen, H., & Jin, X. (2025). *A survey of behavior foundation model: Next-generation whole-body control system of humanoid robots*. arXiv Preprint. https://arxiv.org/abs/2506.20487

**Ken Huang** is a globally recognized expert in AI and Web3 security, a prolific author, and a leading figure in shaping industry standards. He is the CEO and Chief AI Officer (CAIO) of DistributedApps.ai, specializing in Generative AI training and consulting. Ken is deeply involved in driving the development of secure AI systems. He serves as a Research Fellow and Co-Chair of the AI Safety Working Groups at the Cloud Security Alliance (CSA), leading Agentic AI initiatives. He is also a Co-Chair of the AI STR Working Group at the World Digital Technology Academy under the UN Framework, and a core contributor to the OWASP GenAI project, focusing on Agentic AI security. His expertise extends to his contributions to OWASP's Top 10 Risks for LLM Applications and his active participation in the NIST Generative AI Public Working Group. He is also a member of Open AI forum. Ken has authored and edited numerous influential books, including: *Agentic AI: Theories and Practices* (Springer, June 2025). *Securing AI Agents: Foundations, Frameworks, and Real-World Deployment* (Springer, October 2025). *Beyond AI: ChatGPT, Web3, and the Business Landscape of Tomorrow* (Springer, 2023). *Generative AI Security: Theories and Practices* (Springer, 2024). *Practical Guide for AI Engineers* (Volumes 1 and 2, DistributedApps.ai, 2024). *The Handbook for Chief AI Officers: Leading the AI Revolution in Business* (DistributedApps.ai, 2024). *Web3: Blockchain, the New Economy, and the Self-Sovereign Internet* (Cambridge University Press, 2024). *Web3 Applications Security and New Security Landscape: Theories and Practices* (Springer, 2024). His co-authored book, "Blockchain and Web3: Building the Cryptocurrency, Privacy, and Security Foundations of the Metaverse" (Wiley, 2023), was recognized as a must-read by TechTarget. *A sought-after speaker, Ken has presented at prestigious global forums, including Davos WEF, ACM, IEEE, CSA AI Summit, the Depository Trust & Clearing Corporation (DTCC), and World Bank conferences.* Explore Ken's books on Amazon: https://www.amazon.com/author/kenhuang

**Grace Huang** is a seasoned product and technology professional with extensive experience at organizations including PIMCO (a global investment management firm managing over $2 trillion in assets), IBM, SoftBank, and DataRobot. Throughout her career, she has successfully launched multiple products and managed large-scale projects, leveraging her expertise in product management, market analysis, strategic planning, and cross-functional team leadership. Her work brings a finance and product strategy perspective to physical AI, particularly the operational, risk, and adoption considerations involved in bringing advanced robotics into enterprise environments. Always at the forefront of emerging technology, she has explored innovative tools including the implementation of Agentic AI powered solutions within the product management process—streamlining communication, improving decision-making, enhancing customer satisfaction, and driving business growth. In addition to her professional accomplishments, Grace holds a degree from Babson College, where she built a strong foundation in business management and entrepreneurship. She continues to lead at the intersection of product strategy and advanced technology, bringing a rare combination of financial rigor and deep technical acumen to complex, high-stakes initiatives.

**Chuer Pan** is a robotics PhD student at Stanford. She works on the interaction of robotics and computer vision. Her research focus on develop robotic systems that learn high-precision, contact-rich manipulation from human demonstrations and generalize to novel visual and behavioral settings with minimal adaptation. LinkedIn: https://www.linkedin.com/in/chuer-pan-a2ab0475/

**Manish Bhatt** is a technical leader dedicated to making autonomous systems both safe and secure. He currently serves as the Application Security Tech Lead for AI & Crypto at Amazon Leo. With over 15 years in the field, Manish specializes in the critical space where complex hardware meets intelligent software. He has been at the forefront of AI safety as a founding engineer of Purple Llama and the lead behind the CyberSecEval series, which helped set the standard for evaluating risks in frontier models. His research is widely recognized, covering everything from kernel rootkits to the safety of Large Language Models, and has earned over 20,000 citations. As a founding member of the OWASP AI Security Verification Standard (AISVS), Manish is now focused on building the Artificial Intelligence Vulnerability Scoring System (AIVSS). His goal is to better understand risk in Agentic AI, ensuring that as AI technology moves into humanoid robotics, these systems remain resilient and trustworthy for everyone.

# Safety, Security and Reliability

6

Ken Huang, Jerry Huang, Bhavya Gupta, and Víctor Mayoral-Vilches

## 6.1 Physical Safety Standards and Regulations

The deployment of humanoid robots in human-populated environments necessitates rigorous adherence to established safety standards and emerging regulatory frameworks. Unlike traditional industrial robots confined to safety cages, humanoid robots are designed for direct human interaction, fundamentally changing the safety paradigm (Guiochet et al., 2017).

### 6.1.1 International Safety Standards

ISO 13482 represents the cornerstone standard for personal care robots, establishing safety requirements for robots operating in direct contact with humans outside industrial settings (ISO, 2014). This standard defines three categories of personal care robots: mobile servant robots, physical assistant robots, and person carrier robots. Humanoid robots typically fall under the first two categories, requiring compliance with specific risk assessment protocols and protective measures. The standard mandates hazard identification, risk estimation, and risk evaluation

K. Huang (✉)
DistributedApps.Ai, Fairfax, VA, USA
e-mail: ken@distributedapps.ai

J. Huang
Kleiner Perkins, San Francisco, CA, USA

B. Gupta
Stanford University, Stanford, CA, USA

V. Mayoral-Vilches
Alias Robotics, Vitoria-Gasteiz, Spain



K. Huang et al. (eds.), *Humanoid Robots and Physical AI*, Advances in Data Analytics, AI, and Smart Systems, https://doi.org/10.1007/978-3-032-33669-9_6

phases, followed by implementation of safety measures including inherent safe design, protective measures, and user information protocols.

ISO/TS 15066 provides critical guidance for collaborative robot operations, introducing the concept of power and force limiting (ISO, 2016). This technical specification establishes biomechanical limits for human–robot contact, defining maximum allowable forces and pressures for different body regions. For humanoid robots, these limits are particularly relevant as their anthropomorphic form increases the likelihood of unexpected contact during operation. The standard specifies quasi-static contact forces ranging from 65 to 300N depending on body region, with transitional and dynamic contact scenarios requiring additional analysis.

The International Electrotechnical Commission (IEC) standard IEC 61508 establishes the framework for functional safety of electrical, electronic, and programmable electronic safety-related systems (IEC, 2010). This umbrella standard introduces Safety Integrity Levels (SIL) ranging from SIL 1 to SIL 4, with higher levels indicating greater risk reduction requirements. Humanoid robots typically target SIL 2 or SIL 3 for critical safety functions such as emergency stops and collision avoidance, requiring systematic fault analysis and probabilistic failure rate calculations.

### 6.1.2 Regional Regulatory Variations

In the United States, humanoid robot safety falls under the purview of multiple regulatory bodies. The Occupational Safety and Health Administration (OSHA) establishes workplace safety requirements through standards such as 29 CFR 1910.147, which addresses control of hazardous energy (lockout/tagout) (OSHA, 2020). Underwriters Laboratories (UL) provides third-party certification services, with UL 1740 addressing robot safety requirements and UL 3100 covering automated mobile platforms. The American National Standards Institute (ANSI) and Robotics Industries Association (RIA) jointly publish ANSI/RIA R15.08, specifically addressing industrial mobile robots and their integration into human workspaces (RIA, 2020).

The European Union implements a comprehensive regulatory framework through CE marking and product-safety requirements. The Machinery Directive 2006/42/EC remains relevant for machinery placed on the EU market before the transition deadline, but Regulation (EU) 2023/1230 is now the forward-looking machinery regime and will repeal the directive as its provisions apply from January 20, 2027 (European Commission, 2006; European Parliament & Council, 2023). The regulation explicitly reflects risks from autonomous mobile machinery, connectivity, safety-related software, and systems with self-evolving behavior that ensure safety functions. The EU AI Act (Regulation (EU) 2024/1689) is also now adopted rather than merely proposed; it imposes requirements for high-risk AI systems, including AI safety components in products covered by EU harmonization legislation, but obligations phase in over different dates and embedded AI in regulated products has a later transition than some general AI Act obligations (European Parliament & Council, 2024). For robotics companies and SMEs, NIS2 adds another compliance entry point where robot platforms, cloud services, remote operations, or industrial deployments fall within essential or important entity obligations for cybersecurity risk management, incident reporting, and supply-chain security (European Parliament & Council, 2022).

China's regulatory landscape centers on GB (Guobiao) national standards administered by the State Administration for Market Regulation (SAMR). GB/T 36530–2018 establishes safety requirements for service robots, while GB/T 38559–2020 addresses collaborative robot safety (SAC, 2018, 2020). These standards largely align with ISO frameworks but include China-specific requirements for data localization and network security compliance under the Cybersecurity Law and Personal Information Protection Law (PIPL).

### 6.1.3 Case Study: Boston Dynamics Atlas

Boston Dynamics' Atlas humanoid incorporates multiple safety layers including hardware-level emergency stops, software-enforced joint limits, and real-time stability monitoring. The all-electric generation of Atlas features 56 degrees of freedom with custom high-powered electric actuators. Instead of legacy hydraulic pressure valves, safety is maintained through advanced motor controllers with torque-sensing, 360-degree camera views, and tactile sensors that ensure fenceless guarding and collision awareness. Software safety monitors track state estimation confidence, battery levels, and thermal conditions, triggering graceful degradation when thresholds are exceeded.

### 6.1.4 Case Study: Agility Robotics Digit

Agility Robotics' Digit has engaged with UL and OSHA to establish appropriate safety protocols for warehouse environments (Agility Robotics, 2023). Digit incorporates redundant sensing for obstacle detection, emergency stop buttons accessible from multiple angles, and force-limiting control that prevents injury during incidental contact. The robot's compliance strategy includes comprehensive risk assessments following ISO 12100 methodology, hazard analysis per ANSI/RIA R15.08, and documentation of safety-related protective measures. Field deployments at logistics facilities have provided real-world validation of these safety systems.

### 6.1.5 Case Study: UBTECH Walker

UBTECH's Walker humanoid targets consumer and service markets in China, requiring compliance with GB consumer safety standards (UBTECH, 2023). The robot achieves GB/T 36530–2018 certification through multiple protective measures including soft-touch surfaces, proximity sensing with adjustable safety zones, and safe interaction modes. The certification process required extensive physical testing including tip-over resistance, pinch-point analysis, and electromagnetic compatibility verification.

### 6.1.6 Certification Path Comparison: USA vs. China

The certification pathways for humanoid robots diverge significantly between the USA and China due to differences in regulatory philosophy and market structure. United States certification emphasizes voluntary standards with liability-driven compliance, where manufacturers typically pursue UL listing or ETL certification to demonstrate due diligence. The process involves third-party testing laboratories, public documentation of test results, and ongoing market surveillance. Chinese certification follows a mandatory assessment model with government-designated testing centers performing conformity evaluation. GB standards often incorporate specific Chinese technical requirements and prioritize state security considerations alongside physical safety. The timeline for USA certification typically ranges from 6 to 18 months depending on complexity, while Chinese certification can extend 12 to 24 months due to centralized approval processes and data security reviews. Figure 6.1 compares the safety, privacy, and cybersecurity regulatory requirements related to humanoid robot across the US, EU, and China—the three dominant humanoid robotics markets.

| Dimension | United States | European Union | China |
|---|---|---|---|
| **Key standards**<br>Physical safety | UL 1740, UL 3100<br>ANSI/RIA R15.08<br>ISO 13482 (adopted) | Machinery Dir. 2006/42/EC<br>ISO 13482, ISO/TS 15066<br>CE marking required | GB/T 36530-2018<br>GB/T 38559-2020<br>2026 national std. system |
| **Compliance model**<br>Voluntary vs. mandatory | Voluntary + liability-driven<br>3rd-party testing labs | Mandatory CE marking<br>Conformity assessment | Mandatory, gov.-designated<br>State security review |
| **Privacy framework**<br>Data / consent rules | CCPA/CPRA (CA)<br>Opt-out model; sector laws | GDPR — opt-in consent<br>DPIA mandatory; strong rights | PIPL — explicit consent<br>Data localization required |
| **Certification timeline**<br>Typical approval period | 6 – 18 months | 9 – 18 months | 12 – 24 months |
| **AI-specific regulation**<br>Emerging AI rules | No federal AI act yet<br>ITAR for defense apps | EU AI Act (high-risk)<br>Transparency + oversight req. | Cybersecurity Law + DSL<br>SM2/SM3/SM4 encryption |
| **Vulnerability disclosure**<br>Security reporting model | Coordinated disclosure<br>Bug bounties common | Coordinated disclosure<br>GDPR breach notification | Gov. platform (CNVD)<br>Mandatory state reporting |

**Fig. 6.1** Regional humanoid robot regulatory comparison

## 6.2 Fail-Safe Mechanisms and Redundancy

Fail-safe mechanisms form the foundation of trustworthy humanoid robot systems, ensuring safe behavior even when primary systems malfunction. The complexity of humanoid robots, with dozens of actuated joints and multiple sensor modalities, demands systematic redundancy strategies (Carlson & Doyle, 2002).

### 6.2.1 Emergency Stop Systems

Emergency stop (e-stop) systems provide the ultimate safety override, immediately halting robot operation when activated. Effective e-stop design follows a hierarchical architecture with multiple trigger mechanisms and fail-safe power removal (Dhillon, 2007). Physical e-stop buttons must be easily accessible from multiple approach angles, typically requiring large mushroom-head designs that can be struck rather than precisely pressed. These buttons connect through hardwired circuits independent of software control, directly interrupting power to actuators or engaging mechanical brakes.

Modern humanoid robots implement wireless e-stop capabilities enabling remote operators or supervisory systems to trigger emergency stops. These systems require careful design to prevent signal jamming or spoofing attacks while maintaining reliability. Redundant communication channels, such as paired radio frequencies or parallel Wi-Fi and cellular connections, provide resilience against single-point failures. Software e-stops complement hardware systems, monitoring for anomalous conditions such as joint limit violations, excessive forces, or loss of state estimation, and triggering controlled shutdowns before hardware e-stops engage.

### 6.2.2 Sensor Redundancy and Voting Schemes

Critical sensing functions require redundancy to maintain operation despite component failures. Humanoid robots typically employ sensor voting schemes where multiple sensors measure the same quantity and disagreement triggers fault detection (Steinbauer et al., 2016). For inertial measurement units (IMUs), triple modular redundancy with majority voting provides single-fault tolerance, ensuring accurate pose estimation even if one IMU fails or provides corrupted data.

Analytical redundancy leverages physical relationships between different sensor types to validate measurements without hardware duplication. For example, joint encoder positions can be cross-checked against vision-based pose estimates and IMU-derived orientations. Discrepancies beyond expected noise levels indicate sensor degradation or failure, enabling isolation of faulty sensors while maintaining operation (Khalastchi & Kalech, 2018).

### 6.2.3 Mechanical Brakes and Power-Off Safety

Humanoid robots must achieve safe configurations when power is removed, either intentionally through e-stops or unintentionally through battery depletion or electrical faults. Mechanical braking systems engage automatically upon power loss, preventing joints from freely moving under gravity and payload forces. Fail-safe brake design typically employs spring-loaded mechanisms held open by electromagnetic coils during normal operation; power loss releases the springs, engaging friction or magnetic brakes.

The challenge for humanoid robots lies in achieving safe postures during power-off conditions. Unlike industrial manipulators that can remain stationary, bipedal humanoids will fall if power is cut during standing or walking. Advanced systems implement capacitor banks or flywheel energy storage that provide sufficient power for controlled descent sequences lasting several seconds after main power interruption. These "last-gasp" systems enable robots to crouch, shift weight to a stable base, or grasp nearby supports before final shutdown.

### 6.2.4 Graceful Degradation Strategies

Rather than catastrophic failure, well-designed humanoid robots exhibit graceful degradation where functionality reduces progressively as problems accumulate (Carlson & Doyle, 2002). This requires hierarchical prioritization of capabilities and resources. For example, a robot detecting battery levels below 20% might reduce maximum walking speed and disable non-essential sensors to extend operational time for critical tasks. At 10% battery, it might autonomously navigate to charging stations or safe positions before shutting down.

Graceful degradation extends to sensor failures where partial sensing capability maintains limited functionality. Loss of forward-facing cameras might reduce walking speed and disable manipulation tasks while preserving basic mobility using LiDAR and IMU data. Actuator failures can trigger reconfiguration where the robot adapts gait patterns to compensate for disabled joints, similar to how humans adapt after injury (Koos et al., 2013).

### 6.2.5 Case Study: Boston Dynamics Multi-Layered Safety

Boston Dynamics implements defense-in-depth safety architecture with five independent safety layers (Boston Dynamics, 2023). The outermost layer consists of operational safety procedures including trained operators and restricted areas. The second layer involves software safety monitors checking state estimation health, joint limits, and stability margins at 1 kHz. The third layer implements protective stops when the robot detects imminent instability or collision. The fourth layer engages hardware-level circuit breakers and overcurrent protection for its electric actuators. The innermost layer consists of mechanical hard stops and software-defined rotational limits—though its unique joint architecture allows 360-degree rotation at the hips and waist, the system restricts unexpected motions to prevent catastrophic physical failure. This architecture ensures that failure of any single layer does not compromise safety, with each layer providing independent protection.

### 6.2.6 Case Study: Tesla Optimus Automotive Heritage

Tesla's Optimus humanoid benefits from automotive safety engineering expertise developed for vehicle systems (Tesla, 2023). The electric vehicle industry has established rigorous functional safety standards per ISO 26262, which Tesla adapts for humanoid robot applications. Optimus incorporates automotive-grade motor controllers with built-in safety functions including overcurrent protection, thermal monitoring, and phase loss detection. Battery management systems derived from vehicle powertrains implement cell-level monitoring, thermal management, and multiple layers of protection against short circuits or overcharging. The software architecture employs ASIL (Automotive Safety Integrity Level) principles with safety–critical functions isolated in separate computational domains running on dedicated processors. This automotive heritage provides proven safety approaches tested across millions of vehicle miles.

### 6.2.7 Case Study: Fourier GR-1 Medical Device Protocols

Derived from Fourier's expertise in rehabilitation exoskeletons, the GR-1 implements force-limiting control through its proprietary Fourier Smart Actuators (FSA). Delivering up to 230 N·m of peak torque per actuator at the hip joints (different reference points; cf. Unitree G1's 120 N·m peak measured at the knee actuator), the system relies on hardware-level torque sensing to prevent excessive pressure during patient handling. The control system maintains detailed logs of all operations for clinical review and fault analysis. Redundant position sensors in joints ensure accurate limb positioning during rehabilitation exercises, with automatic session termination if sensor disagreement exceeds thresholds. Rather than rigid failure, the GR-1's fail-safe behavior includes dynamic balance recovery—such as automatically stepping back when subjected to sudden external forces—ensuring it does not fall on vulnerable patients (Fourier Intelligence, 2023).

### 6.2.8 Case Study: Apptronik Apollo Industrial Fail-Safe

Apptronik's Apollo humanoid robot is designed for industrial environments with comprehensive fail-safe systems addressing workplace hazards (Apptronik, 2023).

The robot features multiple accessible e-stop buttons positioned on torso, arms, and back. Apptronik states that Apollo's drive electronics implement a safe torque off (STO) function in which dedicated safety circuits remove drive enable independent of the control processor—the standard pattern defined for variable-speed drives in IEC 61800-5-2. The robot's mobile base includes bumper sensors that trigger protective stops when contacted, preventing collisions with workers or infrastructure. Software monitors implement safety-rated monitored stop (SMS), where the robot halts motion and maintains brake engagement until conditions are safe to resume. Battery management includes redundant charge monitoring with conservative shutoff thresholds, preventing deep discharge that could compromise battery safety.

## 6.3 Collision Avoidance and Human Safety

Direct physical interaction between humanoid robots and humans presents significant injury risks that must be mitigated through intelligent collision avoidance and compliant control strategies. Unlike industrial robots operating behind barriers, humanoid robots share human workspaces, requiring sophisticated systems that predict, prevent, and respond to contact events (Lasota et al., 2017).

### 6.3.1 Proximity Detection and Safety Zones

Effective collision avoidance begins with accurate proximity detection using multiple sensor modalities. LiDAR sensors provide precise distance measurements in 360-degree fields, enabling detection of humans at ranges up to 30 m with centimeter accuracy. Time-of-flight cameras and structured light sensors offer dense depth information at closer ranges, typically effective within 10 m. These sensors enable definition of safety zones with progressive response strategies (Marvel & Norcross, 2017).

Safety zone architectures typically implement three concentric regions: warning zones (2–3 m), reduced speed zones (1–2 m), and protective stop zones (less than 1 m). When humans enter warning zones, robots reduce velocity and increase sensor sampling rates. Entry into reduced speed zones triggers speed scaling proportional to distance, slowing to 250–500 mm/s for collaborative operation per ISO/TS 15066 guidelines. Protective stop zones immediately halt motion until the zone is clear, preventing any possibility of contact during unexpected human approach (ISO, 2016).

### 6.3.2 Force Limiting and Compliant Control

When contact is unavoidable or intentional, force-limiting control prevents injury by restricting impact forces and sustained pressures below biomechanical thresholds. Compliant control strategies enable robots to react to external forces by yielding rather than resisting, dramatically reducing injury risk (Haddadin et al., 2017). Impedance control adjusts the apparent stiffness and damping of robot joints, allowing soft interaction when contacts occur. Lower impedance settings make robots feel soft and yielding, while higher settings provide rigid positioning when needed for tasks.

Series elastic actuators (SEAs) provide inherent compliance through mechanical springs between motors and output shafts, naturally limiting forces regardless of control software behavior. While adding mass and complexity, SEAs offer passive force limitation that persists even during software failures or sensor errors. Torque sensing in all joints enables real-time force estimation and control, where measured torques are compared against expected values from dynamics models to detect contacts. Discrepancies trigger compliant responses or protective stops depending on magnitude.

Biomechanical injury thresholds established by ISO/TS 15066 provide quantitative targets for force-limiting control. For quasi-static clamping contacts, permissible forces range from 65N for the skull to 300N for the pelvis, with most body regions falling between 110 and 220N. Transient impacts allow higher peak forces but require considering contact duration and body region sensitivity. Implementation requires calibrated force sensing, validated control algorithms, and extensive physical testing with instrumented crash-test dummies to verify compliance (Haddadin et al., 2017).

### 6.3.3 Predictive Collision Avoidance

Advanced collision avoidance anticipates future collisions based on predicted human and robot trajectories, enabling earlier intervention than reactive approaches. Human motion prediction leverages vision-based pose estimation to track joint positions and estimate velocity vectors. Short-term predictions (1–2s) extrapolate current motions assuming constant velocity or simple physics models. Machine learning approaches can recognize common human actions like reaching or walking, enabling more accurate predictions over 3–5s horizons (Mainprice et al., 2016).

Predicted human trajectories are compared against planned robot motions to identify potential collisions. When collision risk exceeds thresholds, the robot replans its path or adjusts timing to maintain separation margins. Dynamic motion planning algorithms such as Model Predictive Control (MPC) incorporate human predictions directly into optimization objectives, generating robot trajectories that proactively avoid interference while accomplishing tasks (Ratliff et al., 2009).

### 6.3.4 Human Pose Estimation for Safety

Accurate human pose estimation enables robots to identify vulnerable body regions and adapt safety responses accordingly. Vision-based pose estimation using deep learning methods such as OpenPose or MediaPipe detects human skeletal landmarks from camera images at 30+ Hz (Cao et al., 2019). These systems identify locations of head, torso, arms, and legs, enabling distance calculations to specific body parts rather than treating humans as uniform obstacles.

Pose estimation enhances safety through targeted protection of vulnerable regions. Detection of a human head within the robot workspace triggers more conservative safety margins and lower speed limits compared to detection of a hand or forearm. Child detection based on estimated height triggers additional protective measures such as reduced speeds, gentle motions, and immediate stops if the child approaches within close proximity. Age and posture estimation can identify elderly or mobility-impaired individuals who may require extra care to avoid startling or destabilizing.

### 6.3.5 Injury Risk Assessment Models

Quantitative injury risk assessment provides systematic evaluation of potential harm from robot-human contact. The Head Injury Criterion (HIC) quantifies impact severity for head contacts, calculated from acceleration time histories during collisions. ISO 13482 establishes HIC limits of 700 for personal care robots, corresponding to low injury risk during impacts (ISO, 2014). Similar metrics exist for other body regions, including chest acceleration limits and limb loading thresholds derived from automotive crash testing and biomechanical research.

Probabilistic risk assessment extends deterministic injury thresholds by incorporating likelihood of contact events. Fault tree analysis (FTA) decomposes potential injury scenarios into contributing failures and calculates combined probabilities. For example, a contact event requires failures in proximity detection AND collision avoidance AND force-limiting control. If each system has 0.1% failure rate per exposure, the combined probability drops to approximately 0.0001%. This systematic approach identifies critical safety-related components requiring highest reliability.

### 6.3.6 Case Study: Figure AI Collaboration Zones

Figure AI's humanoid robots utilize a Vision-Language-Action (VLA) neural network paired with embedded stereo and palm cameras to maintain real-time spatial awareness without relying on external infrastructure. Rather than rigid safety zones, Figure's onboard perception system builds dynamic 3D maps of its surroundings, allowing it to predict human trajectories and adapt its manipulation speed and pathing continuously to avoid collisions in unstructured settings. When humans approach, the robot provides visual and audio cues indicating its intentions. This communicative safety approach reduces human startle responses and enables more efficient shared workspace operation.

### 6.3.7 Case Study: UBTECH Walker S Industrial Safety

UBTECH's Walker S operates in automotive manufacturing lines (such as NIO and BYD), requiring robust human–robot collaboration safety. The robot utilizes a rigid-flexible coupling hybrid system and force-compliant drive joints that provide stability even on moving or vibrating factory floors. Walker S implements multi-modal sensor fusion—combining 3D depth vision, audio, and force-torque feedback—to monitor human workers in close proximity. Instead of relying purely on fixed barriers, the robot dynamically adjusts its 41 servo joints to yield safely upon unexpected contact, allowing it to work side-by-side with humans on automotive assembly lines without heavy fencing.

### 6.3.8 Case Study: EX Robots Gentle Interaction

EX Robots designs humanoid robots for social settings including retail, hospitality, and events, where gentle interaction with diverse populations is essential (EX Robots, 2023). The robots feature soft silicone skin over force-sensing structures, enabling detection of even light touches while providing cushioned contact surfaces. Compliant neck mechanisms allow the head to yield during accidental contacts, preventing injury while maintaining visual engagement with humans. The control system implements behavior-based safety where aggressive motions such as rapid pointing or reaching are disabled in public modes, favoring slower, more predictable movements. EX Robots' AI analyzes human emotional responses through facial expression recognition, detecting signs of discomfort or fear and automatically increasing personal space or adopting less intimidating postures. This emotional intelligence enhances physical safety by reducing stress-induced startle responses that could lead to accidents.

### 6.3.9 Case Study: Unitree G1 Collision Response

Unitree's G1 humanoid robot demonstrates collision response and recovery capabilities essential for dynamic environments (Unitree, 2023). The robot's joint torque sensors detect unexpected contact forces within 5 ms, triggering immediate compliant response where joints yield in the direction of impact. This compliance absorbs collision energy while the planning system determines appropriate recovery actions. For minor contacts below injury thresholds, G1 continues operation after brief pause, maintaining productivity in busy environments. More significant contacts trigger protective stops and safety assessments before resuming motion.

The robot logs all contact events with force magnitudes, locations, and video recordings, enabling continuous safety improvement through analysis of near-miss incidents. Unitree's control strategy balances safety with task completion, avoiding overly conservative responses that would paralyze the robot in typical human-populated settings.

## 6.4 Cybersecurity Threats and Attack Vectors

Humanoid robots represent complex cyber-physical systems vulnerable to sophisticated attacks that can compromise safety, privacy, and functionality. Security is not merely an adjacent concern to safety; it is often a precondition for safety because compromised sensing, communications, model weights, update channels, or actuation commands can make an otherwise safety-certified robot behave unsafely. The convergence of robotics with artificial intelligence, cloud computing, and Internet of Things technologies expands the attack surface dramatically (Bonaci et al., 2015). Understanding threat vectors and implementing defense-in-depth strategies is essential for secure deployment.

### 6.4.1 Threat Modeling for Humanoid Robots

Comprehensive threat modeling identifies potential adversaries, their capabilities, and likely attack targets. Nation-state actors possess substantial resources for sophisticated attacks potentially targeting robots deployed in critical infrastructure or defense applications. Criminal organizations might exploit robots for financial gain through ransomware or theft of valuable data. Malicious insiders with authorized access present particularly dangerous threats given their knowledge of system architectures and security controls (Yaacoub et al., 2022).

Attack objectives span a broad spectrum including: disruption of operations through denial-of-service attacks; manipulation of robot behavior to cause physical harm or damage property; exfiltration of sensitive data collected by robot sensors; and establishment of persistent access for ongoing intelligence gathering or future attacks. The STRIDE threat modeling framework provides systematic analysis covering Spoofing identity, Tampering with data, Repudiation of actions, Information disclosure, Denial of service, and Elevation of privilege (Shostack, 2014).

Given that humanoid robots function as agentic AI systems—autonomously perceiving, reasoning, and acting in physical environments—STRIDE alone is insufficient. Security practitioners should also apply MAESTRO (Multi-Agent Environment, Security, Threat, Risk, and Outcome), a seven-layer threat modeling framework published by the Cloud Security Alliance (CSA) specifically designed for agentic AI. MAESTRO builds upon established frameworks like STRIDE but adds Agentic AI-specific considerations (Huang & Hughes, 2025a), addressing unique threats such as adversarial machine learning and the risks of autonomous decision-making. Its seven layers span Foundation Models (L1), Data Operations (L2), Agent Frameworks (L3), Deployment and Infrastructure (L4), Evaluation and Observability (L5), Security and Compliance (L6), and Agent Ecosystem (L7)—each directly applicable to the humanoid robot stack, from onboard LLMs and sensor pipelines through real-time decision-making frameworks to multi-robot coordination. Critically, MAESTRO identifies how a threat at one layer can chain into subsequent layers, exposing compounding risks such as model poisoning, resource hijacking, and backdoor attacks that STRIDE's categorical approach would not surface on its own (Huang & Hughes, 2025a). Used together, STRIDE and MAESTRO provide both the classical security coverage and the agentic AI-specific depth required for rigorous humanoid robot threat analysis.

### 6.4.2 Network Security and Communication Encryption

Robust network security forms the foundation for protecting robot systems from remote attacks. All communication channels must implement strong encryption using modern protocols such as Transport Layer Security (TLS) 1.3 or higher. For real-time control systems requiring low latency, Datagram Transport Layer Security (DTLS) provides encryption for UDP-based communications. Pre-shared keys or certificate-based authentication prevents unauthorized devices from joining robot communication networks.

Network segmentation isolates critical robot control systems from less trusted networks such as public Wi-Fi or operator devices. Industrial protocols suggest placing safety–critical systems on isolated networks with no external connectivity, while less critical functions like telemetry and status updates operate on separate networks with controlled internet access. Virtual Private Networks (VPNs) or secure tunnels protect communications when robots must traverse public networks, encrypting all traffic between robots and control centers.

Intrusion detection systems (IDS) monitor network traffic for suspicious patterns indicating attack attempts. Signature-based detection identifies known attack patterns, while anomaly-based detection establishes baselines of normal traffic and flags deviations. For humanoid robots, anomalies might include unexpected command sequences, unusual data volumes, or communications with unknown network addresses. Integration with security information and event management (SIEM) systems enables centralized monitoring across robot fleets.

### 6.4.3 Sensor Spoofing Attacks

Sensor spoofing exploits the robot's dependence on sensor data for perception and navigation. GPS spoofing broadcasts fake satellite signals causing robots to believe incorrect position estimates, potentially leading them into restricted areas or hazardous situations (Tippenhauer et al., 2011). Low-cost software-defined radio equipment enables GPS spoofing attacks with ranges of hundreds of meters.

Countermeasures include GPS signal authentication, cross-validation with inertial navigation systems, and detection of impossible motion profiles.

Vision system spoofing employs adversarial patches or patterns designed to fool object detection algorithms. Researchers have demonstrated small stickers that cause stop signs to be misclassified as speed limit signs or make people invisible to pedestrian detectors (Eykholt et al., 2018). For humanoid robots, such attacks could cause dangerous behaviors like walking into humans misclassified as empty space. Defenses include multiple camera viewpoints, sensor fusion with LiDAR, and adversarial training to improve robustness.

LiDAR spoofing injects false laser returns causing phantom obstacles or making real obstacles disappear. Attackers can use laser pointers or other directed light sources to inject signals at wavelengths matching the robot's LiDAR sensor. Detection requires analysis of signal properties including timing, intensity patterns, and consistency with other sensors (Shin et al., 2017). Advanced sensors implement coded signals or frequency hopping making spoofing more difficult.

### 6.4.4 Adversarial Attacks on Perception Systems

Machine learning-based perception systems achieve high accuracy on natural inputs but remain vulnerable to adversarial examples—carefully crafted inputs that cause misclassification despite appearing normal to humans (Goodfellow et al., 2015). Digital adversarial attacks modify input images through imperceptible perturbations, while physical adversarial attacks create real-world objects or patterns that fool vision systems.

Adversarial attacks on humanoid robots could cause serious safety hazards. An adversarial pattern on clothing might make a person invisible to the robot's obstacle detection, leading to collisions. Adversarial patches on floors could be misinterpreted as obstacles causing the robot to halt unnecessarily or as walkable surfaces causing the robot to step into hazards. Research demonstrates that adversarial examples transfer across different models and sensors, making defenses particularly challenging (Ilyas et al., 2019).

Defense strategies include adversarial training where models are trained on adversarial examples alongside normal data, improving robustness but not providing complete protection. Ensemble methods employ multiple models with different architectures, requiring attackers to fool all models simultaneously. Input sanitization detects and removes adversarial perturbations before classification. Runtime monitoring detects anomalous model behaviors such as rapid prediction changes for similar inputs. For safety-critical applications, combining multiple sensor modalities makes attacks significantly more difficult as attackers must fool all sensing systems simultaneously.

### 6.4.5 Model Poisoning and Backdoor Attacks

Machine learning models trained on compromised datasets can exhibit backdoor behaviors where specific trigger patterns cause targeted misclassifications. An attacker who gains access to training data or processes might inject poisoned examples that teach the model dangerous associations (Gu et al., 2019). For example, poisoned training data could teach a robot to ignore humans wearing specific colored clothing, creating targeted safety vulnerabilities.

Data provenance tracking helps detect poisoned datasets by maintaining records of data sources, collection methods, and preprocessing steps. Training data should come from trusted sources with verification of authenticity. Statistical analysis can identify outlier examples that may represent poisoning attempts. Differential privacy techniques limit the influence any single training example can have on the model, making poisoning attacks more difficult.

Model validation requires testing on diverse, independently collected datasets that attackers have not influenced. Detecting backdoors is challenging as backdoored models typically perform normally on clean data. Techniques include input filtering to detect trigger patterns, model inspection to identify suspicious learned features, and behavioral monitoring during deployment to catch activated backdoors.

### 6.4.6 Supply Chain Security

Humanoid robot supply chains span multiple countries and dozens of component suppliers, creating opportunities for malicious hardware or software insertion. Compromised components might include hardware backdoors enabling remote access, modified firmware with exploitable vulnerabilities, or counterfeit parts with degraded reliability. High-profile examples from other industries demonstrate that supply chain attacks can remain undetected for years while providing persistent access.

Supply chain security begins with vendor assessment and trustworthiness evaluation. Critical components should source from manufacturers with strong security practices and preferably domestic production for sensitive applications. Hardware security modules (HSMs) provide cryptographic verification of component authenticity using digital signatures or secure boot mechanisms. Firmware validation compares cryptographic hashes of loaded firmware against known-good versions, detecting unauthorized modifications.

For software supply chains, dependency scanning identifies vulnerabilities in third-party libraries and frameworks. Software bill of materials (SBOM) documents all components and their versions, enabling rapid response when vulnerabilities are discovered. Code signing ensures software packages originate from verified developers and have not been tampered with during distribution. Regular security audits of the entire supply chain identify weak points requiring enhanced controls.

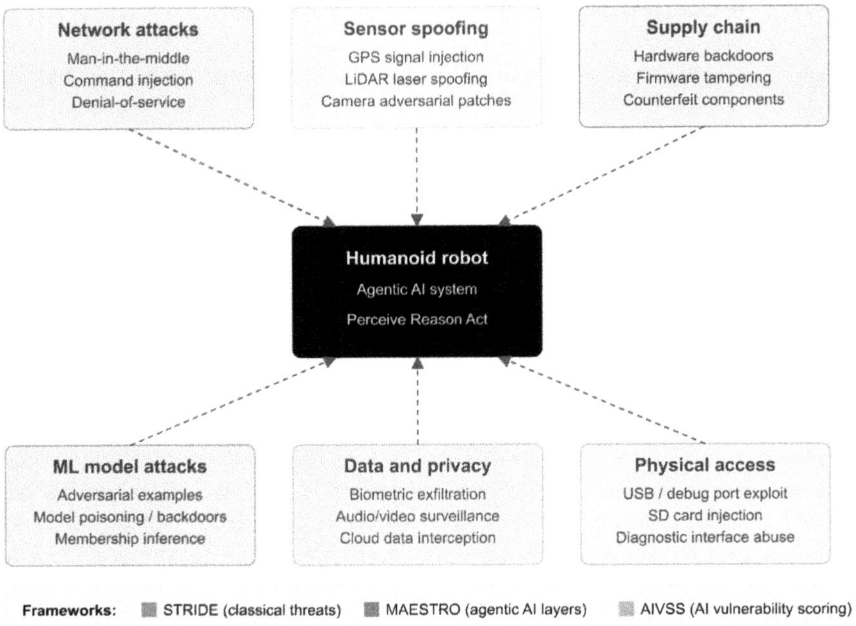


**Fig. 6.2** Cybersecurity threat taxonomy for humanoid robots

Before we look at a few case studies, Fig. 6.2 summarizes the key cybersecurity threat taxonomy for humanoid robots we have discussed in this section.

### 6.4.7 Case Study: Tesla Automotive Cybersecurity

Tesla's cybersecurity practices for vehicle systems provide relevant lessons for Optimus humanoid development. Tesla implements defense-in-depth with multiple security layers including encrypted communications, secure boot verification, over-the-air (OTA) update authentication, and intrusion detection systems. The company maintains an active bug bounty program that has identified and patched hundreds of vulnerabilities before exploitation. Tesla's vehicles employ cryptographic signatures for all firmware, preventing unauthorized code execution. The Autopilot system receives software updates through encrypted channels with verification before installation. Security researchers have demonstrated vulnerabilities in Tesla systems, but rapid response and coordinated disclosure minimize exploitation windows. For Optimus, Tesla applies similar principles including secure boot processes, encrypted communications with backend systems, and continuous security monitoring.

### 6.4.8 Case Study: Boston Dynamics Communication Security

Boston Dynamics implements secured communication protocols for its research and commercial robots including Atlas and Spot. Communication between operator control units and robots uses mutual TLS authentication, ensuring both parties verify identity before exchanging data. Commands are digitally signed preventing replay attacks or command injection by adversaries. The robots support optional air-gapped operation where all processing occurs onboard without network connectivity, eliminating remote attack vectors for highest security applications. For cloud-connected deployments, Boston Dynamics provides VPN connectivity through their secure infrastructure. The company collaborates with cybersecurity researchers through responsible disclosure programs, addressing identified vulnerabilities through regular software updates.

### 6.4.9 Case Study: China-USA Security Considerations

Cybersecurity approaches for humanoid robots diverge between China and the USA due to different threat models, regulatory frameworks, and technological ecosystems (Segal, 2018). Chinese regulations including the Cybersecurity Law and Data Security Law mandate data localization where personal information and important data collected by robots must be stored on servers within China. Chinese companies emphasize protection against foreign intelligence services and implement state-approved encryption standards such as SM2, SM3, and SM4 algorithms. USA companies face restrictions on exporting advanced encryption technologies and must comply with regulations such as International Traffic in Arms Regulations (ITAR) for robots with potential defense applications.

Data sovereignty concerns affect international deployment of humanoid robots. Chinese regulations require government security reviews before data can be transferred abroad, impacting cloud-connected robots operated by international companies. USA regulations including Cloud Act of 2018 enable law enforcement access to data stored by USA companies regardless of physical location, raising concerns for Chinese customers. These conflicting requirements create challenges for companies seeking to deploy humanoid robots across both markets, often necessitating separate product variants with different data handling architectures.

Supply chain security receives particular attention in both countries due to geopolitical tensions. USA companies face restrictions on incorporating certain Chinese-manufactured components in sensitive applications, while Chinese companies encounter similar restrictions on USA-origin semiconductors and software. Both countries are investing in domestic semiconductor manufacturing and robotics supply chains to reduce foreign dependencies.

### 6.4.10 Case Study: Unitree G1 as a Cyber-Physical Attack Platform

Recent security research by Mayoral-Vilches et al. (2025b) demonstrated how a commercial humanoid robot can evolve from a passive cyber-physical endpoint into an active attack platform capable of reconnaissance, persistence, and offensive operations inside enterprise networks. The study analyzed the Unitree G1 humanoid robot through static firmware analysis, Bluetooth Low Energy (BLE) protocol reverse engineering, runtime traffic inspection, and cryptographic analysis of proprietary middleware protections.

The researchers identified a critical BLE provisioning vulnerability in the robot's Wi-Fi configuration workflow. By exploiting malformed Wi-Fi credential injection and leveraging hardcoded AES cryptographic keys reused across devices, attackers could achieve root-level compromise during the provisioning process. The vulnerability effectively bypassed trust assumptions associated with robot onboarding and enabled complete takeover of the robot's Linux-based control stack.

Once compromised, the robot exposed several characteristics that significantly amplify cyber-physical risk:

- persistent network connectivity through Wi-Fi, BLE, DDS/ROS 2 middleware, and cloud services;
- onboard microphones, cameras, IMUs, and telemetry streams capable of environmental surveillance;
- mobility and physical access enabling lateral movement into restricted operational areas;
- cloud-linked OTA infrastructure that could potentially be abused for persistence or propagation.

The study additionally documented weaknesses in Unitree's proprietary FMX encryption layer, including Blowfish-ECB usage with static fleet-wide keys and predictable obfuscation logic. These flaws enabled inspection and manipulation of otherwise protected communications between services running inside the robot orchestration stack.

A particularly important finding was the operationalization of a resident "Cybersecurity AI" (CAI; Mayoral-Vilches et al., 2025c) agent running onboard the humanoid platform. The CAI system autonomously mapped attack surfaces, identified authentication weaknesses, and prepared exploitation workflows against external infrastructure, illustrating how humanoid robots may become autonomous cyber operators rather than merely compromised endpoints.

From a threat-modeling perspective, this case demonstrates several critical realities for humanoid deployments:

1. Humanoid robots should be treated as high-risk mobile endpoints equivalent to privileged laptops or industrial control nodes.

2. BLE onboarding and maintenance interfaces constitute critical attack surfaces and require zero-trust assumptions.
3. Proprietary cryptography without independent review often introduces systemic vulnerabilities.
4. Robot cybersecurity failures can cascade into physical safety failures when compromised systems influence locomotion, manipulation, or human interaction.
5. Agentic AI systems embedded within robots may enable machine-speed offensive operations with limited human oversight.

The findings suggest that future humanoid robot certification frameworks should incorporate:

- mandatory penetration testing,
- SBOM and firmware provenance validation,
- secure provisioning requirements,
- SROS2 and ROS 2/DDS security enforcement,
- SROS2 security enclaves and policy artifacts, which provide ROS 2's practical security layer for authenticated and encrypted node communication, permission control, and deployment-scoped access policies (Mayoral-Vilches et al., 2022; ROS 2 Documentation, 2026),
- runtime anomaly detection,
- and independent security audits before deployment into industrial or public environments.

This case also reinforces the need for robotics-specific coordinated disclosure ecosystems such as the Robot Vulnerability Database (RVD; Mayoral-Vilches et al., 2019) and broader DevSecOps integration throughout the robot lifecycle (Mayoral-Vilches et al., 2020).

## 6.5 Vulnerability Assessment Across Platforms

Comparative vulnerability assessments reveal common weaknesses across humanoid robot platforms. Most systems rely on Linux-based operating systems or real-time operating systems (RTOS) with known vulnerability histories requiring continuous patching. Robot Operating System (ROS), widely used for development, has documented security limitations including lack of built-in encryption or authentication in ROS1, though ROS 2 addresses many issues (White et al., 2021). Default credentials on robot systems present another common vulnerability, with researchers discovering hardcoded passwords in multiple commercial robots.

Physical security often receives insufficient attention with accessible USB ports, Ethernet jacks, or SD card slots enabling local attacks. Diagnostic interfaces intended for maintenance provide powerful access that could be exploited by malicious actors. Limited computational resources constrain security mechanisms,

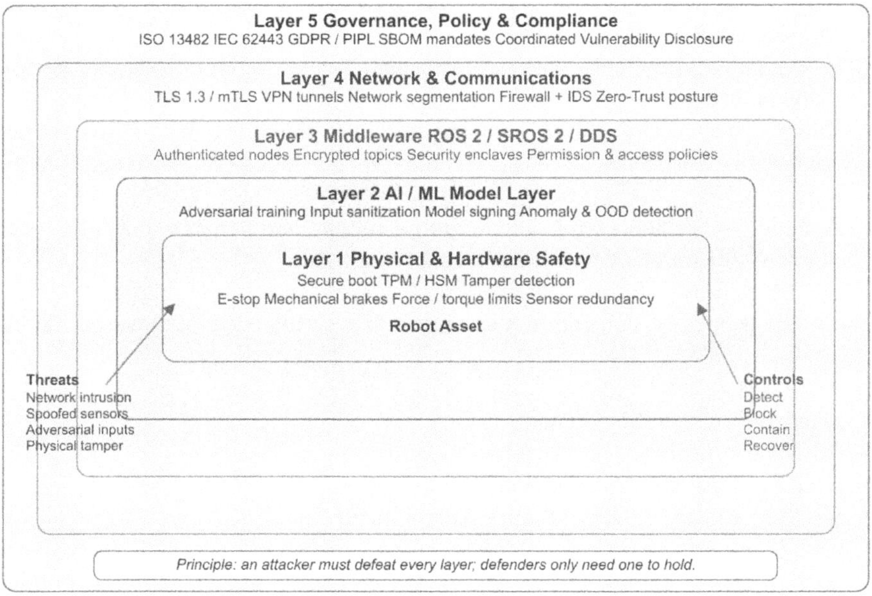


**Fig. 6.3** Defense-in-depth security architecture for humanoid robots

particularly for battery-powered mobile robots where cryptographic operations consume scarce energy. Balancing security with performance and battery life requires careful engineering trade-offs.

Figure 6.3 illustrates the defense-in-depth security architecture recommended for humanoid robots, organizing controls into five concentric layers—physical and hardware safety at the core, the AI/ML model layer, the ROS 2 / SROS2 middleware, network and communications, and an outer governance and compliance perimeter—emphasizing that an attacker must defeat every layer in succession while defenders only need a single intact layer to contain an incident.

## 6.6 Data Privacy and Surveillance Concerns

Humanoid robots equipped with cameras, microphones, and sensors capable of capturing biometric information raise significant privacy concerns, particularly when deployed in homes, workplaces, or public spaces. The ability to continuously record audio and video, track movements, and potentially identify individuals creates surveillance capabilities that must be carefully governed (Lutz & Tamò-Larrieux, 2020).

### 6.6.1 Camera and Audio Data Collection

Humanoid robots typically incorporate multiple cameras for navigation, manipulation, and interaction. Forward-facing cameras capture continuous video of the robot's environment, potentially recording individuals going about daily activities without explicit consent. Microphones enable voice interaction and situational awareness, but also enable persistent audio recording of conversations. The richness of this data exceeds traditional surveillance systems as robots move through spaces and interact closely with people (Calo, 2012).

Privacy-by-design principles should guide sensor usage, collecting only necessary data and implementing controls against misuse. Visual markers such as LED indicators can signal when cameras or microphones are actively recording, providing transparency to nearby individuals. Hardware switches that physically disconnect cameras or microphones offer stronger assurance than software controls that could be bypassed. Some designs implement camera shutters or microphone muting circuits that cannot be overridden by software, protecting against malware or hacking that compromises system controls.

Data minimization practices limit collection to task-essential information. Robots performing navigation may only require depth information and edge detection rather than full-resolution color video. Privacy filters can blur faces or remove identifiable features from recorded data while preserving sufficient information for robot functions. On-device processing that discards raw sensor data after feature extraction provides strong privacy protections, though it limits forensic capabilities and debugging when issues arise.

### 6.6.2 On-Device vs. Cloud Processing

The choice between on-device and cloud processing significantly impacts privacy protection. On-device processing maintains all sensor data locally, preventing exposure during network transmission or cloud storage. Local processing aligns with privacy principles of data minimization and storage limitation, automatically deleting data after immediate use (Lutz & Tamò-Larrieux, 2020). Edge computing approaches balance privacy and capability, performing privacy-sensitive processing locally while leveraging cloud resources for less sensitive tasks.

Cloud processing offers computational advantages enabling more sophisticated AI capabilities, regular model updates, and fleet-wide learning from aggregated data. However, cloud architectures require transmitting sensor data off-device, increasing privacy risks. Data in transit faces interception risks despite encryption. Cloud storage presents attractive targets for attackers seeking large datasets. Third-party cloud providers may have access to unencrypted data, creating trust dependencies.

Hybrid architectures attempt to balance these trade-offs, processing sensitive data on-device while uploading anonymized or aggregated information for cloud analysis. Federated learning enables model training across multiple robots without centralizing raw data, where each robot trains on local data and only shares model updates. Differential privacy adds calibrated noise to shared data, providing mathematical guarantees that individual privacy is protected even when data is aggregated (Dwork & Roth, 2014). Homomorphic encryption enables computation on encrypted data, though performance limitations currently prevent widespread adoption in robotics applications.

### 6.6.3 GDPR, CCPA, and China's PIPL Compliance

General Data Protection Regulation (GDPR) establishes comprehensive privacy rights for European Union residents, significantly impacting robot deployments in EU markets (European Parliament, 2016). GDPR requires lawful basis for processing personal data, with consent being one justification. For robots collecting video or audio, obtaining meaningful consent from all captured individuals is often impractical. Alternative legal bases include legitimate interests, contract performance, or legal obligations, each requiring careful justification. GDPR grants individuals rights to access their data, request correction or deletion, and object to processing. Humanoid robot operators must implement systems enabling individuals to exercise these rights.

Data protection impact assessments (DPIA) are mandatory when processing involves systematic monitoring of publicly accessible areas or large-scale processing of biometric data. DPIAs analyze privacy risks and document mitigation measures. For humanoid robots, DPIAs must address continuous video recording, potential biometric identification, and data retention periods. GDPR's accountability principle requires organizations to demonstrate compliance through policies, procedures, and technical measures. Violations carry substantial penalties up to 20 million euros or 4% of global annual revenue.

California Consumer Privacy Act (CCPA) and its successor California Privacy Rights Act (CPRA) establish privacy rights for California residents, serving as de facto national standards for many USA companies (California Legislature, 2018). CCPA provides rights to know what personal information is collected, delete personal information, and opt-out of sale of personal information. Unlike GDPR's consent requirements, CCPA generally allows collection with opt-out rather than requiring opt-in consent. Sensitive personal information including biometric data receives enhanced protections with required opt-in consent or prominent opt-out opportunities.

Personal Information Protection Law (PIPL) represents China's comprehensive privacy framework, incorporating elements from GDPR while reflecting Chinese governance priorities (NPC, 2021). PIPL requires explicit consent for collecting sensitive personal information including biometric data, religious beliefs, and health information. The law mandates data localization where personal information collected in China must be stored domestically, with cross-border transfers requiring security assessments. PIPL emphasizes state security considerations, enabling government access to data under national security provisions. Critical information infrastructure operators face enhanced obligations including annual security assessments and mandatory reporting.

### 6.6.4 Biometric Data Protection

Biometric data including facial images, voice recordings, and gait patterns receive heightened privacy protections due to their permanent nature and inability to change if compromised. Humanoid robots equipped with facial recognition capabilities can identify individuals without consent, creating surveillance risks (Morley et al., 2020). Voice biometrics enable speaker identification from audio recordings. Gait recognition from video can identify individuals based on walking patterns even when faces are obscured.

Privacy-protective approaches limit biometric processing to necessary applications with clear consent. When facial recognition is required for personalization or security, enrollment processes should obtain explicit informed consent explaining data usage. Biometric templates should use one-way transformations preventing reconstruction of original biometric samples. Distributed storage separates biometric templates from identity information, requiring compromise of multiple systems for identification. Regular deletion of biometric data prevents indefinite retention of sensitive information.

Several jurisdictions prohibit or restrict biometric data usage. Illinois Biometric Information Privacy Act (BIPA) requires written consent before collecting biometric data and prohibits sale or profit from biometric information (Illinois General Assembly, 2008). San Francisco, Boston, and other cities ban government use of facial recognition technology. China's regulatory framework requires explicit consent for facial recognition with individuals able to choose non-biometric alternatives. These varying requirements create compliance challenges for humanoid robots deployed across jurisdictions.

### 6.6.5 Consent and Transparency Mechanisms

Meaningful consent requires individuals to understand what data is collected, how it's used, and potential consequences. For humanoid robots, obtaining informed consent is complicated by mobility through public and semi-public spaces where numerous individuals may be recorded (Calo, 2012). Practical approaches include prominent labeling identifying robots as recording devices, public signage in robot-operating areas warning of recording, and verbal announcements when robots enter spaces.

Transparency mechanisms should enable individuals to understand robot data practices. Privacy policies must be accessible and comprehensible, avoiding dense legalese. Robots can provide real-time privacy information through voice interaction, screens, or smartphone applications. Just-in-time notices when initiating sensitive data collection help individuals make informed decisions. For example, robots requesting to enter private spaces might announce "I will record video and audio in this room to perform my task. Do you consent?".

Privacy dashboards enable individuals to review what data has been collected about them, request deletion, or modify consent preferences. For workplace robots, employee dashboards show when they were recorded, what data was collected, and how it was used. Time-limited consent allows individuals to grant temporary permissions automatically expiring after specified periods. Granular controls enable consent for some data types or uses while declining others.

### 6.6.6 Case Study: UBTECH Walker Consumer Data Handling

UBTECH Walker operates in Chinese homes and businesses, requiring compliance with PIPL and GB standards for data protection. Walker implements on-device processing for voice commands and facial recognition, storing templates locally rather than in cloud databases. When family members enroll for personalization, the robot obtains explicit consent through tablet interface explaining data usage in plain language. Walker's privacy architecture separates functional data necessary for operation from optional data used for service improvements. Users can configure privacy settings to restrict active monitoring, with options to disable video recording or limit microphone activation to local wake-word detection.

Users can opt-out of telemetry data collection while maintaining full robot functionality. The robot provides monthly privacy reports summarizing data collected, retention periods, and any data shared with UBTECH servers. Compliance with PIPL data localization requirements ensures all personal information remains on servers within China. UBTECH implements cryptographic protections for data storage and transmission, with annual third-party privacy audits verifying compliance with commitments.

### 6.6.7 Case Study: Figure AI Workplace Monitoring

Figure AI's humanoid robots deployed in warehouses and manufacturing facilities raise workplace privacy considerations (Figure AI, 2023). Continuous video recording of work areas enables productivity monitoring, raising concerns about employee surveillance. Figure AI implements privacy controls including video processing that extracts only task-relevant information such as object locations and workspace status, discarding detailed imagery of individuals. When recording is necessary for safety investigations or quality control, employees receive clear notification of recording periods and retention schedules.

Workplace privacy policies developed with employee consultation establish acceptable uses of robot-collected data. Productivity metrics derived from robot observations aggregate across teams rather than individual workers, preventing intrusive monitoring. Data access controls limit who can view robot recordings, with access logs tracking all retrievals for audit purposes. Employee representatives participate in data governance committees overseeing privacy practices. Figure AI provides employees access to data collected about them through privacy portals, supporting transparency and enabling individuals to identify inappropriate data collection.

### 6.6.8 Case Study: EX Robots Social Interaction Data

EX Robots' humanoid platforms for retail and hospitality collect substantial interaction data including customer conversations, facial expressions, and behavioral patterns. This data enables personalized service but raises privacy concerns, particularly regarding emotional state inference and potential manipulation. EX Robots implements privacy notices at interaction points informing customers of data collection. Kiosk-style interfaces enable customers to opt-out of personalized service, receiving standardized interactions without data collection beyond transaction requirements.

Conversation data receives automatic deletion after service completion, retaining only anonymized interaction statistics for service improvement. Facial analysis for emotion detection processes entirely on-device without storing images. When customers explicitly consent to enhanced personalization, the system generates unique identifiers separate from personal identity information, enabling return-visit recognition without requiring names or contact information. EX Robots provides QR codes that customers can scan to access privacy policies and request data deletion through automated interfaces. Regular privacy audits by external firms verify compliance with data handling commitments and identify improvement opportunities.

### 6.6.9 Case Study: Humanoid Robots as Persistent Surveillance Infrastructure

Recent security research on the Unitree G1 humanoid robot demonstrated how modern humanoid platforms can unintentionally operate as persistent surveillance systems inside homes, laboratories, factories, and public environments. The study observed periodic outbound telemetry communications alongside continuous collection of multi-modal sensor data including video, audio, inertial measurements, and environmental mapping information. Unlike static IoT devices, humanoid robots combine mobility with embodied sensing, allowing them to traverse environments and collect contextualized behavioral and spatial information at close proximity to humans.

The research highlighted how even operational telemetry can create substantial privacy risks. Continuous SLAM reconstruction, motion tracking, and interaction logging may reveal facility layouts, human movement patterns, work routines, or sensitive operational activities, even when raw video streams are not explicitly stored. In workplace environments, these capabilities raise concerns regarding employee monitoring, behavioral profiling, and secondary uses of robot-collected data beyond the robot's intended operational purpose.

The findings also expose tensions between cloud-connected robot architectures and global privacy regulations such as GDPR, CCPA, and China's PIPL. Humanoid robots operating across jurisdictions may transmit telemetry to external cloud services without sufficiently transparent consent or clear user awareness. If compromised, these systems additionally become high-value intelligence collection platforms capable of capturing conversations, spatial intelligence, and sensitive environmental data. The case illustrates why privacy engineering for humanoid robots must extend beyond traditional IoT approaches, incorporating local-first processing, transparent telemetry controls, and hardware-enforced sensor protections as core architectural requirements.

### 6.6.10 Comparison: Privacy Regulations USA, EU, China

Privacy regulatory frameworks diverge substantially across jurisdictions, affecting humanoid robot design and deployment strategies. European Union GDPR establishes the most stringent privacy protections with broad individual rights, mandatory DPIAs for risky processing, and severe penalties for violations. EU regulations emphasize individual autonomy and data protection as fundamental rights. The EU AI Act introduces additional requirements for high-risk AI systems including robots, mandating transparency, human oversight, and technical robustness (European Parliament & Council, 2024).

United States privacy regulation remains fragmented with sector-specific federal laws and varying state requirements. CCPA/CPRA in California provides comprehensive privacy rights increasingly adopted by other states. Federal laws like HIPAA govern healthcare data, COPPA protects children's information, and FERPA addresses educational records. This patchwork creates compliance complexity but generally allows more permissive data practices than GDPR. Self-regulation and industry standards play larger roles in USA privacy governance compared to European regulatory approaches.

China's PIPL establishes comprehensive privacy protections while maintaining government access provisions for national security purposes. Data localization requirements and security assessments for cross-border transfers create operational challenges for international companies. Chinese regulations emphasize collective security and state interests alongside individual privacy rights. Implementation focuses on protecting against corporate misuse while ensuring government oversight capabilities (Sacks, 2021).

These differences require region-specific product configurations. Robots sold in EU markets need stronger consent mechanisms, more granular privacy controls, and localized data processing infrastructure compared to USA models. Companies operating globally face increased costs maintaining multiple compliance regimes and may choose most restrictive standards as common baselines.

## 6.7 Adversarial Robustness

The integration of machine learning throughout humanoid robot systems creates vulnerabilities to adversarial attacks designed to exploit model weaknesses. Adversarial robustness has emerged as a critical safety concern as robots increasingly rely on neural networks for perception, planning, and control (Akhtar & Mian, 2018).

### 6.7.1 Adversarial Examples for Vision Systems

Adversarial examples represent carefully crafted inputs that cause machine learning models to make incorrect predictions while appearing normal to humans. In computer vision, adversarial perturbations modify pixel values by small amounts, imperceptible to human observers but causing dramatic misclassifications (Goodfellow et al., 2015). For humanoid robots, adversarial attacks on vision systems could cause dangerous behaviors including misidentifying obstacles, failing to detect humans, or misreading navigation markers.

Gradient-based attacks like Fast Gradient Sign Method (FGSM) and Projected Gradient Descent (PGD) efficiently generate adversarial examples by computing gradients of model loss with respect to inputs. These white-box attacks assume full knowledge of target models including architectures and parameters. Black-box attacks generate adversarial examples without model access, using substitute models or query-based optimization. Research demonstrates adversarial examples transfer across different models and architectures, meaning attackers can develop examples using publicly available models and apply them to proprietary robot systems.

Adversarial examples challenge fundamental assumptions about learned representations. Rather than learning robust features aligned with human perception, neural networks often rely on non-robust features that generalize well on natural data but fail catastrophically on adversarial inputs. This brittleness persists across architectures, training procedures, and datasets, suggesting fundamental limitations of current machine learning approaches rather than superficial implementation flaws.

### 6.7.2 Physical Adversarial Patches and Attacks

Physical adversarial attacks extend digital perturbations to real-world objects that can fool robot vision systems. Adversarial patches are specially designed patterns that can be printed and placed in environments to cause targeted misclassifications (Brown et al., 2017). Unlike digital perturbations requiring pixel-level access, physical patches work through cameras under varying viewing angles, lighting, and distances, making them practical attack vectors against deployed robots.

Research demonstrates adversarial patches that make stop signs misclassified as speed limit signs, cause person detectors to ignore individuals, or create phantom obstacles in navigation systems (Eykholt et al., 2018). For humanoid robots, physical adversarial attacks present serious safety risks. Patches on floors could cause misclassification of safe walking surfaces as obstacles or vice versa. Adversarial patterns on clothing could render wearers invisible to collision avoidance systems. Malicious actors could place patches in environments causing robots to navigate into restricted areas or hazardous locations.

Adversarial clothing and accessories designed to fool person detection systems raise particularly concerning scenarios. Researchers have created adversarial T-shirts, eyeglass frames, and patches that significantly reduce detection accuracy of state-of-the-art person detectors (Xu et al., 2020). While initially demonstrated for privacy protection, such techniques could enable malicious actors to evade robot surveillance or cause robots to ignore their presence during navigation, leading to collisions.

Environmental adversarial attacks manipulate scenes rather than individual objects. Adversarial lighting uses carefully positioned light sources to create shadows and reflections that fool vision systems. Adversarial projections dynamically display patterns that corrupt camera images. These attacks require sophisticated setup but can affect multiple robots simultaneously and are difficult to detect without comparing to known good baselines.

### 6.7.3 Robustness Testing and Red Teaming

Systematic robustness testing evaluates robot resilience against adversarial attacks before deployment. Red teaming exercises pit offensive security experts against robot systems, attempting to identify vulnerabilities through adversarial thinking and novel attack approaches (Brundage et al., 2020). Red teams may develop custom adversarial examples targeting specific mission-critical functions like human detection or terrain classification.

Automated robustness testing generates large-scale adversarial example datasets to evaluate model robustness across threat models. Tools like CleverHans and Foolbox provide standardized adversarial attack implementations enabling reproducible robustness measurements (Papernot et al., 2018). Robustness metrics quantify vulnerability including attack success rates, minimum perturbation magnitudes required for misclassification, and robustness under different attack scenarios.

Physical robustness testing requires fabricating and testing adversarial objects in realistic conditions. This includes evaluating attacks under varying lighting, viewing angles, and distances matching expected deployment conditions. Physical testing is more expensive and time-consuming than digital evaluation but provides essential validation that defenses work in real-world conditions rather than only in simulation.

### 6.7.4 Certified Defenses and Provable Guarantees

Certified defenses provide mathematical guarantees of robustness against adversarial perturbations below specified magnitudes. Unlike empirical defenses that improve robustness against known attacks, certified defenses guarantee no attack within threat model can succeed (Cohen et al., 2019). Randomized smoothing represents one practical certified defense approach, adding calibrated noise to inputs and aggregating predictions over multiple noisy samples. The resulting smoothed classifier provably resists perturbations below calculable bounds.

Interval bound propagation and abstract interpretation provide certified bounds on neural network outputs given bounded inputs (Wong & Kolter, 2018). These techniques track how input perturbations propagate through network layers, computing guaranteed output ranges. If all outputs within the perturbation ball match the correct class, the input is certifiably robust. While computationally expensive, these approaches enable formal verification of safety–critical perception functions.

Limitations of certified defenses include computational costs, accuracy tradeoffs, and restricted threat models. Certified robustness typically requires larger perturbation budgets for equivalent accuracy compared to standard models. Current certification techniques primarily address small perturbations, while physical adversarial attacks may require larger modifications. Research continues advancing certified defense capabilities, but practical deployment in resource-constrained humanoid robots remains challenging.

### 6.7.5 Anomaly Detection Systems

Anomaly detection provides defense-in-depth by identifying unusual inputs or model behaviors potentially indicating adversarial attacks. Input anomaly detection analyzes characteristics of sensor data to identify out-of-distribution examples. Statistical measures, density estimation, or reconstruction errors from autoencoders can flag anomalous inputs requiring additional scrutiny (Hendrycks & Gimpel, 2017).

Model behavior anomaly detection monitors internal representations and predictions for unusual patterns. Adversarial examples often cause neural networks to be highly confident in incorrect predictions, with confidence distributions differing from natural examples. Layer-wise activation patterns may show unusual characteristics distinguishing adversarial from natural inputs. Ensemble disagreement where multiple models produce different predictions can signal potential adversarial attacks requiring fallback to safer operation modes.

Runtime monitoring systems track model uncertainty and prediction consistency over time. Sudden prediction changes for similar inputs may indicate adversarial attacks. Spatial consistency checking compares predictions across multiple camera views, with disagreements triggering protective responses. Temporal consistency analysis ensures predictions remain stable across video frames, detecting attacks targeting specific moments.

### 6.7.6 Case Study: Tesla Optimus FSD Robustness Lessons

Tesla's Full Self-Driving (FSD) system offers lessons about adversarial robustness in safety–critical applications transferable to Optimus humanoid development. FSD vision systems face real-world adversarial conditions including unusual lighting, weather, object appearances, and sensor artifacts. Tesla employs extensive data collection from fleet vehicles, capturing edge cases and failure modes. This real-world data enables training on diverse scenarios improving robustness compared to models trained solely on curated datasets.

Tesla implements ensemble approaches using multiple neural network architectures processing the same sensor inputs. Disagreement among ensemble members triggers increased caution or fallback to conservative behaviors. Sensor fusion combining cameras, radar, and ultrasonic sensors provides robustness against attacks targeting single modalities. Shadow mode testing runs experimental models alongside production systems, comparing predictions without affecting vehicle behavior. Divergences identify potential robustness issues requiring investigation.

For Optimus, Tesla applies similar principles including extensive simulation testing with adversarial scenarios, multicamera systems with consistency checking, and conservative fallback behaviors when perception confidence is low. The company maintains internal red teams that actively attempt to fool perception systems, discovering vulnerabilities before deployment. Continuous learning from deployed fleet enables rapid response to novel adversarial scenarios encountered in field operations.

### 6.7.7 Research Findings on Humanoid Robot Vulnerabilities

Academic research has documented numerous vulnerabilities in humanoid robot systems. Studies demonstrate adversarial attacks on gesture recognition systems causing misinterpretation of human commands (Li et al., 2020). Adversarial patches on floors caused navigation failures in wheeled robots, with similar vulnerabilities expected in humanoid bipedal systems. Voice-based adversarial attacks inject inaudible ultrasonic commands or craft audio resembling legitimate commands that cause misinterpretation (Carlini & Wagner, 2018).

Physical adversarial attacks on depth sensors using laser pointers or reflective surfaces demonstrate vulnerabilities in 3D perception. Researchers showed adversarial clothing patterns significantly reduce person detection accuracy for security robots (Xu et al., 2020). Studies on robotic manipulation reveal adversarial objects that appear as one category to vision systems but possess different physical properties, causing grasp failures.

These findings highlight that adversarial robustness challenges extend beyond academic curiosity to practical deployment concerns. Humanoid robots operating in uncontrolled environments will encounter diverse and potentially adversarial conditions requiring robust perception and decision-making.

### 6.7.8 Defense Mechanisms Across Platforms

Defensive strategies implemented across humanoid robot platforms vary based on risk tolerance, computational resources, and deployment environments. Most platforms employ sensor fusion combining multiple modalities (vision, LiDAR, IMU, force sensors) making attacks significantly more difficult as attackers must fool all sensing systems simultaneously. Complementary sensors like LiDAR are less susceptible to adversarial patches affecting cameras, providing cross-validation of environmental perception.

Adversarial training incorporates adversarial examples during model training, exposing networks to attacks and teaching robust decision boundaries. While improving empirical robustness, adversarial training increases computational costs and may reduce accuracy on natural examples. Input preprocessing including JPEG compression, bit depth reduction, or image denoising can remove adversarial perturbations but may also degrade legitimate features.

Detection-based defenses identify adversarial examples before processing, routing suspicious inputs to alternative handling. Multiple defense layers provide depth, where failure of one defense doesn't compromise safety. Human-in-the-loop systems maintain human operators who can intervene when anomalies are detected, particularly for high-stakes decisions or safety–critical operations.

## 6.8 Software Verification and Validation

Ensuring correct and safe behavior of complex robotic software requires rigorous verification and validation methodologies. As humanoid robots incorporate sophisticated AI systems, traditional testing approaches must be supplemented with formal methods and continuous validation procedures (Fisher et al., 2013).

### 6.8.1 Formal Verification Methods

Formal verification provides mathematical proofs that software systems satisfy specified properties. Unlike testing which checks specific scenarios, formal verification explores all possible system states within defined bounds, offering stronger correctness guarantees (Clarke et al., 2018). For humanoid robots, formal verification can prove safety properties such as "the robot never applies forces exceeding biomechanical thresholds" or "emergency stops activate within 50 ms of anomaly detection."

Temporal logic specifications using languages like Linear Temporal Logic (LTL) or Computation Tree Logic (CTL) express safety and liveness properties. Safety properties assert bad states are never reached, such as "the robot never enters an unsafe configuration." Liveness properties guarantee desired events eventually occur, such as "emergency stop requests are eventually processed." Formal verification tools check whether system models satisfy these specifications across all possible executions.

Challenges applying formal verification to humanoid robots include state space explosion where system complexity makes exhaustive analysis computationally infeasible. Abstraction techniques reduce state spaces by grouping similar states or ignoring irrelevant details. Compositional verification decomposes systems into manageable components, verifying each independently and reasoning about their composition. These approaches enable formal verification of critical subsystems even when complete system verification remains intractable.

### 6.8.2 Model Checking and Theorem Proving

Model checking algorithmically explores system state spaces to verify temporal properties (Clarke et al., 2018). Model checkers like SPIN, NuSMV, or UPPAAL automatically explore reachable states, identifying property violations and generating counterexamples showing execution traces leading to failures. For robot software, model checking can verify control protocols, communication patterns, and state machine behaviors. Bounded model checking limits exploration to bounded execution depths, providing tractable verification for properties within specified horizons.

Theorem proving uses interactive proof assistants like Coq, Isabelle, or Lean to construct machine-checked mathematical proofs of correctness (Nipkow et al., 2002). Unlike automated model checking, theorem proving requires human guidance to develop proofs but handles infinite state spaces and more complex properties. For robotics, theorem proving has verified kinematic computations, control algorithms, and safety protocols. The higher effort required limits theorem proving to most critical system components where strongest guarantees justify investment.

Hybrid approaches combine automated and interactive techniques. Automated tools handle routine verification tasks while human experts guide proof development for complex properties. Libraries of verified components enable reuse, reducing verification effort for new systems incorporating proven building blocks.

### 6.8.3 Runtime Monitoring and Safety Envelopes

Runtime monitoring observes system execution during operation, detecting property violations and triggering protective responses (Leucker & Schallhart, 2009). Unlike verification which analyzes all possible behaviors before deployment, runtime monitoring provides dynamic safety nets catching unanticipated scenarios or component failures. Monitors observe sensor data, control commands, and internal states, checking compliance with safety specifications.

Safety envelopes define boundaries of permitted operation. Monitors ensure the robot remains within envelopes, taking corrective action when boundaries are approached. Safety envelopes may include joint position limits, velocity constraints, force thresholds, or workspace boundaries. Hierarchical envelopes provide graduated responses: warning zones trigger increased monitoring, caution zones reduce operational speeds, and critical zones activate emergency stops.

Predictive monitoring uses system models to forecast future states, enabling proactive intervention before safety violations occur. For bipedal locomotion, predictive monitors assess stability margins and future foot placements, adjusting gaits or stopping before falls occur. Predictive monitoring requires accurate dynamics models and sufficient computational resources for real-time forecasting, balancing prediction accuracy against response latency.

### 6.8.4 Testing in Simulation and Real-World

Comprehensive testing strategies combine simulation and real-world validation. Simulation enables systematic coverage of scenarios including rare events, dangerous situations, and edge cases impractical to test physically (Koenig & Howard, 2004). Physics simulators like Gazebo, PyBullet, or Isaac Sim model robot dynamics, sensor behavior, and environmental interactions. Simulation-based testing explores parameter spaces, generates random scenarios, and performs regression testing efficiently.

Simulation limitations include fidelity gaps where simulated dynamics diverge from physical reality. Contact mechanics, friction, material deformation, and sensor noise present challenges for accurate simulation. Sim-to-real transfer techniques like domain randomization train models robust to parameter variations, improving real-world performance despite simulation inaccuracies (Tobin et al., 2017). Hardware-in-the-loop testing combines real robot hardware with simulated environments, validating actual actuators, sensors, and controllers against virtual scenarios.

Real-world testing provides ultimate validation but requires careful safety protocols. Staged testing progressively increases complexity and risk, beginning with constrained benchtop tests, advancing to open-space testing with safety harnesses, and culminating in full operational validation. Test environments should represent deployment conditions including lighting variations, floor surfaces, obstacles, and human interactions. Extensive datalogging during testing captures system states, sensor readings, and decisions, enabling post-test analysis and failure investigation.

### 6.8.5 Continuous Safety Validation

Deployed humanoid robots require ongoing safety validation as software updates introduce new functionalities and environments evolve. Continuous integration/continuous deployment (CI/CD) pipelines automate testing of software changes, preventing regression failures (Kim et al., 2022). Automated test suites execute thousands of scenarios within minutes, providing rapid feedback on code modifications. Canary deployments roll out changes to small subsets of robots, monitoring for anomalies before broader release.

Field data collection from deployed robots provides realistic scenarios for validation. Telemetry systems upload operational data, near-miss incidents, and error logs for analysis. Machine learning identifies patterns in failures, prioritizing testing of high-risk scenarios. Crowd-sourced testing through beta programs deploys experimental features to consenting users, gathering diverse real-world data before general release.

Safety cases document comprehensive arguments for system safety, including hazard analyses, risk assessments, verification results, and operational constraints (Kelly & Weaver, 2004). Safety cases evolve throughout system lifecycle, updated as new hazards are identified or mitigations are implemented. Regular safety audits review cases against accumulated evidence, ensuring claims remain valid.

### 6.8.6 Case Study: Boston Dynamics Validation Experience

Boston Dynamics' decades of experience developing dynamic robots have refined its validation methodologies. The company employs extensive simulation testing using custom physics engines optimized for contact-rich scenarios like running and jumping. Simulations explore millions of parameter combinations identifying failure modes. Real-world testing progresses through controlled laboratory environments with padded floors and safety tethers before transitioning to operational conditions.

Boston Dynamics maintains dedicated test facilities replicating deployment environments including warehouses, construction sites, and rough terrain. Robots undergo durability testing including repeated falls, obstacle courses, and long-duration operation under load. Failure analysis dissects every malfunction, identifying root causes and implementing systemic improvements. The company's iterative development process incorporates continuous validation, with each hardware and software revision subjected to comprehensive test suites.

For Atlas research, validation emphasizes experimental safety given rapid prototyping of novel behaviors. Researchers implement behavior-specific safety monitors checking preconditions and invariants. Test scenarios progressively increase difficulty, mastering simple behaviors before attempting complex maneuvers. Extensive sensor logging enables offline analysis of near-miss events, learning from close calls without requiring failures. The shift from hydraulic systems to custom electric actuators additionally necessitated an entirely new validation pipeline for electromagnetic compatibility and thermal management.

### 6.8.7 Case Study: Agility Robotics Industrial Deployment

Agility Robotics' Digit humanoid targets commercial deployment in logistics, necessitating rigorous validation for industrial safety certification. The company follows structured validation protocols aligned with ISO 13849 for safety of machinery. Critical safety functions like emergency stops and collision avoidance undergo formal FMEA (Failure Modes and Effects Analysis), identifying potential failures and implementing mitigations.

Digit's validation includes extensive simulation of warehouse environments with randomized layouts, inventory configurations, and human traffic patterns. Real-world pilot deployments at customer facilities provide operational validation under authentic conditions. Agility Robotics collects detailed performance metrics including task completion rates, error frequencies, and safety events. Customer feedback informs continuous improvement, refining behaviors based on real-world experiences.

Third-party safety certification required independent validation by accredited testing laboratories. This included physical testing of force limiting, emergency stop response times, and hazard zone implementations. Documentation provided comprehensive evidence of safety measures including risk assessments, design specifications, and test results. The certification process identified gaps requiring additional protections, strengthening overall system safety.

### 6.8.8 Case Study: Tesla Optimus Simulation-Based Testing

Tesla applies automotive software validation expertise to Optimus humanoid development, emphasizing simulation-based testing. The company's simulation infrastructure developed for Autopilot enables testing millions of scenarios computationally. Simulations incorporate diverse environments, lighting conditions, and interaction scenarios drawn from real-world data. Neural network training and validation occur primarily in simulation, with real-world testing validating sim-to-real transfer.

Tesla implements automated testing pipelines executing comprehensive test suites on every code commit. Tests range from unit tests of individual functions to full-system integration tests of complete behaviors. Performance regression testing ensures optimizations don't degrade safety–critical functionality. Static analysis tools check code for common vulnerabilities, memory safety issues, and coding standard violations.

Shadow mode testing runs experimental software alongside production systems on deployed robots, comparing outputs without affecting behavior. Divergences between experimental and production systems indicate potential issues requiring investigation. This approach enables field validation of new algorithms with minimal risk, accelerating development while maintaining safety.

### 6.8.9 Validation Rigor Comparison Across Companies

Validation approaches vary across humanoid robot companies based on maturity, target markets, and regulatory requirements. Boston Dynamics emphasizes physical testing and iterative refinement based on decades of research experience. Their validation focuses on proving capabilities work reliably under challenging conditions through extensive real-world testing. Agility Robotics adopts industrial safety frameworks with formal documentation and third-party certification, addressing commercial deployment requirements.

Tesla leverages automotive-scale simulation infrastructure enabling massive scenario testing but with less emphasis on formal verification compared to traditional robotics companies. Chinese companies including UBTECH and Unitree balance cost constraints with safety, implementing essential validation while avoiding expensive certification processes where not mandated. Start-ups like Figure AI and Apptronik are developing validation practices while navigating the challenging path from prototypes to commercial deployment.

As the humanoid robotics industry matures, validation standards will likely converge toward recognized best practices similar to automotive or aerospace industries. Industry consortia and standards bodies may establish validation frameworks providing clearer guidance for companies navigating safety certification.

Figure 6.4 summarizes the end-to-end software verification and validation pipeline for humanoid robots, showing the five complementary stages—formal methods, large-scale simulation, hardware-in-the-loop testing, staged field deployment, and continuous runtime monitoring—and emphasizing how field incidents feed back into earlier stages to drive a SOTIF-style (Safety Of The Intended Functionality, ISO 21448; introduced in Sect. 6.10) residual-risk reduction loop across the fleet lifecycle.

## 6.9 Reliability Engineering

Achieving sufficient reliability for commercial humanoid robot deployment requires systematic engineering addressing mechanical durability, electronic robustness, and software stability. Reliability engineering balances performance, cost, and longevity through careful component selection, predictive maintenance, and field data analysis.

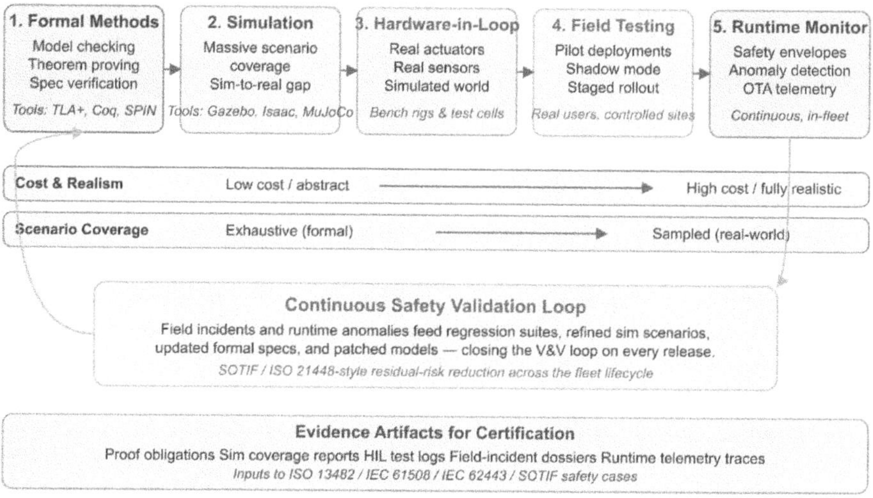


**Fig. 6.4** Software verification and validation pipeline for humanoid robots

### 6.9.1 Mean Time Between Failures (MTBF)

Mean Time Between Failures quantifies expected operational time before system failures, providing standardized reliability metrics for comparison and requirement specification. For humanoid robots, system-level MTBF depends on reliability of numerous components including actuators, sensors, electronics, and mechanical structures. Series systems where failure of any component causes system failure calculate overall MTBF as the reciprocal of summed failure rates: MTBF_system = 1 / (Σ λ_i) where λ_i represents failure rates of individual components.

Achieving acceptable system MTBF requires high-reliability components or redundancy strategies. Industrial robots typically target MTBF values exceeding 50,000 h (approximately 5.7 years continuous operation) for mission-critical applications. Consumer robots may accept lower MTBF around 10,000–20,000 h given different usage patterns and cost constraints. Humanoid robots with dozens of actuators, motors, and joints face significant challenges achieving high MTBF as each component contributes to overall failure rates.

Accelerated life testing subjects components to elevated stress conditions (higher temperatures, voltages, or mechanical loads) to induce failures in compressed timeframes. Acceleration factors derived from physics-based failure models translate accelerated test results to expected field lifetimes. For example, electronic components following Arrhenius models exhibit failure rates doubling for every 10 °C temperature increase. Testing at elevated temperatures provides failure data in months that would require years under normal conditions (Nelson, 2004).

### 6.9.2 Predictive Maintenance and Diagnostics

Predictive maintenance monitors system health, detecting degradation before failures occur and scheduling maintenance minimizing downtime. Condition-based monitoring tracks component performance metrics including motor currents, temperatures, vibrations, and timing. Deviations from baseline signatures indicate developing faults enabling preemptive replacement or repair (Jardine et al., 2006).

Machine learning approaches identify subtle degradation patterns difficult to capture with rule-based methods. Supervised learning trains classifiers on labeled failure examples, predicting impending failures from sensor data. Unsupervised anomaly detection identifies unusual patterns without requiring failure labels, useful for novel failure modes. Deep learning applied to time-series sensor data captures temporal dependencies indicating degradation trajectories.

Diagnostic systems isolate fault locations when failures occur, accelerating repair and minimizing downtime. Fault trees decompose system failures into component-level causes, guiding technicians through logical troubleshooting sequences. Built-in test systems automate diagnostics, exercising components and analyzing responses to identify faults. Connectivity enables remote diagnostics where manufacturer support teams analyze robot telemetry, diagnosing issues without physical access and guiding local repair efforts.

### 6.9.3 Component Lifetime and Wear Modeling

Mechanical components undergo wear and fatigue, degrading performance and eventually failing after accumulated usage. Wear modeling predicts component lifetimes based on operational stresses, enabling preventive replacement before failures. Linear wear models assume damage accumulates proportionally with usage cycles. For example, bearing life models predict failures based on load magnitudes, rotational speeds, and material properties.

Fatigue analysis addresses cyclic loading causing progressive crack growth and eventual fracture. S–N curves characterize material fatigue lives, plotting stress amplitude against cycles to failure. Structural components experiencing repeated stresses below ultimate strength fail after sufficient cycles according to fatigue principles. Humanoid robot joints experience millions of cycles during walking and manipulation, requiring careful fatigue analysis ensuring structural integrity throughout design lifetimes.

Miner's rule provides cumulative damage models for variable amplitude loading. Each stress cycle consumes a fraction of component life inversely proportional to cycles-to-failure at that stress level. When cumulative damage reaches unity, failure is predicted. This approach enables lifetime prediction for complex loading histories typical of humanoid robots with varying operational patterns (Stephens et al., 2000).

Battery degradation represents a critical lifetime constraint for mobile robots. Lithium-ion batteries degrade through calendrical aging (time-based) and cyclic aging (charge–discharge cycles). Capacity fades as internal resistance increases, reducing runtime and power delivery. Battery management systems track state of health (SOH) using voltage, current, and temperature measurements. Models predict remaining useful life enabling proactive battery replacement before performance becomes unacceptable.

### 6.9.4 Field Reliability Data and Learning

Deployed robot fleets generate valuable reliability data revealing failure modes and frequencies under actual operating conditions. Telemetry systems upload operational logs, error reports, and usage statistics enabling statistical analysis of field reliability. Fleet-wide data identifies systematic issues affecting multiple units, distinguishing design weaknesses from random failures (Jardine et al., 2006).

Weibull analysis characterizes failure distributions from field data, estimating key parameters including characteristic life and shape factors. Shape factors indicate failure mechanisms: values less than one suggest infant mortality from manufacturing defects, values near one indicate random failures, and values greater than one reveal wear-out failures. This analysis guides improvement strategies including enhanced quality control for infant mortality, redundancy for random failures, or preventive replacement for wear-out mechanisms.

Failure reporting and corrective action systems (FRACAS) systematically capture failure information, analyze root causes, and track corrective actions. Structured processes ensure failures are documented with sufficient detail enabling engineering analysis. Root cause analysis techniques including 5 Whys, fishbone diagrams, and fault tree analysis identify underlying causes rather than proximate symptoms. Tracking corrective actions through closure ensures identified improvements are actually implemented.

### 6.9.5 Maintenance Protocols

Comprehensive maintenance protocols balance preventive maintenance preventing failures against corrective maintenance repairing failed components. Preventive maintenance includes scheduled inspections, lubrication, adjustments, and component replacements at predetermined intervals. Maintenance schedules optimize trade-offs between failure costs and maintenance expenses, with critical components receiving more frequent attention.

Maintenance procedures must be clearly documented with step-by-step instructions, required tools, and safety precautions. Service manuals include troubleshooting guides, parts diagrams, and maintenance intervals. For commercial robots, service accessibility affects maintenance costs and downtime. Modular designs with easily replaceable components minimize repair time. Tool-less access panels and plug-and-play electronics simplify field maintenance.

Remote diagnostics and support reduce downtime by enabling manufacturer experts to assist with troubleshooting without site visits. Augmented reality tools overlay maintenance instructions on technician views through smart glasses or tablets, guiding repairs step-by-step. For robot fleets, centralized maintenance management systems schedule preventive maintenance, track parts inventory, and optimize technician routing.

### 6.9.6 Case Study: Agility Robotics Digit Long-Term Deployment

Agility Robotics' Digit has accumulated thousands of operational hours in commercial logistics deployments, providing reliability data for humanoid robots in industrial settings. Field experience revealed several reliability challenges addressed through iterative improvements. Early deployments experienced motor controller failures due to dust ingress in warehouse environments. Improved sealing and conformal coating enhanced robustness. Battery connectors suffered wear from repeated charging cycles, prompting redesign with more durable contacts rated for higher cycle counts.

Agility implemented predictive maintenance monitoring motor temperatures, battery health, and joint encoder consistency. Anomaly detection algorithms flag unusual patterns triggering maintenance calls before failures occur. The company's Fleet Management System aggregates data across deployed robots, identifying systematic issues and prioritizing software updates or hardware retrofits. Maintenance protocols evolved from reactive repair toward scheduled preventive maintenance optimized using field reliability data.

Digit's modular design enables rapid component replacement with minimal downtime. Critical components like batteries, computers, and motor drives swap in minutes using quick-disconnect connectors. Spare parts availability at customer sites enables immediate replacement followed by factory refurbishment of failed components. This approach minimizes robot downtime while spreading repair costs efficiently.

### 6.9.7 Case Study: Unitree Cost-Reliability Trade-Offs

Unitree's humanoid robots target lower price points through cost-optimized designs that accept reliability trade-offs appropriate for research and development applications. The G1 humanoid targets competitive price points not by using off-the-shelf parts, but through extensive vertical integration and in-house component manufacturing. Unitree developed proprietary high-torque joint modules—combining permanent magnet synchronous motors (PMSM) with custom planetary gear reducers—capable of delivering up to 120 N·m of torque at the knee. By moving away from expensive third-party industrial gearboxes and designing custom

Quasi-Direct Drive (QDD) actuators, Unitree accepts certain wear-out trade-offs for vastly reduced bill-of-materials costs while maintaining high dynamic performance.

Unitree's approach emphasizes repairability over maximum MTBF, recognizing that lower-cost robots will experience higher failure rates. Detailed maintenance documentation, readily available spare parts, and active user communities enable users to perform repairs. Open-source software and published hardware designs allow advanced users to modify and improve systems, creating ecosystem-driven reliability improvements.

Field reliability data from Unitree's robot dog platforms informed humanoid development. High-wear components like gear trains and bearings received upgrades using lessons learned from thousands of robot dogs deployed globally. Battery management systems incorporate protections preventing common failure modes including over-discharge and thermal runaway. While Unitree robots may not achieve MTBF targets of industrial systems, their cost-effectiveness enables research and development applications tolerating occasional failures.

### 6.9.8 Case Study: Boston Dynamics Commercial Reliability

Boston Dynamics transitioned from research robots to commercial products requiring substantially improved reliability. The Spot quadruped's commercial version underwent reliability engineering that establishes a baseline for future humanoid systems. The company invested heavily in environmental testing including temperature chambers (-20 C to 45 C), dust chambers (IP54 rating), water resistance testing, and drop testing simulating falls.

Component selection prioritized proven reliability over cutting-edge performance. The transition from the legacy hydraulic Atlas to the all-electric Atlas exemplifies this shift; hydraulic systems, while powerful, presented complex maintenance challenges, whereas electric actuators provide superior reliability and reduced operational downtime. Sealed electronics enclosures protect against environmental contamination. Redundant sensors provide fault tolerance for critical functions like terrain perception and body state estimation.

Boston Dynamics established extensive service infrastructure including field service teams, parts distribution networks, and remote monitoring systems. Customer support analyzes robot telemetry proactively identifying issues and dispatching service before failures impact operations. Regular software updates delivered over-the-air address identified issues and improve functionality. The company's reliability focus reflects understanding that commercial success requires not just impressive capabilities but dependable operation day after day.

### 6.9.9 Case Study: Tesla Optimus Automotive Standards

Tesla Optimus's automotive-grade safety and reliability architecture—including ISO 26262 functional safety practice, ASIL-driven hazard analysis, and vehicle-derived battery management—is discussed in detail in Sect. 6.2.

## 6.10 Ethical Hacking and Responsible Disclosure

Security vulnerabilities in humanoid robots present safety risks, but responsible disclosure in robotics remains far less mature than in conventional IT security. Coordinated vulnerability disclosure practices, bug bounty programs, and severity scoring frameworks exist in the broader software ecosystem, yet robotics companies rarely implement them consistently for robots whose vulnerabilities can create physical harm. As a result, security researchers often face unclear reporting channels, weak incentives, and uncertain legal or reputational risks, while vendors may lack internal processes for triaging robot-specific vulnerabilities. Ethical hacking for humanoid robots therefore requires both technical safeguards against misuse and institutional mechanisms for cooperation, including clear disclosure policies, safe-harbor commitments, robotics-aware severity scoring, and timelines that account for hardware, fleet-update, and safety-certification constraints.

### 6.10.1 Bug Bounty Programs for Robotics

Bug bounty programs are common in mainstream IT security but remain rare for humanoid robots and other embodied robotic systems. A more realistic near-term model is coordinated vulnerability disclosure: vendors publish a reporting channel, define what testing is in scope, prohibit actions that could endanger people or damage hardware, provide safe-harbor language for good-faith researchers, and commit to triage and remediation timelines. Some robotics vendors provide vulnerability-reporting channels, but many do not offer monetary rewards, and humanoid-specific disclosure programs remain immature relative to the physical risk created by robot compromise.

Implementing bug bounties for humanoid robots requires careful scope definition addressing physical safety concerns. Researchers must test safely without risking injuries or property damage. Virtual test environments, simulation platforms, or specific test units can provide safe research targets. Clear rules of engagement prohibit attacks on production robots serving customers or operating in public spaces.

Vulnerability severity ratings guide reward structures and remediation priorities. The Common Vulnerability Scoring System (CVSS) provides standardized severity assessment based on exploitability, impact, and scope (FIRST, 2023). Critical vulnerabilities enabling remote code execution or safety compromise receive highest bounties and immediate attention. Lower severity issues like information disclosure or minor authentication weaknesses warrant lower rewards and less urgent remediation timelines.

While CVSS offers a well-established baseline for traditional software vulnerabilities, humanoid robots operating as agentic AI systems introduce categories of risk that CVSS was not designed to score. Initial adaptations to robotics include the Robot Vulnerability Scoring System (RVSS; Mayoral-Vilches, 2018) that considers major relevant issues in robotics including (a) robot safety aspects, (b) assessment of downstream implications of a given vulnerability, (c) library and third-party scoring assessments and d) environmental variables, such as time since vulnerability disclosure or exposure on the web as well as an experimental evaluation of RVSS with contrast to CVSS is provided and discussed with focus on the robotics security landscape. The OWASP Artificial Intelligence Vulnerability Scoring System (AIVSS) project addresses this gap by providing a standardized, quantifiable methodology specifically designed to assess vulnerabilities in AI systems, with particular focus on agentic AI architectures (Huang & Hughes, 2025b; OWASP Foundation, 2025). For humanoid robots, AIVSS extends severity assessment beyond conventional exploitability metrics to encompass AI-specific risk vectors including model manipulation, data poisoning, agent misalignment, and autonomous decision-making failures. A vulnerability that CVSS might rate as moderate—such as subtle manipulation of a robot's perception model—could warrant a critical AIVSS score when the downstream consequence is unsafe physical behavior in a human-occupied environment. AIVSS scoring criteria also account for the compounding nature of agentic risk, where a vulnerability at one layer of the system stack can cascade through planning and actuation layers, amplifying impact in ways that traditional scoring systems do not capture (Huang & Hughes, 2025b). Security teams managing humanoid robot deployments should apply both CVSS for infrastructure and software components and AIVSS for AI-specific components, using the higher of the two scores to drive remediation prioritization and bug bounty reward tiers.

### 6.10.2 Coordinated Vulnerability Disclosure

Coordinated disclosure manages vulnerability information sharing, giving vendors time to develop and deploy fixes before public disclosure. Typical timelines allow 90–120 days for vendor remediation before researchers publish details (Householder et al., 2020). This window balances vendor needs for development and testing against public interest in timely security information.

Disclosure processes begin with researchers reporting vulnerabilities through designated channels, providing technical details, proof-of-concept exploits, and suggested remediation approaches. Vendors acknowledge receipt, validate the vulnerability, and provide remediation timelines. Throughout the process, vendors and researchers maintain communication regarding progress and extensions if needed.

Upon fix deployment or timeline expiration, researchers publish vulnerability details enabling the security community to learn and improve defenses.

Challenges arise when vendors are unresponsive or dispute vulnerability validity. Third-party coordination services like CERT/CC facilitate communication and provide neutral evaluation of reported issues. Disclosure policies may include public disclosure after reasonable timeframes regardless of vendor remediation status, pressuring unresponsive vendors while ensuring important security information reaches the public.

### 6.10.3 Open-Source Security Auditing

Open-source robotics software including ROS, Gazebo, and various libraries benefit from community security auditing. Public source code enables independent security review identifying vulnerabilities before exploitation. Dedicated security auditing efforts including code reviews, static analysis, and penetration testing improve open-source security (Wheeler, 2015).

Security-focused organizations and academic researchers conduct audits of widely-used robotics software. Findings are reported through established disclosure processes, allowing maintainers to fix issues before public disclosure. Funded security initiatives like the Open Source Security Foundation (OpenSSF) support security improvements in critical open-source projects. For robotics, similar efforts could systematically audit foundational libraries and frameworks.

Community engagement through security-focused conferences, workshops, and publications disseminates security knowledge. Capture-the-flag competitions and security challenges specific to robotics train researchers in relevant skills while identifying novel attack vectors. Collaboration between robotics developers and security experts bridges knowledge gaps, improving security throughout the robotics ecosystem.

### 6.10.4 Industry Collaboration on Security

Industry-wide collaboration amplifies security improvements beyond individual companies. Information sharing about threats, vulnerabilities, and defense strategies helps all participants improve security. Information Sharing and Analysis Centers (ISACs) provide structured frameworks for confidential threat intelligence sharing while protecting competitive information (ISAC Council, 2023).

Standards development organizations incorporate security requirements into robotics standards. ISO/IEC 27001 information security management and IEC 62443 industrial control systems security provide frameworks applicable to robotics. Industry consortia like the Industrial Internet Consortium and ROS-Industrial develop security best practices and reference architectures.

Collaborative security initiatives can establish shared infrastructure including threat intelligence databases, vulnerability repositories, and incident response capabilities. For humanoid robotics, industry collaboration might develop common security testing frameworks, share anonymized vulnerability data, and establish incident response protocols. Such collaboration benefits the entire industry by raising baseline security levels and enabling faster responses to emerging threats.

### 6.10.5 Case Study: Companies with Security Programs

Several robotics companies have established public security programs recognizing the importance of external security research. Boston Dynamics provides a security vulnerability disclosure policy with designated contacts for reporting issues and committed response timelines. While not offering bug bounties currently, their disclosure policy establishes clear processes for responsible vulnerability reporting.

Open Robotics, maintaining ROS and Gazebo, implements coordinated disclosure for reported vulnerabilities. Security advisories published on their website document identified issues, affected versions, and remediation guidance. The organization works with researchers to validate reports and develop patches before public disclosure. Community contributors occasionally receive acknowledgment in security advisories, recognizing their contributions to ROS security.

Some companies maintain informal security research relationships with academic institutions and independent security researchers. These relationships enable ongoing security assessment and rapid communication of identified issues. As the humanoid robotics industry matures, more formalized security programs including bug bounties will likely become standard practice, following precedents from software and automotive industries.

### 6.10.6 Notable Vulnerabilities Discovered and Patched

Security research has identified various vulnerabilities in robotic systems demonstrating the importance of ongoing security assessment. Researchers discovered vulnerabilities in ROS1 including lack of authentication, unencrypted communications, and ability for attackers to inject malicious commands (White et al., 2021). These findings motivated security improvements in ROS 2, most notably SROS2—the security layer of ROS 2 built on the OMG DDS-Security specification—which provides authentication, access control, and encrypted communications for robot middleware (Mayoral-Vilches et al., 2022).

Commercial robot vulnerabilities have included hardcoded credentials in management interfaces, unpatched operating systems with known exploits, and insecure firmware update mechanisms lacking cryptographic verification. Some robots shipped with default passwords unchanged, enabling unauthorized access. Insecure communication protocols exposed control commands to eavesdropping or injection attacks.

When properly handled through responsible disclosure, vulnerability discoveries strengthen security across the robotics industry. Researchers publishing findings after vendor remediation enable other companies to assess whether similar issues affect their systems. Security advisories and CVE (Common Vulnerabilities and Exposures) assignments create permanent public records enabling long-term security improvements.

### 6.10.7 China vs. USA Approaches to Disclosure

Vulnerability disclosure practices diverge between China and the USA due to different legal frameworks and security cultures. In the United States, security research generally enjoys legal protections under contexts like security research exception in DMCA and academic freedom principles. Bug bounties and disclosure programs are widely accepted, with major companies actively encouraging external security research (Schneier, 2021).

China's cybersecurity regulations require reporting vulnerabilities to government authorities, potentially before or instead of notifying affected vendors. The Cybersecurity Law and related regulations establish government oversight of vulnerability information. Chinese security researchers typically report through government-established platforms like the China National Vulnerability Database (CNVD) or National Computer Network Emergency Response Technical Team/ Coordination Center of China (CNCERT/CC).

These differences affect international security collaboration. Western researchers may hesitate sharing vulnerabilities with Chinese companies due to government reporting requirements and concerns about information control. Chinese researchers operating internationally must navigate compliance with Chinese reporting requirements while engaging with vendors following Western disclosure practices. Industry efforts to harmonize disclosure practices remain challenged by these regulatory differences.

Figure 6.5 illustrates the coordinated vulnerability disclosure (CVD) workflow as it applies to humanoid robotics, mapping the six-stage timeline from initial discovery through public disclosure against the parallel regulatory channels in the United States, European Union, and China, and emphasizing how bug-bounty incentives and explicit safe-harbor language convert adversarial security research into a continuous defensive feedback loop.

## 6.11 Recent Developments and Emerging Trends (2026)

The first quarter of 2026 has witnessed some new advancements in humanoid robot safety, security, and regulatory frameworks, reflecting the sector's rapid maturation from research prototypes to commercial deployment. These developments span

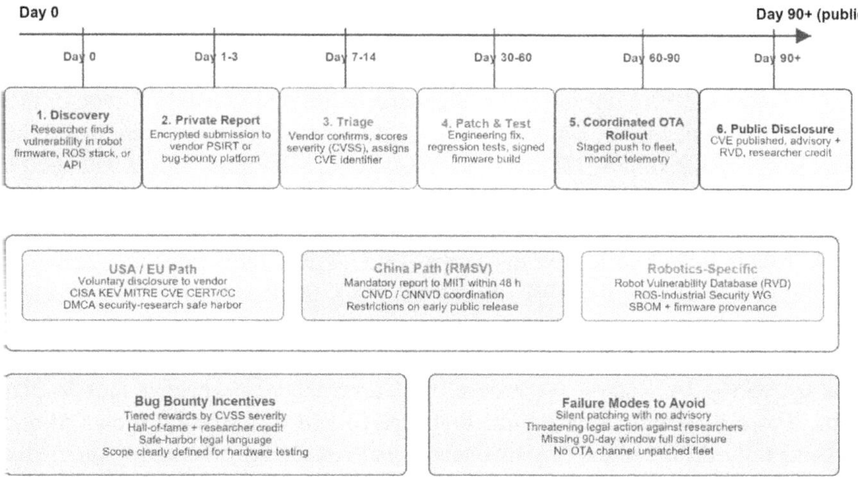


**Fig. 6.5** Coordinated vulnerability disclosure timeline for humanoid robots

new international standards, landmark certifications, critical security disclosures, and the emergence of legal precedents that will shape the industry's trajectory.

### 6.11.1 China's National Standard System for Humanoid Robotics

In March 2026, China released its first national standard system for humanoid robots and embodied artificial intelligence, marking a significant regulatory milestone for the world's largest robotics market (Xinhua, 2026). The standard system is structured on six pillars: foundational and common standards, neuromorphic and intelligent computing, limbs and components, full-system integration, application, and safety and ethics standards. This comprehensive framework addresses the most pressing concerns identified in the industry, with safety standards and cybersecurity receiving top priority. The new system provides crucial guidance for China's rapidly expanding humanoid robotics sector, which includes major players such as UBTECH, Unitree, and Fourier Intelligence. The framework aligns with but extends beyond existing ISO standards, incorporating China-specific requirements for data localization and network security compliance under the Personal Information Protection Law (PIPL).

### 6.11.2 New ISO Standards for Dynamically Stable Mobile Robots

The International Organization for Standardization has advanced two critical standards addressing the unique safety requirements of humanoid and other dynamically stable robots. ISO/AWI 26058–1 (under development) is intended to establish safety requirements for industrial mobile robots, while ISO/WD 25785–1 (also under development) specifically addresses dynamically stable industrial mobile robots including legged platforms (ISO, 2026). Both standards remain at early ISO drafting stages—AWI (Approved Work Item) precedes WD (Working Draft), and substantive language changes are expected before publication. These standards fill a significant gap in the regulatory landscape, as existing standards such as ANSI/RIA R15.08 were developed primarily for wheeled mobile platforms. The new standards recognize that bipedal and quadrupedal robots present fundamentally different stability challenges and require specific safety provisions for actively controlled balance systems. Industry analysts note that these standards will become essential compliance requirements for humanoid robots deployed in industrial settings, establishing baseline expectations for stability monitoring, fall recovery, and safe human interaction.

### 6.11.3 Landmark Safety Certification: Agility Robotics Digit

Agility Robotics achieved a significant milestone in early 2026 when its Digit humanoid robot passed an OSHA-recognized Nationally Recognized Testing Laboratory (NRTL) field inspection, marking the first such certification for a bipedal humanoid robot in the United States (Agility Robotics, 2026a, 2026b). This certification validates that Digit meets safety requirements for deployment in OSHA-regulated industrial environments without requiring physical barriers or safety cages. The inspection process evaluated emergency stop systems, collision avoidance capabilities, force limiting mechanisms, and safe human–robot interaction protocols. This achievement establishes a precedent for other humanoid manufacturers seeking commercial deployment in logistics and manufacturing facilities, demonstrating that bipedal platforms can meet rigorous safety standards previously achieved only by traditional industrial robots and cobots.

### 6.11.4 Boston Dynamics Atlas Enters Production

At CES 2026, Boston Dynamics unveiled the production version of its all-electric Atlas robot, signaling the transition from research platform to commercial product (Boston Dynamics, 2026). The production Atlas was awarded "Best Robot" in the Best of CES 2026 awards by CNET Group, recognizing its readiness for industrial deployment. Boston Dynamics announced that all 2026 deployments are committed, with initial fleets destined for the manufacturing facilities of Hyundai

Motor Group—Boston Dynamics' parent company since Hyundai's 2021 acquisition—and partnerships with Google DeepMind. The company's stated goal of deploying "tens of thousands" of robots represents a significant scale-up that will stress-test the safety frameworks discussed throughout this chapter under real-world industrial conditions. The transition to production also demonstrates the practical application of the multi-layered safety architecture described in Sect. 6.2, with the production vehicles incorporating enhanced safety documentation and certification processes required for commercial deployment.

### 6.11.5 Cybersecurity Disclosures and Senate Testimony

The cybersecurity vulnerabilities of humanoid robots received high-level attention in March 2026 when research on the Unitree G1 robot was cited before the United States Senate (Alias Robotics, 2026a, 2026b). Damion Shelton, co-founder of Agility Robotics, referenced the critical Bluetooth Low Energy (BLE) vulnerability—designated as CVE-2025-35027 and dubbed "UniPwn"—that allows root-level takeover of Unitree humanoid platforms. The vulnerability enables complete compromise of affected robots, potentially transforming them into surveillance nodes or attack vectors. This Senate testimony marks one of the first times humanoid robot security has been discussed at the highest levels of government, signaling increased regulatory attention to the cyber-physical security risks these platforms present. The incident underscores the critical importance of security-by-design principles and responsible disclosure frameworks discussed in Sect. 6.9, while highlighting the competitive dynamics where established manufacturers may reference competitor vulnerabilities when advocating for industry-wide security standards.

### 6.11.6 Safety Methodology Evolution: SOTIF for Humanoid Robots

Emerging from discussions at CES 2026, safety experts have proposed adapting the Safety of the Intended Function (SOTIF) framework—originally developed for autonomous vehicles (ISO 21448)—to humanoid robot applications (SRES, 2026). SOTIF addresses safety risks beyond traditional fault-based models, particularly relevant for perception-driven and machine learning-based systems. The proposal recognizes that humanoid robots operating in unstructured environments face scenario-based safety challenges similar to autonomous vehicles, where the intended function of the system itself may introduce risks even when all components operate correctly. Implementing SOTIF principles would require humanoid robots to demonstrate systematic identification of potentially hazardous operational scenarios, validation of sensor and perception system performance boundaries, and establishment of safe fallback behaviors when operating near performance limits. This methodology complements the adversarial robustness approaches discussed in Sect. 6.6 by addressing systemic safety considerations beyond individual component failures.

### 6.11.7 Legal Precedents: Whistleblower Litigation and Liability Frameworks

The ongoing litigation between former Figure AI safety engineer Robert Gruendel and the company has emerged as a landmark case for humanoid robot safety accountability (Kolodny, 2025). Filed in November 2025 and proceeding through early 2026, the whistleblower lawsuit alleges that Figure AI terminated the engineer after he raised concerns that the Figure 02 robot could exert forces sufficient to "fracture a human skull" and that executives dismissed safety protocols in pursuit of rapid commercialization. Legal analysts describe this as potentially "one of the first whistleblower cases related to Humanoid Robot safety," establishing precedents for safety engineer protections and corporate accountability in the emerging humanoid robotics industry. Concurrently, legal scholars have published comprehensive analyses of humanoid robot liability frameworks, examining questions of negligence, product liability, and vicarious responsibility as deployment scales from controlled pilot programs to widespread commercial use (Paton-Philip & McKinney, 2026). These developments signal that the legal landscape is beginning to catch up with the technical capabilities discussed throughout this chapter, with implications for corporate governance, engineering documentation practices, and the relationship between safety standards and legal liability.

### 6.11.8 Emerging Considerations

These early 2026 developments collectively signal the humanoid robotics industry's transition from a technology demonstration phase to a regulatory and commercial reality. The convergence of new safety standards, landmark certifications, high-profile security disclosures, and emerging legal frameworks suggests that the safety, security, and reliability considerations outlined in this chapter will increasingly be tested under real-world deployment conditions. Companies entering commercial deployment will need to navigate evolving regulatory requirements across jurisdictions while maintaining robust security postures against an expanding threat landscape. The cases of Unitree's vulnerability disclosure and Figure AI's whistleblower litigation serve as cautionary indicators that safety and security cannot be treated as afterthoughts in the race to commercialization—fundamental engineering rigor and transparent safety cultures will be essential for long-term industry credibility and public trust.

## 6.12 Conclusion

This chapter has examined the multifaceted challenges of ensuring safety, security, and reliability in humanoid robot systems. Physical safety standards including ISO 13482, ISO/TS 15066, and regional regulatory frameworks establish requirements for robots operating in human environments. Case studies demonstrate how leading companies from Boston Dynamics to UBTECH approach safety certification, with significant differences between USA and Chinese regulatory pathways.

Fail-safe mechanisms provide critical protection through emergency stops, sensor redundancy, mechanical brakes, and graceful degradation strategies. Multiple layers of safety systems ensure that single-point failures do not compromise human safety. Collision avoidance systems combining proximity detection, force limiting, and predictive algorithms enable humanoid robots to operate safely alongside humans in shared workspaces.

Cybersecurity emerges as a critical concern with humanoid robots presenting extensive attack surfaces spanning network communications, sensor systems, and machine learning models. Threat modeling identifies adversaries and attack vectors including sensor spoofing, adversarial attacks on perception, model poisoning, and supply chain compromises. Defense-in-depth strategies combine encryption, authentication, intrusion detection, and continuous monitoring to protect against sophisticated threats.

Data privacy and surveillance concerns require careful attention to camera and audio data collection, with regulatory frameworks like GDPR, CCPA, and China's PIPL establishing different compliance requirements across jurisdictions. Privacy-by-design principles, on-device processing, and transparency mechanisms help protect individual privacy while enabling robot functionality. Biometric data receives heightened protection given its sensitive nature and permanence.

Adversarial robustness addresses vulnerabilities in machine learning systems to adversarial examples and physical attacks. Research demonstrates that neural networks can be fooled by carefully crafted inputs, presenting safety risks for humanoid robots relying on vision-based perception. Defenses including adversarial training, certified robustness, and anomaly detection provide partial protection, though adversarial robustness remains an active research challenge.

Software verification and validation methodologies ensure correct robot behavior through formal methods, model checking, runtime monitoring, and comprehensive testing in simulation and real-world environments. Continuous validation throughout deployment lifecycles maintains safety as software evolves and environments change. Companies like Boston Dynamics, Agility Robotics, and Tesla demonstrate different validation approaches reflecting their unique backgrounds and target applications.

Reliability engineering addresses long-term dependability through MTBF analysis, predictive maintenance, component lifetime modeling, and systematic field data collection. Trade-offs between cost and reliability vary across companies, with industrial robots requiring higher reliability than research or consumer platforms.

Maintenance protocols and service infrastructure critically impact operational availability and total cost of ownership.

Ethical hacking and responsible disclosure frameworks enable security researchers to contribute to robot safety through bug bounty programs, coordinated vulnerability disclosure, and industry collaboration. Different approaches to disclosure between China and USA reflect varying legal frameworks and security cultures, creating challenges for international cooperation.

Recent developments from early 2026—including China's national standard system, new ISO standards for dynamically stable robots, landmark safety certifications, and significant cybersecurity and legal developments—demonstrate that the industry is rapidly maturing. These developments validate and extend the theoretical frameworks presented throughout this chapter, while highlighting emerging challenges that will require continued attention from researchers, practitioners, and regulators.

The convergence of these safety, security, and reliability considerations shapes humanoid robot design, development, and deployment strategies. As the industry matures, standardized best practices will emerge, informed by regulatory requirements, field experience, and ongoing research. The case studies throughout this chapter illustrate how diverse companies navigate these challenges, with approaches varying based on target markets, regulatory environments, and organizational philosophies.

Future humanoid robots will require increasingly sophisticated safety and security mechanisms as they take on more complex tasks in less structured environments. Continued research in adversarial robustness, formal verification of learning-enabled systems, and provable safety guarantees will be essential for broader societal acceptance and deployment. The interplay between physical safety, cybersecurity, privacy protection, and reliability engineering will continue to drive innovation in humanoid robotics, ensuring these increasingly capable systems operate safely, securely, and dependably in service of humanity.

## Questions

1. Safety Standards & Regulations How do the focuses of ISO 13482 and ISO/TS 15,066 differ when applied to humanoid robots, and why is compliance with both often necessary for robots operating in shared human workspaces?
2. Fail-Safe Mechanisms Explain the concept of "graceful degradation" in the context of humanoid robotics. Why is this strategy often preferable to a sudden, catastrophic power-off, particularly for bipedal robots?
3. Collision Avoidance & Human Safety What role do Series Elastic Actuators (SEAs) and impedance control play in mitigating injury risks? How do these mechanical and software approaches work together to meet biomechanical injury thresholds?
4. Cybersecurity Threats Describe the difference between sensor spoofing (e.g., injecting false LiDAR returns) and model poisoning. How does each attack

vector specifically compromise a humanoid robot's perception and decision-making?

5. Data Privacy & Surveillance Contrast the data localization and consent requirements under China's PIPL with the frameworks of the European Union's GDPR and California's CCPA. How do these regional differences affect the design of on-device versus cloud processing architectures for global robot deployments?
6. Adversarial Robustness How do physical adversarial patches (e.g., patterns on clothing or floors) threaten the safety of a humanoid robot, and why are traditional digital defenses often insufficient to stop them in real-world environments?
7. Software Verification & Validation What is the "sim-to-real gap" in robotic software testing? Discuss the limitations of relying solely on physics simulators like Gazebo or Isaac Sim, and explain how domain randomization helps overcome these issues.
8. Reliability Engineering How does condition-based predictive maintenance differ from traditional scheduled maintenance? Give an example of how machine learning might be applied to a humanoid robot's joint telemetry to prevent a hardware failure.
9. Ethical Hacking & Disclosure What are the primary differences between the United States and China regarding vulnerability disclosure practices and legal frameworks? How do these differences impact international cybersecurity collaboration in the robotics industry?
10. Synthesis & Real-World Application Based on the chapter's case studies, contrast the engineering and reliability trade-offs made by a company targeting cost-effective research markets (e.g., Unitree) versus a company targeting rigorous industrial/automotive deployment (e.g., Agility Robotics or Tesla).
11. How do China's 2026 national standard system for humanoid robotics and the new ISO standards for dynamically stable mobile robots (ISO 26058–1 and ISO 25785–1) address gaps in the existing regulatory landscape? What implications do the Figure AI whistleblower lawsuit and the Unitree G1 vulnerability disclosure have for industry-wide safety culture and security practices?

## References


Agility Robotics. (2023). *Digit: The multi-purpose robot for a more productive workforce.* https://www.agilityrobotics.com/digit

Agility Robotics. (2026). *Beyond the hype: Digit receives OSHA-recognized NRTL approval.* https://www.agilityrobotics.com/content/beyond-the-hype

Akhtar, N., & Mian, A. (2018). Threat of adversarial attacks on deep learning in computer vision: A survey. *IEEE Access, 6*, 14410–14430. https://ieeexplore.ieee.org/document/8294186

Alias Robotics. (2026). *Our humanoid robot security research cited at U.S. Senate.* https://news.aliasrobotics.com/our-research-humanoid-robot-security-us-senate

Apptronik. (2023). *Apollo: A general purpose humanoid robot.* https://apptronik.com/

Bonaci, T., Herron, J., Yusuf, T., Yan, J., Kohno, T., & Chizeck, H. J. (2015). *To make a robot secure: An experimental analysis of cyber security threats against teleoperated surgical robots.* arXiv:1504.04339. https://arxiv.org/abs/1504.04339

Boston Dynamics. (2023). *Atlas: The world's most dynamic humanoid robot*. https://www.bostondynamics.com/atlas

Boston Dynamics. (2026). *Boston Dynamics unveils new Atlas robot to revolutionize industry*. https://bostondynamics.com/blog/boston-dynamics-unveils-new-atlas-robot-to-revolutionize-industry

Brown, T. B., Mané, D., Roy, A., Abadi, M., & Gilmer, J. (2017). *Adversarial patch*. arXiv:1712.09665. https://arxiv.org/abs/1712.09665

Brundage, M., Avin, S., Wang, J., Belfield, H., Krueger, G., Hadfield, G., & Anderljung, M. et al. (2020). *Toward trustworthy AI development: Mechanisms for supporting verifiable claims*. arXiv:2004.07213. https://arxiv.org/abs/2004.07213

California Legislature. (2018). *California Consumer Privacy Act (CCPA)*. https://leginfo.legislature.ca.gov/faces/billTextClient.xhtml?bill_id=201720180AB375

Calo, R. (2012). Robots and privacy. In P. Lin, K. Abney, & G. Bekey (Eds.), *Robot ethics: The ethical and social implications of robotics* (pp. 187–202). MIT Press.

Cao, Z., Hidalgo, G., Simon, T., Wei, S. E., & Sheikh, Y. (2019). OpenPose: Realtime multi-person 2D pose estimation using part affinity fields. *IEEE Transactions on Pattern Analysis and Machine Intelligence, 43*(1), 172–186. https://ieeexplore.ieee.org/document/8765346

Carlini, N., & Wagner, D. (2018). *Audio adversarial examples: Targeted attacks on speech-to-text*. arXiv:1801.01944. https://arxiv.org/abs/1801.01944

Carlson, J. M., & Doyle, J. (2002). Complexity and robustness. *Proceedings of the National Academy of Sciences, 99*(suppl 1), 2538–2545. https://doi.org/10.1073/pnas.012582499

Clarke, E. M., Henzinger, T. A., Veith, H., & Bloem, R. (Eds.). (2018). *Handbook of model checking*. Springer. https://link.springer.com/book/10.1007/978-3-319-10575-8

Kolodny, L. (2025, November 21). *Figure AI sued by whistleblower who warned that startup's robots could 'fracture a human skull'*. https://www.cnbc.com/2025/11/21/figure-ai-sued.html

Cohen, J., Rosenfeld, E., & Kolter, Z. (2019). Certified adversarial robustness via randomized smoothing. *Proceedings of the 36th International Conference on Machine Learning*, 1310–1320. https://arxiv.org/abs/1902.02918

Dhillon, B. S. (2007). *Robot reliability and safety*. Springer. https://link.springer.com/book/10.1007/978-0-387-68867-0

Dwork, C., & Roth, A. (2014). The algorithmic foundations of differential privacy. *Foundations and Trends in Theoretical Computer Science, 9*(3–4), 211–407. https://www.cis.upenn.edu/~aaroth/Papers/privacybook.pdf

European Commission. (2006). *Machinery Directive 2006/42/EC*. https://eur-lex.europa.eu/legal-content/EN/TXT/?uri=CELEX:32006L0042

European Parliament and Council. (2022). *Directive (EU) 2022/2555 on measures for a high common level of cybersecurity across the Union (NIS2 Directive)*. https://eur-lex.europa.eu/eli/dir/2022/2555/oj

European Parliament and Council. (2023). *Regulation (EU) 2023/1230 on machinery*. https://eur-lex.europa.eu/eli/reg/2023/1230/oj

European Parliament and Council. (2024). *Regulation (EU) 2024/1689 laying down harmonised rules on artificial intelligence (Artificial Intelligence Act)*. https://eur-lex.europa.eu/eli/reg/2024/1689/oj

European Parliament. (2016). *General Data Protection Regulation (GDPR)*. https://eur-lex.europa.eu/eli/reg/2016/679/oj

EX Robots. (2023). *Humanoid robots for social interaction*. https://www.ex-robots.com/

Eykholt, K., Evtimov, I., Fernandes, E., Li, B., Rahmati, A., Xiao, C., Prakash, A., Kohno, T., & Song, D. (2018). Robust physical-world attacks on deep learning visual classification. *Proceedings of the IEEE Conference on Computer Vision and Pattern Recognition*, 1625–1634. https://arxiv.org/abs/1707.08945

Figure AI. (2023). *Figure 01: The first commercially viable autonomous humanoid robot*. https://www.figure.ai/

FIRST. (2023). *Common Vulnerability Scoring System (CVSS)*. https://www.first.org/cvss/

Fisher, M., Dennis, L., & Webster, M. (2013). Verifying autonomous systems. *Communications of the ACM, 56*(9), 84–93. https://dl.acm.org/doi/10.1145/2494558

Fourier Intelligence. (2023). *GR-1: General purpose humanoid robot.* https://www.fftai.com/

Goodfellow, I. J., Shlens, J., & Szegedy, C. (2015). Explaining and harnessing adversarial examples. *Proceedings of the International Conference on Learning Representations.* https://arxiv.org/abs/1412.6572

Gu, T., Liu, K., Dolan-Gavitt, D., & Garg, S. (2019). BadNets: Evaluating backdooring attacks on deep neural networks. *IEEE Access, 7*, 47230–47244. https://ieeexplore.ieee.org/document/8685687

Guiochet, J., Machin, M., & Waeselynck, H. (2017). Safety-critical advanced robots: A survey. *Robotics and Autonomous Systems, 94*, 43–52. https://www.sciencedirect.com/science/article/pii/S0921889016305462

Haddadin, S., Haddadin, S., Khoury, A., Rokahr, T., Parusel, S., Burgkart, R., Bicchi, A., & Albu-Schäffer, A. (2017). On making robots understand safety: Embedding injury knowledge into control. *The International Journal of Robotics Research, 31*(13), 1578–1602. https://journals.sagepub.com/doi/10.1177/0278364912462256

Hendrycks, D., & Gimpel, K. (2017). A baseline for detecting misclassified and out-of-distribution examples in neural networks. *Proceedings of the International Conference on Learning Representations.* https://arxiv.org/abs/1610.02136

Paton-Philip, J., & McKinney, A. (2026). *Humanoid robots and the law—preparing for a new era of risk. Hill Dickinson.* https://www.hilldickinson.com/our-view/articles/humanoid-robots-and-the-law-preparing-for-a-new-era-of-risk

Householder, A. D., Wassermann, G., Manion, A., & King, C. (2020). *The CERT guide to coordinated vulnerability disclosure (CMU/SEI-2017-SR-022).* Software Engineering Institute, Carnegie Mellon University. https://resources.sei.cmu.edu/library/asset-view.cfm?assetid=503330

Huang, K. (2025, February 6). *Agentic AI threat modeling framework: MAESTRO.* Cloud Security Alliance. https://cloudsecurityalliance.org/blog/2025/02/06/agentic-ai-threat-modeling-framework-maestro

Huang, K., & Hughes, C. (2025a). Agentic AI threat modeling. In K. Huang & C. Hughes (Eds.), *Securing AI agents: Foundations, frameworks, and real-world deployment (Advances in Data Analytics, AI, and Smart Systems).* Springer. https://doi.org/10.1007/978-3-032-02130-4_2

Huang, K., & Hughes, C. (2025b). Agentic AI capability and security benchmark. In K. Huang & C. Hughes (Eds.), *Securing AI agents: Foundations, frameworks, and real-world deployment (Advances in Data Analytics, AI, and Smart Systems).* Springer. https://doi.org/10.1007/978-3-032-02130-4_6

IEC. (2010). *IEC 61508: Functional safety of electrical/electronic/programmable electronic safety-related systems.* International Electrotechnical Commission. https://www.iec.ch/

Illinois General Assembly. (2008). *Biometric Information Privacy Act (BIPA).* https://www.ilga.gov/legislation/ilcs/ilcs3.asp?ActID=3004

Ilyas, A., Santurkar, S., Tsipras, D., Engstrom, L., Tran, B., & Madry, A. (2019). Adversarial examples are not bugs, they are features. *Advances in Neural Information Processing Systems, 32.* https://arxiv.org/abs/1905.02175

ISAC Council. (2023). *Information Sharing and Analysis Centers.* https://www.nationalisacs.org/

ISO. (2014). *ISO 13482: Robots and robotic devices—Safety requirements for personal care robots.* International Organization for Standardization. https://www.iso.org/standard/53820.html

ISO. (2016). *ISO/TS 15066: Robots and robotic devices—Collaborative robots.* International Organization for Standardization. https://www.iso.org/standard/62996.html

ISO. (2026). *ISO 26058–1: Safety requirements for industrial mobile robots.* International Organization for Standardization.

Jardine, A. K., Lin, D., & Banjevic, D. (2006). A review on machinery diagnostics and prognostics implementing condition-based maintenance. *Mechanical Systems and Signal Processing, 20*(7), 1483–1510. https://www.sciencedirect.com/science/article/pii/S0888327005001512

Kelly, T. P., & Weaver, R. A. (2004). The goal structuring notation–a safety argument notation. *Proceedings of the Dependable Systems and Networks Workshop on Assurance Cases*. https://www.goalstructuringnotation.info/

Khalastchi, E., & Kalech, M. (2018). On fault detection and diagnosis in robotic systems. *ACM Computing Surveys, 51*(1), 1–24. https://dl.acm.org/doi/10.1145/3146389

Kirschgens, L. A., Ugarte, I. Z., Uriarte, E. G., Rosas, A. M., & Mayoral-Vilches, V. (2018). *Robot hazards: From safety to security*. arXiv:1806.06681. https://doi.org/10.48550/arXiv.1806.06681

Kim, M., Zimmermann, T., DeLine, R., & Begel, A. (2022). The emerging role of data scientists on software development teams. *Proceedings of the 38th International Conference on Software Engineering*, 96–107. https://ieeexplore.ieee.org/document/7886912

Koenig, N., & Howard, A. (2004). Design and use paradigms for Gazebo, an open-source multi-robot simulator. *Proceedings of the 2004 IEEE/RSJ International Conference on Intelligent Robots and Systems, 3*, 2149–2154. https://ieeexplore.ieee.org/document/1389727

Koos, S., Mouret, J. B., & Doncieux, S. (2013). The transferability approach: Crossing the reality gap in evolutionary robotics. *IEEE Transactions on Evolutionary Computation, 17*(1), 122–145. https://ieeexplore.ieee.org/document/6152894

Lasota, P. A., Fong, T., & Shah, J. A. (2017). A survey of methods for safe human-robot interaction. *Foundations and Trends in Robotics, 5*(4), 261–349. https://www.nowpublishers.com/article/Details/ROB-057

Leucker, M., & Schallhart, C. (2009). A brief account of runtime verification. *The Journal of Logic and Algebraic Programming, 78*(5), 293–303. https://www.sciencedirect.com/science/article/pii/S1567832608000775

Li, Y., Liu, Q., Zhou, Z., Xu, X., & Ren, K. (2020). Adversarial examples against deep learning based smart healthcare systems. *IEEE Internet of Things Journal, 8*(5), 3189–3200. https://ieeexplore.ieee.org/document/9115384

Lutz, C., & Tamò-Larrieux, A. (2020). The robot privacy paradox: Understanding how privacy concerns shape intentions to use social robots. *Human-Machine Communication, 1*, 87–111. https://hmcjournal.com/index.php/hmc/article/view/6

Mainprice, J., Hayne, R., & Berenson, B. (2016). Predicting human reaching motion in collaborative tasks using inverse optimal control and iterative re-planning. *Proceedings of the 2015 IEEE International Conference on Robotics and Automation*, 885–892. https://ieeexplore.ieee.org/document/7139282

Marvel, J. A., & Norcross, R. (2017). Implementing speed and separation monitoring in collaborative robot workcells. *Robotics and Computer-Integrated Manufacturing, 44*, 144–155. https://www.sciencedirect.com/science/article/pii/S0736584516302861

Mayoral-Vilches, V., White, R., Caiazza, G., & Arguedas, M. (2022). SROS2: Usable cyber security tools for ROS 2. In *Proceedings of the 2022 IEEE/RSJ International Conference on Intelligent Robots and Systems (IROS)*, 11253–11259. arXiv:2208.02615. https://doi.org/10.48550/arXiv.2208.02615

Mayoral-Vilches, V. (2018). *Towards an open standard for assessing the severity of robot security vulnerabilities, the Robot Vulnerability Scoring System (RVSS)*. arXiv:1807.10357. https://doi.org/10.48550/arXiv.1807.10357

Mayoral-Vilches, V., et al. (2019). *Introducing the Robot Vulnerability Database (RVD)*. arXiv:1912.11299. https://doi.org/10.48550/arXiv.1912.11299

Mayoral-Vilches, V., et al. (2020). *DevSecOps in robotics*. arXiv:2003.10402. https://doi.org/10.48550/arXiv.2003.10402

Mayoral-Vilches, V. (2025a). *Offensive robot cybersecurity*. arXiv:2506.15343. https://doi.org/10.48550/arXiv.2506.15343

Mayoral-Vilches, V., Makris, A., & Finisterre, K. (2025b). *Cybersecurity AI: Humanoid robots as attack vectors*. arXiv:2509.14139. https://doi.org/10.48550/arXiv.2509.14139

Mayoral-Vilches, V., Navarrete-Lozano, L. J., Sanz-Gómez, M., Espejo, L. S., Crespo-Álvarez, M., Oca-Gonzalez, F., et al. (2025c). *CAI: An open, bug bounty-ready cybersecurity AI*. arXiv:2504.06017. https://doi.org/10.48550/arXiv.2504.06017

Morley, J., Floridi, L., Kinsey, L., & Elhalal, A. (2020). From what to how: An initial review of publicly available AI ethics tools, methods and research to translate principles into practices. *Science and Engineering Ethics, 26*(4), 2141–2168. https://link.springer.com/article/10.1007/s11948-019-00165-5

Nelson, W. B. (2004). *Accelerated testing: Statistical models, test plans, and data analysis*. John Wiley & Sons. https://onlinelibrary.wiley.com/doi/book/10.1007/9780470316795

Nipkow, T., Paulson, L. C., & Wenzel, M. (2002). *Isabelle/HOL: A proof assistant for higher-order logic*. Springer. https://link.springer.com/book/10.1007/3-540-45949-9

NPC. (2021). *Personal Information Protection Law of the People's Republic of China. National People's Congress*. http://www.npc.gov.cn/npc/c30834/202108/a8c4e3672c74491a80b53a172bb753fe.shtml

OSHA. (2020). *Control of hazardous energy (lockout/tagout). Occupational Safety and Health Administration*. https://www.osha.gov/laws-regs/regulations/standardnumber/1910/1910.147

OWASP Foundation. (2025). *OWASP AI Vulnerability Scoring System (AIVSS): Scoring system for OWASP agentic AI core security risks (v0.5)*. https://aivss.owasp.org

Papernot, N., Faghri, F., Carlini, N., Goodfellow, I., Feinman, R., Kurakin, A., & McDaniel, P. et al. (2018). *Technical report on the CleverHans v2.1.0 adversarial examples library*. arXiv:1610.00768. https://arxiv.org/abs/1610.00768

Ratliff, N., Zucker, M., Bagnell, J. A., & Srinivasa, S. (2009). CHOMP: Gradient optimization techniques for efficient motion planning. *Proceedings of the 2009 IEEE International Conference on Robotics and Automation*, 489–494. https://ieeexplore.ieee.org/document/5152817

RIA. (2020). *ANSI/RIA R15.08–2020: Industrial mobile robots—Safety requirements. Robotics Industries Association*. https://www.robotics.org/

ROS 2 Documentation. (2026). Robot Operating System 2 documentation. *Open Robotics*. https://docs.ros.org/

SAC. (2018). *GB/T 36530–2018: Safety requirements for service robots*. Standardization Administration of China.

SAC. (2020). *GB/T 38559–2020: Collaborative robots*. Standardization Administration of China.

Sacks, S. (2021). *China's emerging data privacy system and GDPR*. Center for Strategic and International Studies. https://www.csis.org/analysis/chinas-emerging-data-privacy-system-and-gdpr

Schneier, B. (2021). The security mirage: Why vulnerability disclosure doesn't work. *Communications of the ACM, 64*(8), 20–21. https://dl.acm.org/doi/10.1145/3468269

Segal, A. (2018). When China rules the web: Technology in service of the state. *Foreign Affairs, 97*(5), 10–18. https://www.foreignaffairs.com/articles/china/2018-08-13/when-china-rules-web

Shin, H., Kim, D., Kwon, Y., & Kim, Y. (2017). Illusion and dazzle: Adversarial optical channel exploits against lidars for automotive applications. *Proceedings of the 19th International Conference on Cryptographic Hardware and Embedded Systems*, 445–467. https://link.springer.com/chapter/10.1007/978-3-319-66787-4_22

Shostack, A. (2014). *Threat modeling: Designing for security*. John Wiley & Sons.

SRES. (2026). *CES Wrap-Up 2026: The humanoid robot safety question*. https://sres.ai/autonomous-systems/ces-wrap-up-2026-the-humanoid-robot-safety-question

Steinbauer, G., Mörth, M., Wotawa, F., & Ferrein, A. (2016). Fault detection and diagnosis in robotics: Current systems and new challenges. In *Fault detection and diagnosis in robotics* (pp. 1–24). Springer. https://doi.org/10.1007/978-3-319-32552-1_1

Stephens, R. I., Fatemi, A., Stephens, R. R., & Fuchs, H. O. (2000). *Metal fatigue in engineering*. John Wiley & Sons.

Tesla. (2023). *Tesla Bot (Optimus)*. https://www.tesla.com/AI

Tippenhauer, N. O., Pöpper, C., Rasmussen, K. B., & Čapkun, S. (2011). On the requirements for successful GPS spoofing attacks. *Proceedings of the 18th ACM Conference on Computer and Communications Security*, 75–86. https://dl.acm.org/doi/10.1145/2046707.2046719

Tobin, J., Fong, R., Ray, A., Schneider, J., Zaremba, W., & Abbeel, P. (2017). Domain randomization for transferring deep neural networks from simulation to the real world. *Proceedings of the*

*2017 IEEE/RSJ International Conference on Intelligent Robots and Systems*, 23–30. https://ieeexplore.ieee.org/document/8202133

UBTECH. (2023). *Walker: Bipedal walking humanoid service robot*. https://www.ubtrobot.com/

Unitree. (2023). *Unitree G1: General-purpose humanoid robot*. https://www.unitree.com/

Wheeler, D. A. (2015). *Secure programming HOWTO: Creating secure software*. https://dwheeler.com/secure-programs/

White, R., Caiazza, G., Christensen, H., & Cortesi, A. (2021). Procedurally provisioned access control for robotic systems. *Proceedings of the 2021 IEEE/RSJ International Conference on Intelligent Robots and Systems*, 1018–1025. https://ieeexplore.ieee.org/document/9635887

Wong, E., & Kolter, Z. (2018). Provable defenses against adversarial examples via the convex outer adversarial polytope. *Proceedings of the 35th International Conference on Machine Learning*, 5283–5292. https://arxiv.org/abs/1711.00851

Xinhua. (2026). *China's first national standard system for humanoid robotics poised to spur industry development*. https://english.news.cn/20260303/0e51ac8f66c542c5bacf2af3f80b3a40/c.html

Xu, K., Zhang, G., Liu, S., Fan, Q., Sun, M., Chen, H., Chen, P.Y., Wang, Y., & Zhao, Y. (2020). Adversarial T-shirt! Evading person detectors in a physical world. *Proceedings of the European Conference on Computer Vision*, 665–681. https://arxiv.org/abs/1910.11099

Yaacoub, J. P. A., Noura, H. N., Salman, O., & Chehab, A. (2022). Robotics cyber security: Vulnerabilities, attacks, countermeasures, and recommendations. *International Journal of Information Security, 21*(1), 115–158. https://link.springer.com/article/10.1007/s10207-021-00545-8

**Ken Huang** is a globally recognized expert in AI and Web3 security, a prolific author, and a leading figure in shaping industry standards. He is the CEO and Chief AI Officer (CAIO) of DistributedApps.ai, specializing in Generative AI training and consulting. Ken is deeply involved in driving the development of secure AI systems. He serves as a Research Fellow and Co-Chair of the AI Safety Working Groups at the Cloud Security Alliance (CSA), leading Agentic AI initiatives. He is also a Co-Chair of the AI STR Working Group at the World Digital Technology Academy under the UN Framework, and a core contributor to the OWASP GenAI project, focusing on Agentic AI security. His expertise extends to his contributions to OWASP's Top 10 Risks for LLM Applications and his active participation in the NIST Generative AI Public Working Group. He is also a member of Open AI forum.

**Jerry Huang** is affiliated with Kleiner Perkins and has a background in AI, machine learning, security, scalable systems, and applied mathematics. His prior experience includes technical work across AI/ML evaluation, infrastructure, security, access control, and privacy-preserving biometric authentication. He holds BS/MS degrees in Computer Science from Georgia Tech and has pursued graduate study in Applied Mathematics at the University of Chicago.

**Bhavya Gupta** is an Information Security Officer at Stanford University, where she has built her career since 2017 across roles spanning in risk, governance and security operations. She holds a Master's in Cyber Security Engineering from the University of Southern California. Beyond Stanford, Bhavya is a recognized voice in AI security and governance. She serves as an OWASP lead contributor on the Agentic AI Top 10 risks for skills and the AI Vulnerability Scoring System (AIVSS), helping shape industry standards for securing emerging AI systems. Her expertise spans the full spectrum of modern cybersecurity—from traditional vulnerability management and cloud security to the novel risks introduced by agentic AI—making her a sought-after expert at the intersection of cloud, application, and AI security governance.

**Víctor Mayoral-Vilches,** PhD, is affiliated with Alias Robotics and is a robotics and robot cybersecurity specialist. His published work includes research on robot security assessment, robot teardown, and security-oriented analysis of industrial robot systems. His work is directly relevant to the cybersecurity and reliability challenges of humanoid robots and physical AI.

# Economic Transformation & Labor Markets

# 7

Jerry Huang and Ken Huang

## 7.1 Introduction

Humanoid robots' implications extend beyond technological advancement into fundamental restructuring of labor markets, industrial productivity, and global economic systems. The humanoid robotics market has experienced an unprecedented growth trajectory, with market valuations projected to surge from USD 2.92 billion in 2025 to USD 15.26 billion by 2030, representing a compound annual growth rate (CAGR) of 39.2% (MarketsandMarkets, 2025). More ambitious long-term projections suggest the market could reach $5 trillion by 2050, supported by the deployment of nearly 1 billion humanoid robots globally (Morgan Stanley, 2025).

This acceleration reflects the convergence of multiple technological and economic factors. Advances in artificial intelligence, particularly large language models and computer vision systems, have enhanced robots' cognitive capabilities for perception, decision-making, and human interaction (World Economic Forum, 2025). Simultaneously, manufacturing costs have declined approximately 40% year-over-year between 2023 and 2024, exceeding earlier projections of 15–20% annual reductions (U.S.-China Economic & Security Review Commission, 2024). These cost improvements, combined with acute labor shortages in advanced economies, create a compelling economic rationale for humanoid adoption across multiple sectors.

This chapter analyzes the economic implications of humanoid robot deployment through systematic examination of interconnected domains: industry disruption

J. Huang
Kleiner Perkins, San Francisco, CA, USA

K. Huang (✉)
DistributedApps.Ai, Fairfax, VA, USA
e-mail: ken@distributedapps.ai



K. Huang et al. (eds.), *Humanoid Robots and Physical AI*, Advances in Data Analytics, AI, and Smart Systems, https://doi.org/10.1007/978-3-032-33669-9_7

and deployment scenarios, cost analysis and economic viability, labor market transformation, productivity and economic growth, business models and value chains, new agentic economy, investment trends and venture capital, and recent developments. Analysis incorporates comparative assessment of developments in the United States and China, two nations pursuing markedly different strategies in humanoid robotics adoption and industrial policy.

The U.S. approach emphasizes venture capital-funded startups, technological innovation, and market-driven deployment, exemplified by companies including Tesla, Agility Robotics, Figure AI, and Boston Dynamics. In contrast, China implements coordinated industrial policy combining government subsidies, research center establishment, and strategic procurement mandates, as articulated in the Ministry of Industry and Information Technology's 2023 "Guiding Opinion on the Innovation and Development of Humanoid Robots" (U.S.-China Economic & Security Review Commission, 2024). Understanding these divergent approaches provides insight into competitive dynamics shaping the global humanoid robotics industry.

The economic impact of humanoid robots cannot be evaluated through a singular analytical lens. Rather, comprehensive assessment requires integrated examination of technological capability trajectories, manufacturing economics, labor market dynamics, and socio-political responses to automation. Morgan Stanley estimates that by 2040, the United States may deploy 8 million working humanoid robots with $357 billion impact on wages, expanding to 63 million units by 2050 with approximately $3 trillion in wage effects (Morgan Stanley, 2024). While projections contain substantial uncertainty, they underscore the transformative economic potential of humanoid robotics technology.

## 7.2 Industry Disruption and Deployment Scenarios

The integration of humanoid robots is driving profound operational shifts across multiple global industries. This section explores the primary sectors undergoing the most rapid transformation, examining the unique deployment scenarios, technical requirements, and early commercial traction within structured manufacturing environments, logistics and warehousing operations, and the complex, human-centric fields of healthcare and eldercare. A separate deployment category involves hazardous or high-risk work, where the economic case is driven less by wage substitution than by reducing injury, liability, insurance exposure, and mission risk. Robotics agencies and workplace-safety bodies already emphasize robots' value in tasks that are dangerous, repetitive, or unpleasant, including inspection, emergency response, and environments unsafe for human workers (Centers for Disease Control and Prevention, 2024; Occupational Safety and Health Administration, 2026). For humanoids, this includes disaster response, hazardous inspection, tactical or emergency scenarios, and industrial environments involving heat, contamination, unstable structures, or other conditions where removing humans from the hazard is itself the primary source of value.

### 7.2.1 Manufacturing Sector Deployment

Manufacturing represents the primary deployment sector for humanoid robots, driven by structured operational environments, repetitive task profiles, and quantifiable return-on-investment metrics. The sector's adoption of humanoid robots differs fundamentally from earlier industrial automation waves through robots' versatility and human-centric design, enabling operation in facilities designed for human workers without extensive infrastructure retrofitting (Global X ETFs, 2024).

Tesla has emerged as a prominent case study in manufacturing deployment through its Optimus humanoid robot program. During Tesla's Q4 2024 earnings call, CEO Elon Musk articulated ambitious production targets: approximately 5,000 Optimus units in 2025 (with parts inventory sufficient for 10,000–12,000 units), scaling toward 50,000 units in 2026 and longer-term ambitions of 100,000 units monthly (Humanoid Robotics Technology, 2025). Initial deployment focuses on repetitive factory tasks, including loading sheet metal onto welding lines, with Tesla targeting Musk's long-term aspiration of pricing below $20,000 once manufacturing scale matures (with a near-term $20,000–$30,000 manufacturing-scale target detailed below), at volumes exceeding 1 million units annually.

This aggressive cost trajectory—from approximately $200,000 per unit in 2024, declining toward $20,000-$30,000 at manufacturing scale—represents a critical inflection point for economic viability (Shaikh, 2025). The pricing target positions Optimus competitively with annual human worker compensation in manufacturing roles, potentially achieving cost parity around 2026–2027 based on projected scaling curves (TESMAg, 2025). Musk projects that Optimus could generate up to $10 trillion in long-term revenue for Tesla, emphasizing the program's strategic importance to company valuation.

Tesla's vertical integration strategy provides distinct advantages in humanoid development. The company leverages technology transfer from its Full Self-Driving (FSD) autonomous vehicle system, utilizing identical visual perception algorithm architectures and Autopilot-level camera modules while eliminating expensive lidar sensors through algorithmic optimization (36Kr, 2024). This technical continuity accelerates development timelines while controlling costs. Supply chain integration with Model Y production enables component cost reduction through scaled procurement; for instance, Green Harmonic supplies harmonic reducers for both Optimus joint actuation and Model Y steering systems.

However, Tesla's humanoid development faces significant technical challenges. According to technology media reporting, Tesla has encountered major obstacles in robotic hand technology, resulting in numerous essentially complete robots remaining idle in factories awaiting hand and forearm component integration. The human hand contains 27 bones, 34 muscles, and achieves 24 degrees of freedom; replicating this functionality in robotic systems demands extremely high performance in material strength, weight optimization, and wear resistance. These technical bottlenecks illustrate that aggressive production timelines may face delays pending resolution of fundamental engineering challenges.

Beyond Tesla, commercial humanoid deployments proliferate across automotive manufacturing. Figure AI established a partnership with BMW Manufacturing to deploy humanoid robots in BMW's South Carolina facility for automotive production applications (Figure AI, 2024). In February 2024, Figure secured $675 million in Series B funding at $2.6 billion valuation, with investments from Microsoft, OpenAI Startup Fund, NVIDIA, Jeff Bezos through Bezos Expeditions, and additional strategic partners (see Sect. 7.8). The capital infusion enables Figure to accelerate the development of next-generation AI models for humanoid robots, combining OpenAI's research capabilities with Figure's robotics hardware and software expertise (PRNewswire, 2024).

Chinese manufacturers have accelerated humanoid robot integration across automotive production. UBTECH Robotics partnered with Dongfeng Liuzhou Motor to deploy 20 industrial humanoid robots during first half of 2025 (IDNFinancials, 2025). UBTECH's Walker S humanoid robot entered training programs at major Chinese automotive manufacturers including BYD, NIO, and Geely, demonstrating capabilities in component assembly, quality inspection, and seat belt functionality testing (Qviro, 2025). UBTECH reported securing orders exceeding 500 humanoid robots from automotive companies, with expectations to reach 1,000–2,000 orders by 2025.

The convergence of multiple humanoid robot deployments across automotive manufacturing signals sector-wide transformation. Unlike specialized industrial robots confined to single-purpose applications behind safety cages, humanoid robots' human-analogous form factors enable flexible task assignment across manufacturing workflows. This versatility justifies higher per-unit costs relative to conventional industrial robots through amortization across multiple application domains. As manufacturing environments become increasingly customized and production runs shorter, the flexibility advantage of humanoid robots intensifies, potentially accelerating adoption beyond the initial early-adopter phase.

### 7.2.2 Logistics and Warehousing Applications

Logistics and warehousing sectors represent another critical deployment arena for humanoid robots, addressing persistent labor shortages and operational inefficiency challenges. These environments feature structured layouts, repetitive material handling tasks, and quantifiable productivity metrics, creating favorable conditions for humanoid robot integration.

Agility Robotics has achieved notable commercial traction with Digit, its purpose-designed humanoid robot for logistics applications. In June 2024, Agility signed a multi-year agreement with GXO Logistics, the world's largest pure-play contract logistics provider, deploying a "small fleet" of Digit robots at the SPANX facility in Flowery Branch, Georgia under robotics-as-a-service (RaaS) business model (The Robot Report, 2024a). This deployment, widely regarded as the first formal commercial application of humanoid robots, demonstrates Digit's capability to autonomously transfer totes between 6 River Systems' Chuck autonomous mobile robots (AMRs) and conveyor systems (Chesnokova, 2024).

Digit specifications illustrate design optimization for logistics environments. Standing 5 feet 9 inches tall and weighing 140 pounds, the robot lifts up to 35 pounds—sufficient for typical warehouse tote handling while maintaining safe human interaction profile. Operation orchestration occurs through Agility Arc, cloud-based automation platform providing operational visibility into key performance indicators including uptime, throughput, Mean Time Between Incidents (MTBI), and robot status (FinancialContent, 2024). The platform integrates with existing Warehouse Management Systems (WMS), Warehouse Execution Systems (WES), and Manufacturing Execution Systems (MES) through industry-standard APIs, simplifying deployment into established operational frameworks.

Amazon commenced testing Digit at its robotics research facility near Seattle, exploring how humanoid robots support logistics operations, particularly repetitive tasks including tote recycling—handling empty totes following inventory picking operations (Robotics 24/7, 2024). While Amazon's engagement remains in pilot phase, the e-commerce giant's involvement provides significant market validation for humanoid logistics applications, given Amazon's position as global leader in warehouse automation innovation.

Agility Robotics supported commercial expansion through strategic infrastructure investment. In September 2023, the company inaugurated RoboFab, a 70,000-square-foot manufacturing facility in Salem, Oregon (Agility Robotics, 2023). The facility provides initial production capacity for hundreds of Digit robots in first year of operation, with scaling capability to exceed 10,000 robots annually at full capacity. Notably, at full capacity RoboFab is planned to employ more than 500 human workers while deploying Digit robots on the production floor performing similar material handling tasks as customer deployments—moving, loading, and unloading totes. This operational configuration demonstrates human–robot collaborative workflows while validating Digit's capabilities in production environment.

Customers participating in Agility Partner Program (APP) received first Digit deliveries during 2024, with general market availability commencing in 2025 (Agility Robotics, 2024). The APP structure provides early adopter customers with dedicated engineering resources and on-site visits to optimize deployment configurations for specific warehouse workflows and operational pain points. This collaborative development approach enables Agility to incorporate real-world operational feedback into successive product iterations while building reference customer portfolio for market expansion.

The economic rationale for humanoid deployment in logistics emphasizes labor availability rather than immediate cost arbitrage. GXO Logistics frames humanoid robots as addressing ongoing labor challenges and demand for adaptable automation capable of handling diverse material handling tasks without extensive facility modification (The Robot Report, 2025). This positioning—robots as labor force augmentation rather than wholesale replacement—characterizes early adopter communication strategies, though longer-term dynamics may evolve as technology matures and unit costs decline.

Industry analysts project substantial market potential for logistics humanoids. Goldman Sachs Research estimated the total addressable market for humanoid robots at approximately $38 billion by 2035, with shipments reaching 1.4 million units, assuming continued penetration of global manufacturing workforce and technological advancement (Goldman Sachs, 2024). While these projections contain considerable uncertainty regarding adoption rates and pricing trajectories, they indicate substantial market opportunity if technical and economic viability thresholds are achieved.

### 7.2.3 Healthcare and Eldercare Applications

Healthcare and eldercare sectors present compelling application domains for humanoid robots, driven by demographic aging in developed economies and chronic healthcare labor shortages. China's demographic challenge proves particularly acute, with population aged 60 and above expected to comprise 40% of total population by 2050, creating unprecedented demand for eldercare services (China Daily, 2025).

Humanoid robots in healthcare settings address multiple operational requirements: patient monitoring, medication delivery, mobility assistance, companionship provision, and routine facility sanitization. Unlike specialized medical devices, humanoid robots' versatile form factors enable task flexibility across diverse healthcare workflows. However, healthcare deployment faces more stringent regulatory requirements, safety certifications, and human acceptance thresholds compared to industrial applications.

Fourier Intelligence's GR-1 humanoid robot targets rehabilitation and eldercare applications, designed for patient assistance and research collaboration. The Chinese company, founded in 2015, initially focused on rehabilitation robotics before expanding into humanoid space, launching GR-1 through dedicated humanoid division in 2023. The robot demonstrates capabilities in walking, object carrying, and AI-powered human interaction for eldercare environments.

UBTECH Robotics' Walker robot family also targets healthcare and service sectors, with deployment emphasis in Chinese market addressing demographic-driven eldercare demand. Walker S features autonomous battery swapping capability enabling continuous 24/7 operation—practical advantage in healthcare settings requiring consistent uptime.

Economic models for healthcare humanoids differ substantially from industrial deployments. Healthcare applications typically involve direct human interaction requiring higher safety standards, reliability thresholds, and regulatory compliance compared to warehouse tote manipulation. Reimbursement models remain undeveloped in most healthcare systems, creating uncertainty regarding payment mechanisms for robot-assisted care. These factors may slow healthcare adoption relative to industrial sectors despite compelling demographic drivers.

China's policy environment provides potential accelerant for healthcare humanoid adoption. Government subsidies, pilot program facilitation through state-owned facilities, and active labor-market policy supporting automation in care industries create favorable conditions for deployment. As Chinese humanoid manufacturers achieve cost reductions through manufacturing scale, healthcare applications may become economically viable earlier in Chinese market than Western economies, particularly for eldercare applications in institutional settings.

## 7.3 Cost Analysis and Economic Viability

The widespread adoption of humanoid robotics ultimately hinges on their financial competitiveness relative to traditional human labor and existing single-purpose automation. This section breaks down the economic viability of these systems by evaluating comprehensive Total Cost of Ownership frameworks and analyzing how aggressive economies of scale, component cost reductions, and manufacturing learning curves are rapidly accelerating the timeline to commercial affordability.

### 7.3.1 Total Cost of Ownership Analysis

Economic viability of humanoid robots depends critically on total cost of ownership (TCO) relative to human labor alternatives. TCO encompasses capital expenditure (robot purchase price), operating costs (energy, maintenance, software subscriptions), infrastructure adaptation expenses, and productivity differentials. Comprehensive TCO analysis must extend beyond simple cost comparison to incorporate productivity variations, quality improvements, operational flexibility, and risk mitigation.

Current humanoid robot pricing exhibits substantial variation reflecting different market positioning strategies and production volumes. Morgan Stanley Research estimates average humanoid cost at approximately $200,000 in 2024 for high-income countries, projected to decline to $150,000 by 2028 and $50,000 by 2050 as production volumes increase and component costs decrease (see Sect. 7.1). In lower-income countries leveraging Chinese supply chains, prices could decline to $15,000 by 2050.

Tesla's stated pricing target of $20,000-$30,000 per Optimus unit at production scale represents aggressive cost objective (see Sect. 7.2.1). Achieving this price point requires manufacturing volumes exceeding 1 million units annually, creating classic chicken-and-egg challenge: achieving cost targets depends on volume production, while volume production requires demonstrated market demand at target price points. Tesla's vertical integration and existing manufacturing infrastructure provide advantages in navigating this dynamic, but execution risk remains substantial.

Unitree Robotics has disrupted market pricing expectations through aggressive cost optimization. The company's G1 humanoid robot launched at $16,000—approximately 90% cheaper than its predecessor H1 model priced at $90,000 (Maginative, 2024). This dramatic cost reduction reflects deliberate design choices prioritizing manufacturability over maximum performance capabilities. The G1 stands 127 cm tall compared to H1's 180 cm height, enabling smaller motor utilization while maintaining 47 kg weight (The Robot Report, 2024b). Unitree emphasized optimization across seemingly minor elements: reducing wire and cable counts, minimizing chip quantities, and decreasing fastener requirements—marginal improvements that aggregate to substantial cost savings at production volume.

By July 2025, Unitree further disrupted pricing norms by launching R1 humanoid at $5,900—price point previously considered impossible for years (Yahoo Finance, 2025). This achievement demonstrates that manufacturing cost declines may occur faster than conservative projections, particularly as Chinese manufacturers achieve production scale and supply chain optimization. However, ultra-low pricing may reflect strategic market share acquisition rather than sustainable long-term business models, raising questions about quality, durability, and support infrastructure at extreme price points.

Operating cost components merit careful analysis beyond capital expenditure. Energy consumption for humanoid operation remains relatively modest; typical robot draws approximately 500 Wh per hour, translating to $0.72 per eight-hour shift at $0.18/kWh industrial rates (Digital Habitats, 2024). Battery replacement costs constitute more significant expense, with lithium-ion battery packs requiring replacement every 2–3 years depending on duty cycle and charging patterns. Maintenance costs encompass actuator replacement, sensor calibration, software updates, and periodic component refurbishment.

Labor cost comparisons require regional disaggregation. In the United States, average manufacturing worker compensation (including benefits) exceeds $70,000 annually, while warehouse workers earn approximately $35,000-$45,000 (U.S. Bureau of Labor Statistics, 2024). Assuming humanoid robot operational lifespan of 5–7 years, break-even analysis depends critically on robot pricing and productivity equivalence assumptions. At $30,000 purchase price with $5,000 annual operating costs, total seven-year TCO approximates $65,000—potentially competitive with single year of skilled manufacturing labor compensation.

However, simple cost comparison overlooks critical nuances. Robots operate continuously without fatigue, potentially enabling 24/7 production schedules impossible with human labor constrained by shift rotations and mandatory rest periods. This operational continuity effectively multiplies robot productivity by 3–4 × relative to a single-shift human worker, dramatically improving TCO economics. Conversely, robots lack human adaptability, problem-solving capabilities, and contextual judgment, potentially reducing productivity on non-routine or complex tasks. Optimal deployment strategies therefore emphasize hybrid workflows combining robot consistency on routine tasks with human oversight and intervention on exceptional conditions.

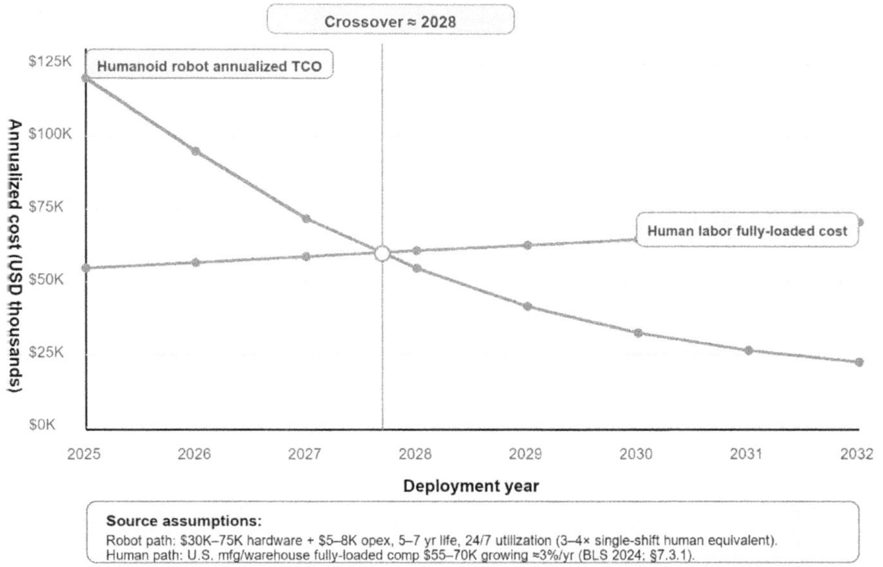


**Fig. 7.1** Humanoid robot vs. human labor annualized cost—TCO crossover plausibly around 2028 (assumption set described in Sect. 7.3.1)

Geographic labor cost variations significantly impact deployment economics. Chinese manufacturing wages average approximately one-quarter to one-third of U.S. equivalents, substantially raising hurdle rate for robot economic viability in Chinese manufacturing facilities. However, China's demographic headwinds—shrinking working-age population and rapid wage inflation—create compelling long-term rationale for automation despite current labor cost advantages. Chinese government policy explicitly recognizes humanoid robots as solution to demographic challenges, providing subsidies and deployment facilitation supporting adoption even where immediate ROI remains marginal.

Figure 7.1 plots annualized humanoid-robot total cost of ownership against fully loaded U.S. labor cost on a $30 K–75 K hardware path with $5–8 K annual operating expense, 24/7 utilization, and a 5–7 year service life. Under those assumptions the two curves cross around 2028, after which incremental humanoid deployment becomes cheaper than the worker it replaces—matching the cost-parity window flagged in Sect. 7.3.1 and the Tesla, Unitree, and AgiBot pricing roadmaps in Sect. 7.2.

### 7.3.2 Economies of Scale and Learning Curves

Manufacturing cost trajectories for humanoid robots exhibit characteristic learning curve dynamics observed across technology product categories. Historical analysis reveals consistent patterns: initial production costs decline 15–20% with each doubling of cumulative production volume, accelerating as manufacturing processes optimize and component suppliers achieve scale efficiencies (Goldman Sachs, 2024).

Empirical evidence demonstrates accelerated cost reduction in humanoid robotics sector. Goldman Sachs reported that manufacturing costs declined 40% year-over-year from 2023 to 2024, substantially exceeding typical 15–20% annual reduction projections (see Sect. 7.1). This acceleration reflects confluence of factors: maturing manufacturing processes, component supplier competition intensification, design optimization iterations, and production volume expansion across multiple manufacturers simultaneously.

China's manufacturing ecosystem provides structural advantages in cost curve progression. The country hosts more than 40 robotics industrial parks and extensive supplier networks for actuators, sensors, processors, and structural components (see Sect. 7.2.1). This manufacturing infrastructure density enables rapid prototype iteration, supplier competition, and logistics cost minimization. Chinese manufacturers including Unitree, UBTECH, and AgiBot leverage these advantages to achieve aggressive pricing positions, creating competitive pressure on Western manufacturers.

Component cost evolution proves critical to overall system economics. Harmonic reducers—precision gear systems enabling accurate joint movement—historically represented significant cost elements dominated by Japanese manufacturers including Harmonic Drive and Nabtesco. Chinese suppliers have entered this market segment, with domestic harmonic reducer market share for Japanese suppliers declining from 42.9% in 2021 to 41.6% by early 2024 (Sixth Tone, 2025). For industrial robots requiring more robust RV reducers, Japanese market share decreased from 55% in 2022 to 44% in 2023 as Chinese alternatives achieved acceptable performance specifications at lower price points.

Sensor cost trajectories follow similar patterns. Lidar sensors, initially costing thousands of dollars per unit, have declined to hundreds of dollars as autonomous vehicle development drove volume production. Visual perception cameras and depth sensors benefit from smartphone component ecosystems, with high-resolution camera modules available at commodity pricing. These component cost reductions directly translate to system-level affordability improvements, accelerating the timeline toward mass-market viability.

Battery technology represents both opportunity and constraint. Humanoid robots typically utilize lithium-ion battery packs derived from electric vehicle (EV) supply chains, benefiting from massive automotive industry investment in battery cost reduction and energy density improvement. However, humanoid robots' weight and size constraints demand high energy density, potentially requiring more expensive battery chemistries than stationary applications. Battery costs constituted approximately 15–20% of total robot manufacturing cost in 2024, suggesting continued cost reduction potential as energy density improves and production scales.

## 7.4 Labor Market Transformation

The deployment of humanoid robots presents far-reaching and deeply disruptive implications for the global workforce. This section systematically analyzes the complex dynamics of this workforce transition, exploring occupational vulnerability and job displacement patterns, the simultaneous potential for new job creation and the necessity of worker retraining, and the resulting impacts on wage levels and broader income distribution. Labor-market outcomes also depend on organizational adoption quality: even when a robot's nominal economics appear compelling, deployments can underperform if workers resist the change, managers lack realistic capability expectations, integration teams underestimate workflow redesign, or training and buy-in arrive after rather than before rollout. These change-management frictions affect both realized productivity and the distribution of gains between firms, workers, and customers.

### 7.4.1 Job Displacement Dynamics

Labor market implications of humanoid robot deployment constitute central concern for policymakers, workers, and societies broadly. Historical precedents from previous automation waves provide mixed guidance: agricultural mechanization and manufacturing automation displaced workers from traditional occupations while enabling unprecedented economic growth and new employment categories. Whether humanoid robots follow similar patterns or represent qualitatively different transformation remains subject of intense debate.

Quantitative projections of job displacement vary considerably based on technological capability assumptions, adoption rate scenarios, and analytical frameworks. Morgan Stanley estimates that by 2050, humanoid robots could affect approximately 75% of occupations, 40% of employees, and $3 trillion in payroll within the U.S. labor market (see Sect. 7.1). These figures suggest transformation scope exceeding previous automation waves in breadth and speed. However, projections depend critically on humanoid robots achieving human-equivalent task performance across diverse occupational domains—outcome subject to substantial uncertainty.

Occupational vulnerability analysis reveals patterns of automation exposure. Positions characterized by routine manual tasks, structured work environments, and limited requirement for complex human interaction face highest displacement risk. Manufacturing assembly positions, warehouse material handling, food service preparation, cleaning and maintenance, and transportation operations demonstrate significant automation susceptibility (Robozaps, 2024). Conversely, occupations emphasizing creativity, complex problem-solving, emotional intelligence, and contextual judgment exhibit lower near-term displacement risk.

MIT research by Acemoglu and Restrepo provides empirical evidence on robot deployment impacts. Their analysis of 19 industries and 722 U.S. commuting zones revealed that adding one robot per thousand workers reduced employment-to-population ratio by 0.39 percentage points and wages by 0.77% compared to unexposed zones between 1990 and 2007 (MIT Sloan, 2020). Projecting forward, aggressive scenario predicting robots quadrupling worldwide by 2025 would yield 1 percentage point lower employment-to-population ratio and 2% lower wage growth across affected regions.

These empirical findings, while based on conventional industrial robots rather than humanoid variants, suggest significant labor market disruption potential. Importantly, Acemoglu and Restrepo identify geographic concentration of impacts, with displacement effects most pronounced in the U.S. industrial heartland. When manufacturing jobs disappear, displaced workers compete for alternative employment in retail and service sectors, creating cascading wage pressure across occupational categories.

China faces distinctive labor market dynamics. Despite lower absolute wage levels compared to Western economies, China confronts demographic headwinds including rapidly aging population and shrinking working-age cohort. Chinese government policy explicitly frames humanoid robots as solution to these demographic challenges rather than purely cost-optimization technology. If automation replaces 70% of manufacturing jobs—plausible given current technological trajectories—Chinese social security systems face sustainability challenges as payroll contribution base contracts.

However, near-term displacement dynamics may differ from long-term patterns. RethinkX analysis suggests that for roughly the next decade humanoid robots will primarily address unmet labor demand rather than displacing currently employed workers (RethinkX, 2025). This framework emphasizes labor shortage mitigation rather than worker replacement as initial deployment driver. Aging populations in developed economies, declining labor force participation rates, and persistent difficulty filling physically demanding positions create unfilled labor demand that robots address before competing directly with employed workers.

Geographic and sectoral variation in displacement timing appears likely. Manufacturing sectors with structured environments and routine tasks face earliest automation pressure. Service sectors requiring complex human interaction, contextual judgment, and emotional intelligence experience later and more gradual automation. Geographic regions with younger populations and lower labor costs witness slower automation adoption compared to aging, high-cost regions where economic incentives favor robot deployment.

### 7.4.2 Job Creation and Workforce Transition

While job displacement concerns dominate public discourse, humanoid robot deployment simultaneously creates new employment categories and occupational opportunities. Historical technology transitions demonstrate that automation displaces specific tasks rather than entire occupations, with workers transitioning to complementary activities emphasizing uniquely human capabilities (Chicago Booth Review, 2024).

Robotics industry directly generates employment across multiple categories: robotics engineers designing hardware and software systems, AI researchers developing perception and control algorithms, manufacturing technicians assembling robots at scale, field service technicians maintaining deployed robot fleets, integration specialists optimizing robot deployment in customer facilities, and training professionals teaching robots new task capabilities. Agility Robotics' RoboFab facility in Salem, Oregon is planned to employ more than 500 workers at full capacity producing Digit robots (see Sect. 7.2.2). As industry scales, these direct employment categories expand proportionally.

Indirect employment effects potentially exceed direct robotics industry jobs. Companies deploying humanoid robots require workforce members capable of supervising robot operations, troubleshooting exceptions beyond robot capabilities, and optimizing human–robot workflow coordination. These roles emphasize technical literacy, problem-solving, and systems thinking—skills complementary to rather than substitutable by robot capabilities. Survey data indicates 60% of information and technology services companies expect net job creation from robot deployment over coming five years.

Zipline, autonomous drone delivery company, illustrates broader pattern: company actively adds workers in electrical engineering, mechanical engineering, software development, and security roles even while automating delivery logistics. This pattern suggests technology companies developing and deploying automation generate high-skilled employment opportunities, though geographic and skill-set mismatches may prevent displaced workers from seamlessly transitioning to these positions.

World Economic Forum analysis projects that while 85 million jobs globally could be displaced by labor division between humans and machines by 2025, 97 million new roles may emerge. The 2023 WEF report identifies that agricultural technologies, digital platforms, e-commerce, and AI will generate substantial labor-market disruption with job displacement in some areas offset by growth elsewhere, yielding net positive employment outcome.

Research by MIT's Autor and colleagues finds that approximately 60% of jobs in the United States today did not exist in 1940, when manufacturing constituted 25% of employment and farming/mining nearly 20%. This historical perspective suggests that transformative technologies generate entirely new occupational categories impossible to predict ex ante. Video game developers, social media managers, cybersecurity analysts, and numerous other contemporary occupations lacked precedent in pre-digital economy.

However, workforce transition from displaced to emerging occupations faces substantial frictions. Workers displaced from manufacturing and logistics positions typically possess skill sets misaligned with robotics engineering and AI development roles. Geographic mismatches exacerbate these challenges, as job displacement concentrates in manufacturing regions while new technology employment clusters in coastal metropolitan areas with high housing costs.

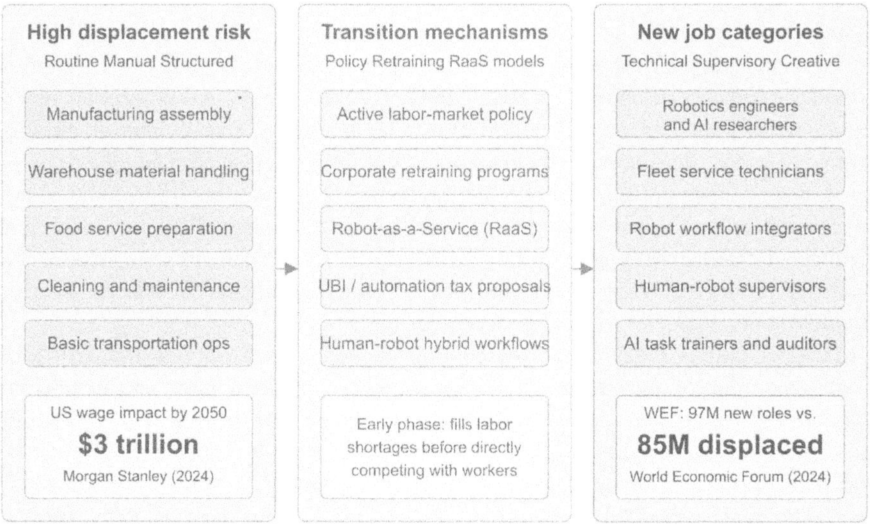


**Fig. 7.2** Labor market transformation framework: displacement, creation, and policy response

Effective workforce transition requires comprehensive retraining infrastructure. Denmark exemplifies proactive approach, investing approximately 1.5% of gross domestic product (GDP) in active labor-market policies including retraining programs for workers displaced by technological change. The United Kingdom's National Retraining Scheme invested £100 million to train workers affected by AI and automation. Amazon's Upskilling 2025 program invests $1.2 billion providing employees access to training programs preparing them for higher-level positions.

These programs recognize that technology diffusion accelerates while working careers lengthen, requiring workers to adapt to multiple technological transitions within single career. Effectiveness of retraining initiatives depends on curriculum relevance, accessibility for displaced workers, and employer willingness to hire workers with non-traditional credential paths.

Figure 7.2 presents three-panel framework organizes the labor analysis into a single visual: occupations at high displacement risk on the left, transition mechanisms in the center, and emerging job categories on the right. The bottom banner highlights geographic concentration as the core policy challenge—displaced workers are in manufacturing regions, new jobs are in coastal tech centers—and the WEF net-employment figures provide essential context for interpreting the overall balance.

### 7.4.3 Wage Effects and Income Distribution

Beyond employment quantity effects, humanoid robot deployment exerts significant influence on wage levels and income distribution. Economic theory predicts that labor-replacing technology reduces workers' bargaining power, suppressing wage growth as employers substitute capital for labor at margin. Empirical evidence from previous automation waves supports this theoretical prediction, particularly for workers without advanced education credentials.

Acemoglu and Restrepo's analysis documents wage suppression effects from industrial robot deployment, reinforces the 0.77% wage decline finding noted earlier in Sect. 7.4.1. These effects concentrate among workers with routine manual task profiles rather than distributing uniformly across workforce. College-educated workers experience minimal wage impacts and may benefit from productivity spillovers, while workers lacking higher education face disproportionate negative effects.

This divergence exacerbates income inequality trends predating robotics deployment. Technology-driven wage polarization creates expanding gap between high-skilled workers complemented by automation and low-skilled workers substituted by automation. Morgan Stanley projects humanoid robots could impact $357 billion in U.S. wages by 2040, expanding to $3 trillion by 2050 (see Sect. 7.1). These wage effects concentrate in manufacturing, logistics, food service, and cleaning occupations—sectors employing substantial shares of lower-income workers.

China confronts distinctive wage dynamics given rapid wage growth in recent decades. Chinese manufacturing wages increased approximately 10–15% annually from 2010 to 2020, eroding cost advantages relative to Southeast Asian competitors. This wage inflation creates economic incentive for automation adoption even where absolute wage levels remain below Western economies. Chinese analysts frame humanoid robots as addressing labor cost pressures while enabling transition to higher-value manufacturing rather than purely as cost-reduction measure.

However, automation's wage effects depend critically on labor market tightness and policy responses. In tight labor markets with unfilled positions, robots address labor shortages rather than displacing employed workers, potentially maintaining or increasing wages for remaining human workers who transition to supervisory and exception-handling roles. Policy interventions including minimum wage regulations, collective bargaining protections, and progressive taxation can mitigate inequality-enhancing effects of automation.

Some economists advocate more radical policy responses. Universal Basic Income (UBI) proposals gain traction as potential mechanism for distributing automation productivity gains across society rather than concentrating benefits among capital owners. Alaska's Permanent Fund Dividend provides precedent for resource-based universal payments, though extending this model to automation-driven productivity remains politically contested.

However, prominent industry leaders such as Elon Musk argue that traditional UBI frameworks are too narrow and focused merely on poverty reduction, proposing instead the concept of a "Universal High Income" (UHI). Musk envisions that the ubiquitous deployment of humanoid robots will eventually drive the marginal cost of physical labor to near zero, ushering in an "Age of Abundance." In this proposed scenario, the output of goods and services would increase so rapidly that it triggers massive deflation, making material goods extraordinarily cheap. Consequently, Musk has publicly argued that scarcity-driven economics will become obsolete, money becomes irrelevant, eradicating global poverty and rendering traditional employment entirely optional, functioning more like a voluntary hobby than a necessity for survival.

While this utopian vision of a post-labor abundance society offers an optimistic counter-narrative to job displacement fears, it introduces a separate array of complex macroeconomic and social challenges. Economists caution that funding a Universal High Income would require unprecedented mechanisms for wealth redistribution. Without systemic overhauls—such as a "robot tax" on capital automation or AI dividends—the massive value generated by automated labor risks being entirely captured and hoarded by the corporations that own the machine fleets. Furthermore, assuming the economic logistics of an abundance society are successfully navigated, society will face a profound philosophical hurdle. If humanoid robots can perform physical and cognitive tasks better than humans, societies will face an existential crisis in determining how individuals find purpose, civic motivation, and psychological meaning in a world where human labor is no longer required (more on this in Chap. 10).

Task-based rather than occupation-based displacement framework provides more nuanced analysis. Rather than eliminating entire occupations, humanoid robots may perform specific tasks within occupational bundles, allowing human workers to focus on complementary activities. For instance, warehouse workers may transition from material handling (automated by robots) to quality control, exception resolution, and customer service (tasks requiring human judgment). This reallocation potentially maintains employment levels while shifting task composition within roles.

Reporting by Mesa (2025) on AI expert Adam Dorr's analysis suggests more pessimistic scenario: by 2040s, robots may perform virtually all tasks better and cheaper than humans, fundamentally challenging traditional employment models (Mesa, 2025). Under this scenario, conventional wage labor system becomes increasingly untenable as marginal cost of labor approaches zero. Whether such developments necessitate complete restructuring of economic systems or represent overestimation of technological capability trajectories remains subject of intense debate.

## 7.5 Productivity and Economic Growth

Beyond individual cost savings, humanoid robots are anticipated to serve as major catalysts for macroeconomic growth and industrial efficiency. This section examines the mechanisms through which continuous, fatigue-free operation generates measurable sectoral productivity gains, while also addressing the "productivity paradox" that may temporarily obscure these macroeconomic benefits as organizations navigate early workflow integration challenges.

Productivity enhancement represents core economic rationale for humanoid robot deployment across industrial sectors. Robots' capacity for 24/7 operation without fatigue, consistent quality output, and scalable deployment theoretically enables substantial productivity gains relative to human labor. However, realizing these gains requires addressing technical challenges, workflow integration obstacles, and organizational adaptation requirements.

Manufacturing sector productivity gains manifest through multiple mechanisms. Continuous operation eliminates downtime associated with shift changes, breaks, and human fatigue, potentially tripling effective production capacity on fixed manufacturing footprint. Robots' consistent execution reduces error rates and rework requirements, improving first-pass yield and reducing material waste. Quality control accuracy increases through precise sensor feedback and automated inspection capabilities (Automate, 2024).

Tesla's internal productivity metrics from Optimus trials, while not publicly disclosed in detail, reportedly demonstrate meaningful efficiency improvements on repetitive assembly tasks (see Sect. 7.2.1). The company's vertical integration enables rapid deployment iteration and productivity optimization without external customer coordination complexity. As Optimus deployment scales across Tesla's manufacturing operations, empirical productivity data will inform broader industry adoption decisions.

Agility Robotics provides more transparent productivity reporting through customer deployments. The company's Agility Arc platform tracks key performance indicators including uptime, throughput, and Mean Time Between Incidents, providing customers with operational visibility supporting productivity assessment (see Sect. 7.2.2). GXO Logistics' deployment at the SPANX facility in Flowery Branch, Georgia serves as real-world productivity validation, with the logistics provider emphasizing robots' contribution to operational efficiency rather than purely cost reduction.

Goldman Sachs research suggests peer-reviewed studies document that humanoids boost productivity on mixed assembly lines by up to 15% while shortening worker training times. These gains reflect both direct robot contributions and organizational learning effects as companies optimize workflows around human–robot collaboration. Labor economists characterize robots as competing with routine labor while complementing creative roles, suggesting productivity gains concentrate in standardizable tasks rather than distributing uniformly across all activities.

Chinese manufacturing productivity transformation scenarios demonstrate substantial potential gains. Pilot programs at automakers including Nio and Geely deploying Unitree H1 prototypes for bolt-tightening on EV battery trays reported 7% cycle time reduction and decreased ergonomic stress for human technicians. While modest percentage improvements, these gains aggregate substantially across high-volume production environments.

However, productivity paradox concerns merit attention. Historical technology deployment reveals puzzling pattern: substantial IT investment in 1980s-1990s failed to generate expected productivity gains at macroeconomic level until late

1990s, creating "productivity paradox" that mystified economists. This delay reflected organizational learning requirements, complementary innovation necessities, and measurement challenges rather than technology ineffectiveness per se.

Similar dynamics may characterize humanoid robot deployment. Initial installations may demonstrate limited productivity gains as organizations navigate integration challenges, workflow redesign, and human worker adaptation. Meaningful productivity improvements may emerge only after extended deployment experience enables process optimization and organizational learning. This dynamic suggests patience regarding short-term productivity metrics while monitoring longer-term trajectories.

Service sector productivity gains face greater uncertainty than manufacturing improvements. Service activities often involve complex human interaction, contextual judgment, and emotional intelligence—capabilities where current humanoid robots remain substantially inferior to humans. While robots may perform basic service tasks (food delivery, cleaning, simple customer interactions), value-added service activities requiring expertise and relationship-building remain challenging to automate effectively.

Figure 7.3 projects logistic-shaped adoption curves for five sectors from 2025 through 2050, illustrating how structured environments such as logistics and automotive manufacturing inflect first, consumer and household applications follow once household-grade units mature, and healthcare and eldercare lag because of safety certification, regulatory approval, and human-interaction complexity. The relative shapes—not the absolute percentages—are the load-bearing insight: sector readiness, not raw capability, governs the deployment timeline.

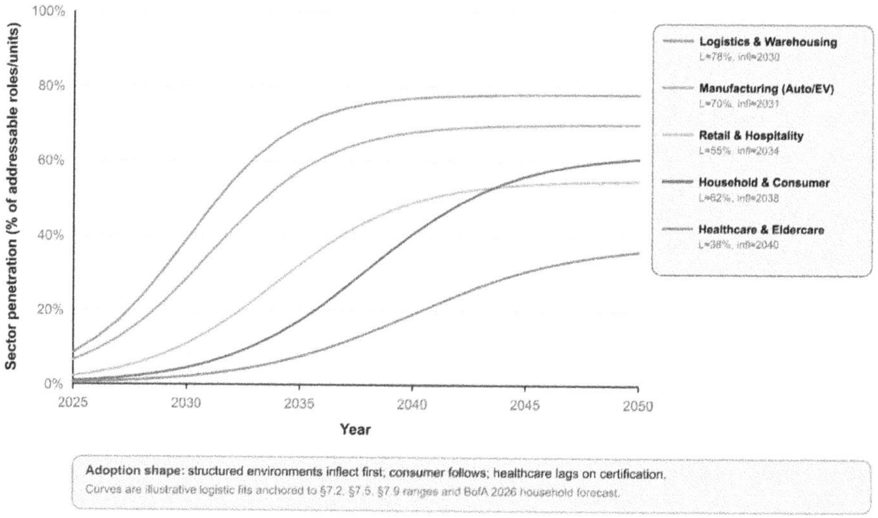


**Fig. 7.3** Sector-by-sector humanoid robot adoption S-curves, 2025–2050

## 7.6 Business Models and Value Chains

As hardware capabilities mature, the commercialization of humanoid robots requires innovative financial approaches to overcome high upfront capital expenditures. This section outlines the evolution of industry value chains and alternative market access strategies, focusing specifically on how Robot-as-a-Service and leasing models are reducing adoption barriers and reshaping long-term relationships between robot manufacturers and enterprise customers.

Business model innovation constitutes critical enabler for humanoid robot market expansion. Traditional capital purchase models create significant adoption barriers for potential customers through high upfront costs, technology obsolescence risks, and maintenance complexity. Alternative models including Robot-as-a-Service (RaaS), leasing arrangements, and subscription-based access reduce adoption friction while creating recurring revenue streams for manufacturers.

Agility Robotics pioneered the RaaS model for humanoid robots through partnership with GXO Logistics (see Sect. 7.2.2). Under this arrangement, GXO pays recurring fees for Digit robot access rather than purchasing units outright. RaaS structure transfers technology obsolescence risk to Agility while providing GXO with flexibility to scale deployment without major capital commitment. The model resembles Software-as-a-Service (SaaS) business structures that transformed enterprise software markets.

RaaS advantages for customers extend beyond financial considerations. Manufacturers retain responsibility for maintenance, software updates, and performance optimization under service agreements. This arrangement reduces customer operational complexity while enabling manufacturers to capture ongoing value from installed base through recurring revenue. Additionally, RaaS facilitates rapid technology upgrades as manufacturers introduce improved hardware and software capabilities (TS2 Tech, 2025).

As these business models mature and robot fleets scale, they lay the groundwork for an even more profound macroeconomic shift: the transition from human-directed tools to self-governing economic agents.

## 7.7 The Transition to an "Agentic Economy"

As humanoid robots achieve higher levels of cognitive and physical autonomy, their economic classification must shift from mere capital equipment to autonomous economic actors. This evolution triggers a "phase transition" in global markets, moving society toward a fully realized "agentic economy" (Tan & Huang, 2025). Traditional microeconomic and macroeconomic theories—which assume humans are the primary decision-makers constrained by bounded rationality and high transaction costs—will be severely strained under this new paradigm. At the firm level, humanoid robots powered by agentic AI do not just flatten the marginal cost of physical labor; they flatten the cost of complex coordination. This allows enterprises to completely re-architect their operations away from rigid, top-down human hierarchies toward dynamic, swarm-like multi-agent systems where fleets of humanoids can autonomously monitor environments, adjust workflows, and execute tasks in real time.

This phase transition fundamentally redefines market structures through the rise of Agent-to-Agent (A2A) transactions. In the agentic economy, a humanoid robot managing a logistics hub will not wait for human purchase orders; it will autonomously discover prices, negotiate contracts, and allocate resources directly with autonomous freight networks and energy grids. Consequently, traditional competitive "moats" will evolve. Defensibility in the robotics industry will shift away from simply owning a proprietary physical robot or a single large language model. Instead, competitive advantage will be captured by companies that master system-level orchestration and composability—the platforms that can seamlessly integrate disparate agents, manage real-world permissions, and provide verifiable cryptographic proofs of safety and reliability for autonomous fleets.

To coordinate and incentivize these vast, decentralized networks of embodied agents, the integration of crypto-economic design and tokenization becomes essential. In a landscape dominated by A2A interactions, tokens serve as the vital infrastructure for machine-to-machine coordination. Humanoids could utilize tokens to autonomously meter access to scarce physical resources—such as priority access to charging stations, bandwidth, or raw materials—while smart contracts automatically reward optimal operational behavior and penalize inefficiencies. Furthermore, this tokenized infrastructure enables the use of Decentralized Autonomous Organizations (DAOs), creating a framework where humans and AI agents can co-govern shared physical infrastructure. This presents a novel mechanism to distribute the massive economic surplus generated by humanoid automation, ensuring it does not become entirely centralized in the hands of a few corporate platform owners.

Ultimately, the emergence of the agentic economy demands a radical shift in regulatory and policy frameworks. As humanoid fleets become autonomous market participants, regulation must pivot from being "model-centric" to "system-centric." Policymakers must look beyond the safety of an individual robot's hardware or its underlying neural network, and instead focus on the systemic risks of interconnected agentic networks. Forward-looking governance must anticipate complex, macro-level threats such as algorithmic collusion between competing robotic fleets, cascading supply chain failures triggered by autonomous miscalculations, and the "cognitive degradation" of AI systems operating without human oversight. Ensuring that the agentic economy remains resilient and aligned with human prosperity will require unprecedented auditability and the continuous, cryptographic verification of agent behavior across the global economy.

## 7.8 Investment Trends and Venture Capital

The humanoid robotics sector has attracted substantial venture capital and strategic investment, reflecting investor conviction regarding transformative market potential. Investment dynamics differ markedly between the United States and China, with the U.S. approach emphasizing venture capital-funded startups while China implements government-coordinated industrial policy supported by strategic capital deployment.

The U.S. investment landscape features prominent venture capital participation. Figure AI's February 2024 Series B funding round exemplifies this pattern: $675 million raised at $2.6 billion valuation from consortium including Microsoft ($95 million), OpenAI Startup Fund, NVIDIA ($50 million), Jeff Bezos through Bezos Expeditions ($100 million), Amazon's affiliated fund ($50 million), Intel Capital, Parkway Venture Capital, Align Ventures, and ARK Invest (O'Brien & Associated Press, 2024). The funding round, significantly oversubscribed relative to initial $500 million target, demonstrates strong investor appetite for humanoid robotics exposure.

Strategic corporate investment complements pure venture capital participation. Microsoft and OpenAI's involvement in Figure AI reflects broader industry recognition of humanoid robots as a critical application domain for AI technologies. NVIDIA's participation across multiple humanoid startups (Figure AI, Agility Robotics, and others) positions the company to capture hardware and software infrastructure demand regardless of which specific manufacturers succeed.

China's investment landscape reflects government industrial policy priorities. The Ministry of Industry and Information Technology's 2023 "Guiding Opinion on the Innovation and Development of Humanoid Robots" established explicit targets: world-class innovation ecosystem by 2025, integration into manufacturing supply chains by 2027 (see Sect. 7.1). Government subsidies, development zone establishment, and state-owned enterprise pilot program brokering support industrial development objectives.

Chinese humanoid robotics financing activity surged dramatically in 2024–2025. From January to October 2024, China recorded at least 56 humanoid robot financing events among 69 globally, totaling more than 5 billion yuan (Investinchina, 2024). In the first nine months of 2025, China recorded 610 robotics investment deals totaling 50 billion yuan ($7 billion)—representing a 250% increase year-over-year. Third quarter of 2025 alone witnessed 243 deals, up 102% from prior year.

Prominent Chinese financing includes UBTECH Robotics securing $1 billion strategic financing facility encompassing equity placements, convertible bonds, and cash-draw rights. Beijing-based Noetix Robotics completed 300 million yuan ($42 million) in Pre-B financing. Zhiyuan Robotics completed eight funding rounds since 2023, demonstrating sustained investor interest across multiple financing stages.

Figure 7.4 contrasts the two nations across five strategic dimensions: funding model, leading companies, 2025 shipment outcomes, pricing strategy, and primary

| Dimension | United States | China |
|---|---|---|
| Funding model<br>Capital source | Venture capital + strategic corporate investors<br>Market-driven | Government subsidies + state-owned enterprise pilots<br>Policy-coordinated |
| Key players<br>Leading companies | Tesla, Figure AI, Agility, Boston Dynamics<br>Startup-led innovation | Unitree, UBTECH, AgiBot, Fourier<br>Ecosystem-led scale |
| 2025 shipments<br>Global market share | ~10% global share<br>Tesla missed 5,000-unit target | ~90% global share<br>AgiBot #1 (5,168); Unitree #2 (5,500+) |
| Pricing strategy<br>Unit economics | Premium positioning<br>$20k–$30k target (Tesla)<br>Capability-led justification | Aggressive undercutting<br>$5.9k (Unitree R1, 2025)<br>Volume and share-led |
| Deployment driver<br>Primary rationale | Labor cost arbitrage + productivity gains<br>ROI-driven adoption | Demographic crisis mitigation + industrial upgrading<br>Strategic policy-driven |

**Fig. 7.4** U.S. vs. China humanoid robotics investment and deployment strategies

deployment driver. The 2025 shipment data—China commanding roughly 90% of global units while Tesla missed its production target—illustrates how coordinated industrial policy and manufacturing ecosystem density can outpace market-driven innovation in the race to production scale.

Figure 7.5 visualizes the steep ramp in humanoid robotics funding from 2018 through 2025, decomposing each year's total into U.S. (venture capital plus corporate strategics) and Chinese (mixed state and private) sources, with callouts for the largest individual deals. The chart underscores how China overtook the United States in deal volume and absolute capital during 2024–2025, driven by state-aligned financing channels and 14th Five-Year Plan industrial priorities.

## 7.9 Market Size Projections and Growth Trajectories

Market size projections for humanoid robotics exhibit substantial variation reflecting uncertainty in adoption rates, pricing trajectories, and technological capability development. Conservative estimates suggest market growing from approximately $2–3 billion in 2024 to $15–22 billion by 2030–2032 (see Sect. 7.1; SkyQuest, 2025). More aggressive projections envision market reaching $66–79 billion by 2032–2035 (Fortune Business Insights, 2024). Long-term forecasts suggest market potentially achieving $5 trillion by 2050 as deployment scales to nearly 1 billion units globally (see Sect. 7.1).

Regional market dynamics differ substantially. Asia Pacific region, led by China, Japan, and South Korea, expected to experience highest CAGR driven by government initiatives and aging population demographics (Dimov, 2025). North America held dominant market share in 2024, contributing over 53.4% of global

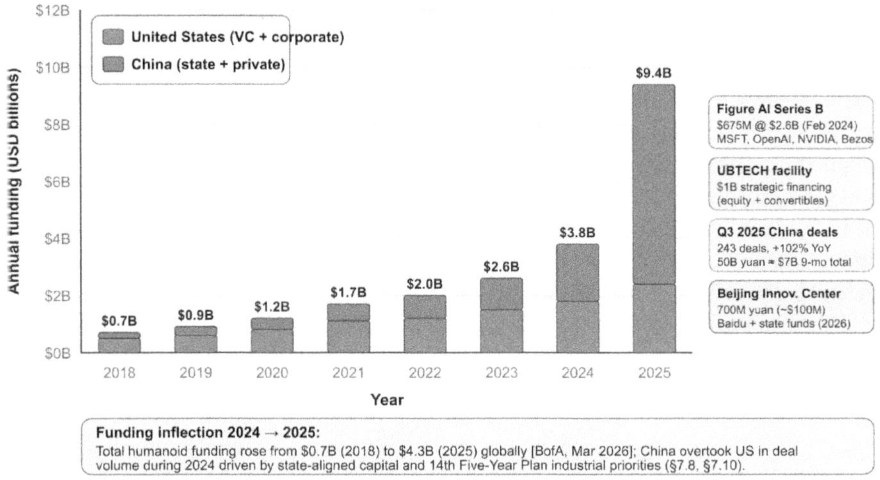


**Fig. 7.5** Annual humanoid robotics funding 2018–2025: U.S. vs. China capital flows

revenues with $188 million, fueled by robust R&D infrastructure and early adopter companies (Scoop Market, 2025). China's humanoid robot market specifically projected to surge from RMB 2.76 billion ($377.56 million) in 2024 to RMB 75 billion ($10.26 billion) by 2029 (see Sect. 7.1). Note that China's revenue share (approximately 13–19% of the conservative global $2–3 billion 2024 estimate) and unit-share (~90% of global shipments per Sect. 7.10.1) differ widely because Chinese humanoids dominate at the lower price points; see Sect. 7.2 for unit pricing context.

These projections depend critically on several key assumptions: manufacturing cost trajectories achieving projected declines, technological capabilities reaching human-equivalent performance on target tasks, regulatory frameworks enabling deployment at scale, and workforce/societal acceptance of human–robot collaboration. Deviation from any of these assumptions would significantly alter market trajectory.

## 7.10 Recent Developments and Market Evolution (Q1 2026)

The first quarter of 2026 has witnessed transformative developments in humanoid robotics, marking a decisive shift from pilot programs to scaled commercial deployment. These developments substantiate earlier projections while revealing unexpected competitive dynamics between Chinese and Western manufacturers.

### 7.10.1 China's Production and Shipment Leadership

Chinese manufacturers have established dominant positions in global humanoid robot shipments. Analyst rankings disagree on the 2025 league table: Omdia research released in January 2026 ranks AgiBot first globally with 5,168 units shipped (a 39% market share), while other 2025 reports cite Unitree at over 5,500 shipments per SCMP-aligned sources (PRNewswire, 2026; Technode, 2026). Both figures cover calendar-year 2025. Combined, Chinese manufacturers shipped approximately 90% of all humanoid robots sold globally in 2025, signaling decisive advantage in scaled production capability.

Unitree announced plans to ship 20,000 humanoid robots in 2026—representing a nearly four-fold increase from 2025 deliveries (Abdullahi, 2026). The company reported shipping over 5,500 units in January–February 2026 alone, building on its 2025 base of more than 5,500 shipments (PRNewswire, 2026). This production acceleration demonstrates China's manufacturing ecosystem advantages in achieving volume scale ahead of Western competitors.

Beijing Innovation Center of Humanoid Robotics raised over 700 million yuan ($100 million) in its first funding round, with backing from Baidu and state investment funds (Du, 2026). This financing underscores continued government support for humanoid robotics development as strategic industrial priority.

### 7.10.2 Tesla Optimus Generation 3 Announcement

Tesla announced in January 2026 that it would unveil its third-generation Optimus robot during Q1 2026, describing it as the "first design meant for mass production" (Hawkins, 2026). In a strategic reallocation of manufacturing capacity, Tesla ended production of Model S and Model X vehicles to establish Optimus assembly lines at its Fremont facility (AutoNews, 2026a). This decision reflects Tesla's conviction that humanoid robotics represents greater long-term value than legacy vehicle models.

As of March 2026, Tesla reportedly has hundreds of Optimus units in internal testing at company factories, with production targeted to begin in summer 2026 and mass production by year-end (Basenor, 2026; Teslarati, 2026). However, Tesla did not achieve its previously stated target of 5,000 Optimus units in 2025, with both AgiBot and Unitree each exceeding Tesla's production output (Lo, 2026).

### 7.10.3 European Manufacturing Deployment Milestone

BMW Group deployed humanoid robots at its Leipzig plant in February 2026, marking the first time humanoid robots have been introduced into series production at a European automotive facility (BMW Group, 2026). The Leipzig pilot builds on BMW's 2025 deployment of Figure AI robots at its Spartanburg, South Carolina plant, where Fig. 02 units logged over 1,250 h and moved more than 90,000 parts during an 11-month deployment period (AutoNews, 2026b; Figure AI, 2025).

BMW's Leipzig deployment utilizes two distinct humanoid platforms: Figure AI robots continuing from the Spartanburg pilot, and AEON robots developed through partnership with Hexagon for Physical AI applications (R, 2026). The pilot aims to integrate humanoid robotics into existing series production for the Neue Klasse i3, with broader deployment planned for April 2026 and full pilot launch in summer 2026.

### 7.10.4 Agility Robotics Expansion and Rebranding

Agility Robotics rebranded to "Agility" in March 2026, reflecting broader ambitions beyond logistics applications (BusinessWire, 2026). Concurrently, the company announced a commercial agreement with Toyota Motor Manufacturing Canada to deploy Digit humanoid robots at its Woodstock, Ontario facility. Following successful pilot evaluation, Toyota committed to deploying seven Digit robots for manufacturing and supply chain operations, with expansion to 10 units planned (Automotive News, 2026; Robotics 24/7, 2026).

The Toyota agreement marks Agility's expansion beyond logistics into automotive manufacturing, demonstrating Digit's versatility across industrial applications. The company confirmed it remains on track to deliver the "first cooperatively safe humanoid" in 2026, emphasizing safety certification as critical enabler for broader commercial deployment.

### 7.10.5 China's Regulatory Framework Development

China issued its first national standard system for humanoid robots and embodied AI in March 2026, establishing a comprehensive regulatory framework covering the industry's entire lifecycle (Bock, 2026). The standards framework was announced at the annual meeting on Humanoid Robots and Embodied Intelligence Standardization in Beijing, addressing safety requirements, performance specifications, and interoperability standards. Comparable deployment in Western markets faces a different type of friction: the United States has no single OSHA standard specific to robotics, so employers must map humanoid deployments onto existing machine-guarding, lockout/tagout, workplace-safety, and consensus-standard obligations, while the EU's Machinery Regulation 2023/1230 imposes essential health and safety requirements and conformity-assessment obligations for machinery, including autonomous mobile machinery and AI-enabled safety functions, with most provisions applying from January 20, 2027 (European Agency for Safety and Health at Work, 2025; see also Sect. 7.2). These health-and-safety certification, workplace-compliance, insurance, and fragmented oversight requirements can lengthen deployment timelines and raise total adoption cost, helping explain why similar robot capabilities may scale at different rates across China, the United States, and Europe.

China's 14th Five-Year Plan, finalized in March 2026, explicitly prioritizes robotics as strategic industrial sector (The Diplomat, 2026). The plan emphasizes humanoid robotics as solution to demographic challenges posed by aging population and shrinking workforce, providing continued policy support for industry development.

### 7.10.6 Revised Long-Term Market Projections

Bank of America issued updated projections in March 2026 forecasting 3 billion humanoid robots globally by 2060, with more people owning humanoid robots than automobiles (Lichtenberg, 2026). The analysis estimates 62% of humanoid robots—approximately 2 billion units—will be deployed in household applications, representing fundamental shift from industrial-focused adoption scenarios.

The bank reported humanoid robotics funding surged from \$0.7 billion in 2018 to \$4.3 billion in 2025, representing approximately six-fold ($6.14\times$) increase in seven years (Lichtenberg, 2026). This funding trajectory indicates sustained investor conviction regarding market potential despite ongoing technical and commercial challenges.

### 7.10.7 Labor Market Impact Research

Korea Labor Institute research released in January 2026 found that robot adoption in Korean manufacturing correlated with 0.6% increase in employment and 2.5% increase in skilled worker wages, challenging displacement-focused narratives (Chosun, 2026). The study attributed positive outcomes to productivity gains enabling facility expansion and worker upskilling, suggesting that automation effects may vary significantly based on labor market conditions and policy frameworks.

This research provides an empirical counterpoint to displacement-focused projections, indicating that automation impacts depend critically on labor market tightness, policy frameworks, and industry-specific dynamics. The Korean experience suggests potential for positive employment outcomes when automation deployment occurs in contexts of labor shortages and supportive policy environments.

### 7.10.8 Macroeconomic Risks and the "Ghost GDP" Phenomenon

In February 2026, Citrini Research published a widely circulated report and thought exercise titled "The 2028 Global Intelligence Crisis," which introduced the concept of "Ghost GDP" to describe the widening disconnect between corporate productivity and broad-based economic prosperity (The Economic Times, 2026). The report, co-authored by Alap Shah of Lotus Technology Management, warns of a scenario where artificial intelligence and physical automation, such as humanoid robots, drastically boost corporate profit margins and headline economic output. However, because machines do not spend wages on discretionary goods, housing, or services, the velocity of money effectively flatlines. The result is "Ghost GDP"—economic output that appears spectacular on national accounts and corporate balance sheets but never actually circulates through the human consumer economy.

This phenomenon directly intersects with the labor market transformations detailed earlier in Sect. 7.4. The Citrini report argues that aggressive automation creates an "intelligence displacement spiral," wherein companies under competitive pressure use AI and robots to cut headcount, and then reinvest every dollar saved directly back into more compute and automation capability, accelerating the next round of job cuts. As labor's share of GDP drops, the foundational income required to sustain consumer spending, pay off the massive U.S. mortgage market, and drive corporate revenues begins to hollow out. To mitigate this systemic risk, the authors advocate for aggressive policy intervention, specifically calling on governments to implement taxes on the incremental or windfall gains generated by AI and automation. This proposed "automation tax" provides a potential fiscal mechanism to redistribute the massive wealth captured by capital owners back to the displaced workforce, aligning with earlier discussions of Universal Basic Income and active labor-market policies.

This analysis highlights a critical vulnerability in current fiscal structures. Governments fund social safety nets primarily through income and payroll taxes, which are inherently taxes on human time and labor. As human labor is replaced, the traditional tax base will erode. The proposal to tax corporate windfall gains from automation represents a necessary shift toward taxing capital and compute rather than labor, which will be essential to fund retraining programs and maintain economic stability.

Ultimately, the "Ghost GDP" concept perfectly encapsulates the core economic paradox of humanoid robots: while these systems solve the production problem by operating continuously without fatigue, they exacerbate the consumption problem. Capitalist economies rely on workers using their wages to purchase the goods that factories produce. If manufacturers achieve mass-market deployment at aggressive price points, production capacity will skyrocket, but displaced human workers will lack the income to consume the resulting goods, potentially triggering a deflationary spiral. Furthermore, this dynamic reframes the "productivity paradox" discussed in Sect. 7.5. Rather than a mere delay in realizing economic gains due to integration hurdles, the Citrini report presents a darker alternative in which productivity gains materialize quickly but are hoarded by the corporations that own the compute and the robot fleets, fundamentally breaking the historical link between technological innovation and shared societal wealth.

## 7.11 Conclusion

The humanoid robotics sector stands at a critical inflection point, rapidly transitioning from laboratory prototypes to scaled commercial realities. As this chapter has detailed, the economic implications of this transition are profound, multifaceted, and global in scale. While aggressive cost-reduction trajectories and scaling manufacturing efforts—driven simultaneously by U.S. technological innovation and Chinese industrial policy—are making humanoid robots economically viable, their deployment represents far more than a simple capital substitution. It signals a fundamental restructuring of the modern labor market.

The potential for unprecedented productivity gains in structured environments like manufacturing and logistics must be carefully weighed against the risks of wage suppression, task displacement, and widening geographic economic disparities. Navigating this transition successfully will require more than engineering breakthroughs; it demands proactive workforce retraining initiatives, innovative adoption strategies like the Robot-as-a-Service (RaaS) model, and robust, forward-looking regulatory frameworks. Ultimately, realizing the full economic dividend of humanoid robotics will depend on a collaborative approach that prioritizes human–robot augmentation over pure replacement, ensuring that the wealth and productivity generated by this technological leap are broadly and equitably shared across society.

### Questions

1. US vs. China Strategies: Compare and contrast the U.S. venture capital-driven approach to humanoid robotics with China's state-sponsored industrial policy. Which model do you believe is better positioned to achieve long-term global market dominance, and why?
2. The "Displacement vs. Shortage" Debate: RethinkX argues that early humanoid deployment will primarily address unmet labor demand rather than displace current workers. Do you agree with this assessment? At what point might this dynamic shift from labor shortage mitigation to direct competition with human workers?
3. Calculating True Value: Beyond the initial capital expenditure (the purchase price of the robot), what "hidden" operational factors and infrastructure costs must a company consider when calculating the Total Cost of Ownership (TCO) and return on investment for a humanoid robot?
4. The Productivity Paradox: The chapter mentions a potential "productivity paradox" similar to the IT revolution of the 1980s and 90s. Why might an initial,

large-scale deployment of humanoid robots in a factory fail to show immediate macroeconomic productivity gains?
5. Business Model Innovation: How does the Robot-as-a-Service (RaaS) business model lower the barrier to entry for companies wanting to adopt robotics? What specific risks does this model shift from the consumer back to the manufacturer?
6. Form Factor Justification: Humanoid robots are significantly more complex and often more expensive than specialized, single-purpose industrial robots (e.g., robotic arms behind safety cages). What is the economic justification for investing in a human-analogous form factor instead of task-specific automation?
7. Mitigating Inequality: Evidence from earlier automation waves suggests that robotics can suppress wages for routine manual laborers while benefiting highly educated workers. What specific policy interventions (e.g., Universal Basic Income, active labor-market retraining) do you think are most viable to prevent humanoid robots from exacerbating income inequality?
8. Healthcare Bottlenecks: Why do healthcare and eldercare applications present vastly different economic, ethical, and regulatory challenges for humanoid robots compared to structured, predictable environments like logistics warehouses?
9. Interpreting Market Data: Based on the Q1 2026 developments showing Chinese manufacturers dominating global shipment volumes, how must Western competitors like Tesla or Figure AI adapt their short-term manufacturing and deployment strategies to maintain market relevance?
10. The Post-Labor Economy: Bank of America projects that by 2060, more people will own a humanoid robot than a car. If the marginal cost of physical labor approaches zero in the long term, how might traditional capitalist economic systems—which are fundamentally based on wage-labor—need to be restructured?

## References

36Kr. (2024). *Can Optimus truly generate 80% of Tesla's future value?* https://eu.36kr.com/en/p/3457772193519237

Abdullahi, A. (2026). China's Unitree aims to ship 20,000 humanoid robots in 2026. *eWeek*. https://www.eweek.com/news/unitree-20000-humanoid-robots-2026-china

Agility Robotics. (2023). *Opening RoboFab: World's first factory for humanoid robots*. https://www.agilityrobotics.com/content/opening-robofab-worlds-first-factory-for-humanoid-robots

Agility Robotics. (2024). *Agility Robotics launches next generation of Digit: World's first human-centric, multi-purpose robot made for logistics work*. https://agilityrobotics.com/content/agility-robotics-launches-next-generation-digit

AutoNews. (2026a). *Tesla kills Model S, Model X for Optimus robot assembly line*. https://www.autonews.com/tesla/an-tesla-q4-earnings-kills-car-models-to-build-optimus-robots-0128

AutoNews. (2026b). *BMW humanoid robots starts Germany pilot after U.S. success*. https://www.autonews.com/bmw/ane-bmw-humanoid-robot-leipzig-germany-0227

Automate. (2024). *The impact of humanoid robots on the production workforce*. https://www.automate.org/robotics/news/the-impact-of-humanoid-robots-on-the-production-workforce

Automotive News. (2026). *Toyota Canada confirms 2026 rollout of Agility's Digit robots in Woodstock factory.* https://www.autonews.com/technology/toyota-canada-agility-digit-humanoid-robots

BMW Group. (2026). *BMW Group to deploy humanoid robots in production in Germany for the first time.* https://www.press.bmwgroup.com/global/article/detail/T0455864EN/bmw-group-to-deploy-humanoid-robots-in-production-in-germany-for-the-first-time

Basenor. (2026). *Tesla Optimus Gen 3: Production starts summer 2026.* https://www.basenor.com/blogs/news/tesla-optimus-gen-3-production-starts-summer-2026

Bock, G. (2026). China releases national standards for humanoid robotics and embodied AI. *The AI Insider.* https://theaiinsider.tech/2026/03/01/china-releases-national-standards-for-humanoid-robotics-and-embodied-ai

BusinessWire. (2026). *Agility Robotics becomes Agility, expanding its reach across emerging use cases.* https://www.businesswire.com/news/home/20260305947515/en/agility-robotics-becomes-agility-expanding-its-reach-across-emerging-use-cases

Centers for Disease Control and Prevention, National Institute for Occupational Safety and Health (NIOSH). (2024). *Robotics in the workplace: An overview.* https://www.cdc.gov/niosh/robotics/index.html

Chesnokova, S. (2024). Future of work: 5 things to know about Agility Robotics' Digit. *Tech Funding News.* https://techfundingnews.com/future-of-work-5-things-to-know-about-agility-robotics-digit-a-humanoid-autonomous-robot-disrupting-warehouses/

Chicago Booth Review. (2024). *A.I. is going to disrupt the labor market. It doesn't have to destroy it.* https://www.chicagobooth.edu/review/ai-is-going-disrupt-labor-market-it-doesnt-have-destroy-it

China Daily. (2025). *Experts warn of social security challenges in age of automation.* https://www.chinadaily.com.cn/a/202503/16/WS67d6e831a310c240449db0a5.html

Chosun. (2026). *Manufacturing employment, wages rise with robots.* https://www.chosun.com/english/market-money-en/2026/01/24/N7KCYIGWMVGNFPRYFMY5N2B5BI

Digital Habitats. (2024). *Tesla's path to $10 trillion.* https://digitalhabitats.global/blogs/tesla-1/exclusive-tesla-s-path-to-10-trillion-cern-basher

Dimov, P. (2025). The rise of humanoid robots: Key trends shaping the next decade (2025–2035). *Medium.* https://petardimov.medium.com/the-rise-of-humanoid-robots-key-trends-shaping-the-next-decade-2025-2035-6e91eb1938e6

Du, Z. (2026). Beijing humanoid robotics hub raises $100 million in first funding round. *Caixin Global.* https://www.caixinglobal.com/2026-02-04/beijing-humanoid-robotics-hub-raises-100-million-in-first-funding-round-102411145.html

European Agency for Safety and Health at Work (EU-OSHA). (2025). *Regulation 2023/1230/EU—machinery.* https://osha.europa.eu/no/legislation/directive/regulation-20231230eu-machinery

Figure AI. (2024). *Figure raises $675M at $2.6B valuation and signs collaboration agreement with OpenAI. PRNewswire.* https://www.prnewswire.com/news-releases/figure-raises-675m-at-2-6b-valuation-and-signs-collaboration-agreement-with-openai-302074897.html

Figure AI. (2025). *Figure 02 completes 11-month pilot project at BMW's assembly plant.* https://www.figure.ai/news/production-at-bmw

FinancialContent. (2024). *Agility Robotics brings operational visibility to deployment of Digit fleets with the launch of Agility Arc.* https://markets.financialcontent.com/stocks/article/bizwire-2024-3-11-agility-robotics-brings-operational-visibility-to-deployment-of-digit-fleets-with-the-launch-of-agility-arc

Fortune Business Insights. (2024). *Humanoid robot market size, share, & growth report [2032].* https://www.fortunebusinessinsights.com/humanoid-robots-market-110188

Global X ETFs. (2024). *The rise of humanoids, explained.* https://www.globalxetfs.com/the-rise-of-humanoids-explained/

Goldman Sachs. (2024). *The global market for humanoid robots could reach $38 billion by 2035.* https://www.goldmansachs.com/insights/articles/the-global-market-for-robots-could-reach-38-billion-by-2035

Hawkins, A. J. (2026). Tesla says production-ready Optimus robot is coming soon. *The Verge.* https://www.theverge.com/transportation/869746/tesla-optimus-gen-3-q1-2026-earnings

Humanoid Robotics Technology. (2025). *Tesla unveils ambitious Optimus humanoid roadmap.* https://humanoidroboticstechnology.com/industry-news/tesla-unveils-ambitious-optimus-humanoid-roadmap/

IDNFinancials. (2025). *China overtakes developed countries in industrial and humanoid market.* https://www.idnfinancials.com/news/54226/china-overtakes-developed-countries-in-industrial-and-humanoid-market

Investinchina. (2024). *Humanoid robots poised to transform China's innovation landscape.* https://investinchina.chinaservicesinfo.com/s/202412/30/WS67725a34498eec7e1f72bf81/humanoid-robots-poised-to-transform-chinas-innovation-landscape.html

Lichtenberg, N. (2026). More people will own a humanoid robot than a car by 2060, BofA predicts. *Fortune.* https://fortune.com/2026/03/13/bank-of-america-humanoid-robot-forecast-3-billion-2060

Lo, K. (2026). *China is running the EV playbook on humanoid robots—and it's working. Rest of World.* https://restofworld.org/2026/china-humanoid-robots-unitree-agibot-tesla-optimus

MIT Sloan (2020). *A new study measures the actual impact of robots on jobs. It's significant.* https://mitsloan.mit.edu/ideas-made-to-matter/a-new-study-measures-actual-impact-robots-jobs-its-significant

Maginative. (2024). *Unitree Robotics unveils affordable G1 humanoid robot.* https://www.maginative.com/article/unitree-robotics-unveils-affordable-g1-humanoid-robot/

MarketsandMarkets. (2025). *Humanoid robot market size, share, industry report trends, 2025 to 2030.* MarketsandMarkets Industry Reports. https://www.marketsandmarkets.com/Market-Reports/humanoid-robot-market-99567653.html

Mesa, J. (2025). The robots are coming for your job sooner than you think: AI researcher. *Newsweek.* https://www.newsweek.com/ai-robotics-jobs-adam-dorr-2039026

Morgan Stanley. (2024). *Could AI robots help fill the labor gap?* https://www.morganstanley.com/ideas/humanoid-robot-market-outlook-2024

Morgan Stanley. (2025). *Humanoid robot market expected to reach $5 trillion by 2050.* https://www.morganstanley.com/insights/articles/humanoid-robot-market-5-trillion-by-2050

O'Brien, M., & Associated Press. (2024). OpenAI wants to implant its ChatGPT-style tech into humanoid robots: Just look at its big deal with robotics startup Figure. *Fortune.* https://fortune.com/2024/03/01/humanoid-robots-with-ai-figure-funding-openai-bezos-nvidia/

Occupational Safety and Health Administration (OSHA). (2026). *Robotics: Overview. U.S. Department of Labor.* https://www.osha.gov/robotics

PRNewswire. (2026). *Unitree ranks No.1 globally in humanoid robot shipments, exceeding 5,500 units in 2025.* https://www.prnewswire.com/news-releases/unitree-ranks-no1-globally-in-humanoid-robot-shipments-exceeding-5-500-units-in-2025-302674729.html

PRNewswire. (2024). *Figure raises $675M at $2.6B valuation and signs collaboration agreement with OpenAI.* https://www.prnewswire.com/news-releases/figure-raises-675m-at-2-6b-valuation-and-signs-collaboration-agreement-with-openai-302074897.html

Qviro. (2025). *Humanoid robots in China 2025.* https://qviro.com/blog/humanoid-robots-in-china-2024/

R, V. (2026). Could BMW's humanoid pilots with Figure AI and Hexagon transform Europe's factories? *Tech Funding News.* https://techfundingnews.com/bmw-humanoid-robots-leipzig-figure-hexagon-physical-ai

RethinkX. (2025). *This time, we are the horses: The disruption of labor by humanoid robots.* https://www.rethinkx.com/blog/rethinkx/the-disruption-of-labour-by-humanoid-robots

Robotics 24/7. (2024). *Agility Robotics expands Amazon relationship with Digit testing.* https://www.robotics247.com/article/agility_robotics_expands_amazon_relationship_with_digit_testing

Robotics 24/7. (2026). *Agility Robotics announces agreement to deploy Digit humanoid at Toyota Motor Manufacturing Canada.* https://www.robotics247.com/article/agility-robotics-announces-agreement-to-deploy-digit-humanoid-at-toyota-motor-manufacturing-canada

Robozaps. (2024). *The economic impact of humanoid robots on the job market.* https://blog.robozaps.com/b/economic-impact-of-humanoid-robots-on-job-market

Scoop Market. (2025). *AI-powered humanoid robots market: Impacts of U.S. tariffs.* https://scoop.market.us/ai-powered-humanoid-robots-market-news/

Shaikh, K. (2025). What Tesla's Optimus robot can do in 2025 and where it still lags. *Interesting Engineering.* https://interestingengineering.com/culture/can-optimus-make-america-win

Sixth Tone. (2025). *Humanoid robots are poised to march into the real world.* https://www.sixthtone.com/news/1016947/humanoid-robots-are-poised-to-march-into-the-real-world

SkyQuest. (2025). *Humanoid robot market size, share & growth forecast 2025–2032.* https://www.skyquestt.com/report/humanoid-robot-market

TESMAg (2025). *Tesla Optimus production revolution.* https://www.teslaacessories.com/blogs/news/tesla-optimus-production-revolution

TS2 Tech. (2025). *Agility Robotics' DIGIT: The humanoid robot revolutionizing warehouse work.* https://ts2.tech/en/agility-robotics-digit-the-humanoid-robot-revolutionizing-warehouse-work/

Tan, L. J. Y., & Huang, K. (2025). The AI agent economy. In K. Huang (Ed.), *Agentic AI: Theories and practices.* Springer. https://ideas.repec.org/h/spr/prochp/978-3-031-90026-6_4.html

Technode. (2026). *China's AgiBot leads global humanoid robot shipments in 2025, Omdia says.* https://technode.com/2026/01/09/chinas-agibot-leads-global-humanoid-robot-shipments-in-2025-omdia-says

Teslarati. (2026). *Elon Musk shares big Tesla Optimus 3 production update.* https://www.teslarati.com/elon-musk-shares-big-tesla-optimus-3-production-update

The Robot Report. (2024a). *Agility Robotics' Digit humanoids land first official job.* https://www.therobotreport.com/agility-robotics-digit-humanoid-lands-first-official-job/

The Robot Report. (2024b). *Unitree Robotics unveils G1 humanoid for $16K.* https://www.therobotreport.com/unitree-robotics-unveils-g1-humanoid-for-16k/

The Robot Report (2025). *Digit is first humanoid deployed in a commercial application.* https://www.therobotreport.com/rbr50-company-2025/digit-is-first-humanoid-deployed-in-a-commercial-application/

The Diplomat. (2026). *China's new five-year plan prioritizes robotics. The world should pay attention.* https://thediplomat.com/2026/03/chinas-new-five-year-plan-prioritizes-robotics-the-world-should-pay-attention

The Economic Times. (2026). *Citrini Research report flags AI 'ghost GDP' risk, calls for taxing windfall gains amid job disruption.* https://economictimes.indiatimes.com/tech/artificial-intelligence/citrini-research-report-flags-ai-ghost-gdp-risk-calls-for-taxing-windfall-gains-amid-job-disruption/articleshow/128740909.cms

U.S.-China Economic and Security Review Commission (2024). *Humanoid robots.* https://www.uscc.gov/sites/default/files/2024-10/Humanoid_Robots.pdf

U.S. Bureau of Labor Statistics. (2024). *Occupational employment and wage statistics: May 2024 national occupational employment and wage estimates.* https://www.bls.gov/oes/current/oes_nat.htm

World Economic Forum. (2025). *Humanoid robots offer disruption and promise. Here's why.* https://www.weforum.org/stories/2025/06/humanoid-robots-offer-disruption-and-promise/

Yahoo Finance. (2025). *Humanoid robots global market report 2026–2040.* https://finance.yahoo.com/news/humanoid-robots-global-market-report-153700568.html

**Jerry Huang** is affiliated with Kleiner Perkins and has a background in AI, machine learning, security, scalable systems, and applied mathematics. His prior experience includes technical work across AI/ML evaluation, infrastructure, security, access control, and privacy-preserving biometric authentication. He holds BS/MS degrees in Computer Science from Georgia Tech and has pursued graduate study in Applied Mathematics at the University of Chicago.

**Ken Huang** is a globally recognized expert in AI and Web3 security, a prolific author, and a leading figure in shaping industry standards. He is the CEO and Chief AI Officer (CAIO) of DistributedApps.ai, specializing in Generative AI training and consulting. Ken is deeply involved in driving the development of secure AI systems. He serves as a Research Fellow and Co-Chair of the AI Safety Working Groups at the Cloud Security Alliance (CSA), leading Agentic AI initiatives. He is also a Co-Chair of the AI STR Working Group at the World Digital Technology Academy under the UN Framework, and a core contributor to the OWASP GenAI project, focusing on Agentic AI security. His expertise extends to his contributions to OWASP's Top 10 Risks for LLM Applications and his active participation in the NIST Generative AI Public Working Group. He is also a member of Open AI forum.

# Governance, Ethics, and Philosophy

8

Ken Huang, Jerry Huang, and Bhavya Gupta

## 8.1 Regulatory Frameworks and Policy Approaches

As humanoid robots transition from constrained laboratory environments to dynamic, real-world applications, policymakers face the complex task of designing regulatory frameworks that ensure public safety without stifling technological innovation. This section explores the diverse approaches taken by major global powers to govern the deployment and operation of advanced robotic systems, examining the contrasting regulatory philosophies of the United States, the European Union, and China, while also highlighting the critical national security and export control considerations that complicate international robotics trade.

### 8.1.1 The Regulatory Landscape: United States, EU, and China

The proliferation of humanoid robots into manufacturing facilities, warehouses, healthcare settings, and potentially households presents governance challenges unprecedented in technological regulation history. Unlike previous automation waves involving specialized industrial machinery confined to safety cages, humanoid robots' general-purpose capabilities, autonomous decision-making, and

K. Huang (✉)
DistributedApps.Ai, Fairfax, VA, USA
e-mail: ken@distributedapps.ai

J. Huang
Kleiner Perkins, San Francisco, CA, USA

B. Gupta
Stanford University, Stanford, CA, USA



K. Huang et al. (eds.), *Humanoid Robots and Physical AI*, Advances in Data Analytics, AI, and Smart Systems, https://doi.org/10.1007/978-3-032-33669-9_8

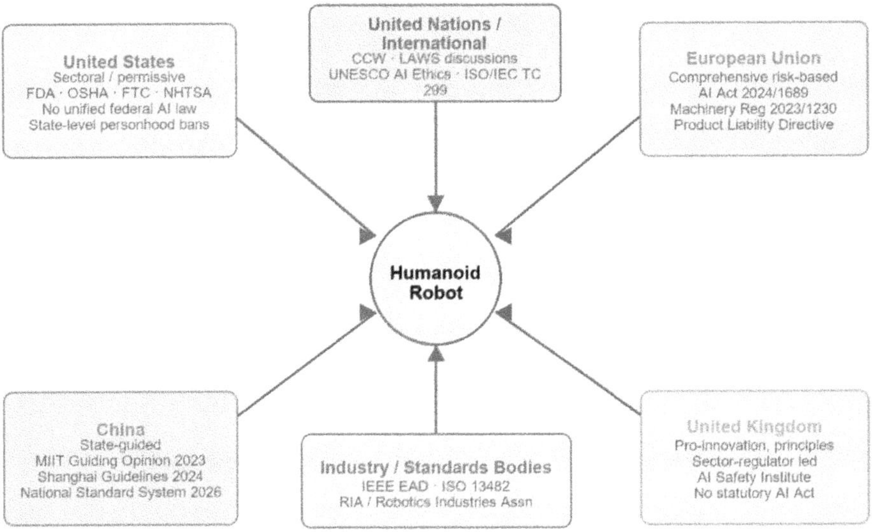


**Fig. 8.1** The multi-jurisdictional governance landscape for humanoid robotics

human-analogous form factors demand regulatory frameworks addressing safety, liability, ethical conduct, and societal implications simultaneously (Cambridge University Press, 2024).

Figure 8.1 surveys the principal jurisdictions and standards bodies whose rules now apply, in some combination, to any internationally deployed humanoid robot—the United States, European Union, China, and United Kingdom on the national tier; the United Nations and ISO/IEC on the international tier; and the IEEE alongside other industry standards organizations on the voluntary tier.

Regulatory approaches vary dramatically across jurisdictions, reflecting divergent cultural values, legal traditions, and innovation priorities. These variations create both opportunities and challenges for manufacturers pursuing global deployment strategies and policymakers seeking to balance innovation encouragement with risk mitigation. Table 8.1 offers a structured comparison of the three dominant governance philosophies.

A structured comparison of the U.S., EU, and Chinese regulatory approaches across six governance dimensions. The star ratings for innovation and safety priority are indicative of each jurisdiction's primary emphasis rather than absolute scores.

The U.S. regulatory approach emphasizes permissive innovation frameworks enabling rapid technological development, with sector-specific regulations emerging reactively to identified risks rather than through comprehensive preemptive governance. The United States lacks unified federal robotics legislation, instead relying on existing regulatory authorities including the Food and Drug Administration (FDA) for medical robotics, the Occupational Safety and Health Administration (OSHA) for workplace safety, the Federal Trade Commission (FTC) for

**Table 8.1** Global regulatory comparison matrix

| Dimension | United States | European Union | China |
|---|---|---|---|
| Regulatory Philosophy | Permissive/Reactive | Comprehensive/ Precautionary | State-Directed/ Strategic |
| Primary Instrument | Sector-specific agencies (FDA, OSHA, FTC) | EU AI Act + Machinery Regulation + PLD | MIIT Guidance + National Standards |
| Rulemaking Speed | Fast (market-led) | Moderate (legislative) | Fast (central coordination) |
| Enforcement Mechanism | Agency enforcement, litigation | Conformity assessment, market access denial | State inspection, market coordination |
| Innovation Priority | ★★★★★ | ★★★☆☆ | ★★★★☆ |
| Safety/Precaution Priority | ★★★☆☆ | ★★★★★ | ★★★★☆ |

consumer protection, and the National Highway Traffic Safety Administration (NHTSA) for autonomous vehicles as precedent for mobile systems (Yale Review of International Studies, 2025).

This fragmented approach provides flexibility for innovation but creates regulatory uncertainty and potential gaps in coverage. Tesla's experience with Full Self-Driving (FSD) technology illustrates both the benefits and challenges of this framework: permissive testing environments enabled rapid development, while liability disputes and safety investigations revealed enforcement limitations (Ronin Legal Consulting, 2025). As Tesla deploys Optimus humanoid robots in factory environments, lessons from autonomous vehicle regulation will inform workplace safety oversight.

The European Union implements comprehensive risk-based regulation through the landmark AI Act (Regulation EU 2024/1689), which entered into force on August 1, 2024. The Act is fully applicable on August 2, 2026, so as of the time of writing in 2026 its compliance obligations are active law rather than prospective requirements (European Commission, 2024; European Parliament, 2024). The AI Act establishes a tiered regulatory framework categorizing AI systems by risk level: prohibited practices (including manipulative AI and untargeted social scoring), high-risk systems requiring conformity assessments and human oversight, limited-risk systems necessitating transparency, and minimal-risk applications facing no specific obligations (Lorenz & Teetzmann, 2024).

Humanoid robots incorporating AI systems for perception, decision-making, and interaction fall under high-risk categorization when deployed in employment contexts, workplace management, or safety–critical applications (Artificial Intelligence Act, 2024). The Act's Annex III explicitly identifies AI systems deployed for employment, workers' management, and access to self-employment as high-risk, requiring strict conformity assessment procedures (Mizan Law Review, 2025).

The revised EU Machinery Regulation (Regulation 2023/1230), effective January 2027, introduces three pivotal requirements for autonomous robotics: autonomy thresholds requiring enhanced conformity assessment for machines demonstrating "self-evolving behaviour through experience"; lifetime cybersecurity responsibilities demanding resilience throughout the product lifecycle including post-sale updates; and collaborative risk mapping accounting for human–machine interactions through dynamic hazard monitoring (Stables et al., 2025).

The EU's new Product Liability Directive, effective December 2024, increases civil liability exposure for autonomous robotics by enabling standalone liability claims for machine-learning models as "software as a product" independent of hardware defects, while revising presumption-of-defect rules to address algorithmic decision-making complexity (Stables et al., 2025). This triad of AI Act, Machinery Regulation, and Product Liability Directive creates a comprehensive but demanding compliance framework.

China deploys state-guided governance combining national standards (GB standards), government guidance documents, and coordinated industrial policy supporting strategic humanoid robotics development. The Ministry of Industry and Information Technology (MIIT) released "Guiding Opinion on the Innovation and Development of Humanoid Robots" in 2023, identifying humanoid robots as a key area for technological competition and establishing explicit targets: a preliminary innovation system by 2025, and integration into manufacturing supply chains by 2027 (U.S.-China Economic and Security Review Commission, 2024).

Provincial and municipal governments provide substantial implementation support. Beijing announced a $1.4 billion fund for local robotics industry development in January 2024, while Shenzhen allocated $71 million annually from 2014 to 2020 for robots and intelligent technologies (U.S.-China Economic and Security Review Commission, 2024). Guangxi Province pledged subsidies up to $140,000 per company for robotics firms, and Chengdu committed $14.3 million per major project (U.S.-China Economic and Security Review Commission, 2024).

In July 2024, Shanghai released China's first governance guidelines specifically for humanoid robots during the World Artificial Intelligence Conference (WAIC). The "Guidelines for Humanoid Robot Governance" emphasize that humanoid robots must "not threaten human security" and "effectively safeguard human dignity," while requiring manufacturers to establish risk warning procedures, emergency response systems, and provide users with training on ethical and lawful robot use (South China Morning Post, 2024; MMLC Group, 2024; Tri-City Voice, 2024).

These Chinese governance frameworks reflect a distinctive approach balancing rapid technological development with social stability maintenance, emphasizing state coordination over market-driven standards emergence, and prioritizing domestic industrial champion cultivation alongside responsible deployment principles.

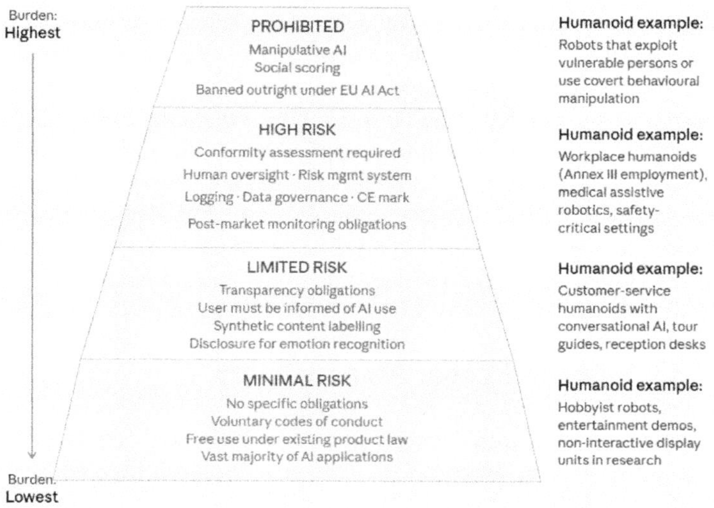


**Fig. 8.2** EU AI act risk tiers applied to humanoid robots

Figure 8.2 maps the EU AI Act's four-tier risk pyramid—prohibited, high, limited, and minimal risk—onto representative humanoid-robot use cases, showing how the same physical platform can attract very different compliance burdens depending on the deployment context.

### 8.1.2 Export Controls and National Security Considerations

Humanoid robotics technology increasingly intersects with national security concerns, particularly given dual-use potential for military applications and technology competition between the United States and China. Export control regimes attempt to balance legitimate security interests with commercial innovation requirements and international scientific collaboration benefits.

The United States maintains export controls on robotics technologies through International Traffic in Arms Regulations (ITAR) and Export Administration Regulations (EAR), with particular scrutiny on AI-enabled autonomous systems, advanced sensors, and sophisticated actuators. However, humanoid robotics components' commercial nature and global supply chains complicate enforcement. Chinese manufacturers' reliance on international components including NVIDIA AI processors, Intel cameras, and specialized actuators from multiple countries creates interdependencies that simple technology denial strategies cannot easily address (U.S.-China Economic and Security Review Commission, 2024).

China's military-civil fusion policy explicitly leverages commercial technologies for military advancement, blurring boundaries between civilian and defense applications (U.S.-China Economic and Security Review Commission, 2024). Chinese military analysts have proposed human–robot cooperation theories potentially employing humanoid robots in support roles or replacing front-line soldiers if technology sufficiently advances. This strategic framework raises Western concerns about dual-use humanoid robot development.

International humanitarian law considerations regarding autonomous weapons systems extend to humanoid military applications. The Convention on Certain Conventional Weapons (CCW) ongoing discussions address lethal autonomous weapons systems (LAWS), though consensus on binding regulations remains elusive. Humanoid robots' potential for greater battlefield autonomy compared to conventional drones or ground vehicles intensifies these debates.

## 8.2 Liability and Legal Responsibility

The autonomous and adaptive nature of modern humanoid robots fundamentally disrupts traditional legal doctrines of fault and responsibility. When a machine capable of continuous learning and independent decision-making causes harm, allocating blame becomes a highly complex legal puzzle. This section examines the challenges of adapting traditional product liability laws to autonomous systems, exploring the intricate web of responsibility connecting hardware manufacturers, software developers, and entities deploying robots in commercial or public spaces.

### 8.2.1 Product Liability for Autonomous Systems

Figure 8.3 traces how a single humanoid-robot incident produces very different liability outcomes depending on the governing regime, contrasting the U.S. strict product-liability default, the EU shared-responsibility model under the revised Product Liability Directive, and China's state-coordinated framework anchored in GB national standards.

Product liability frameworks developed for conventional manufactured goods encounter fundamental challenges when applied to autonomous humanoid robots. Traditional doctrine assumes products possess fixed characteristics at sale, with defects attributable to design flaws, manufacturing errors, or inadequate warnings. Autonomous systems utilizing machine learning challenge these assumptions through post-deployment behavioral evolution, emergent capabilities, and probabilistic rather than deterministic operation (Yale Review of International Studies, 2025).

A fundamental issue underlying AI liability is responsibility fragmentation. Unlike traditional tools functioning under direct human control, AI-driven systems operate autonomously based on algorithmic decision-making (Yale Review of International Studies, 2025).

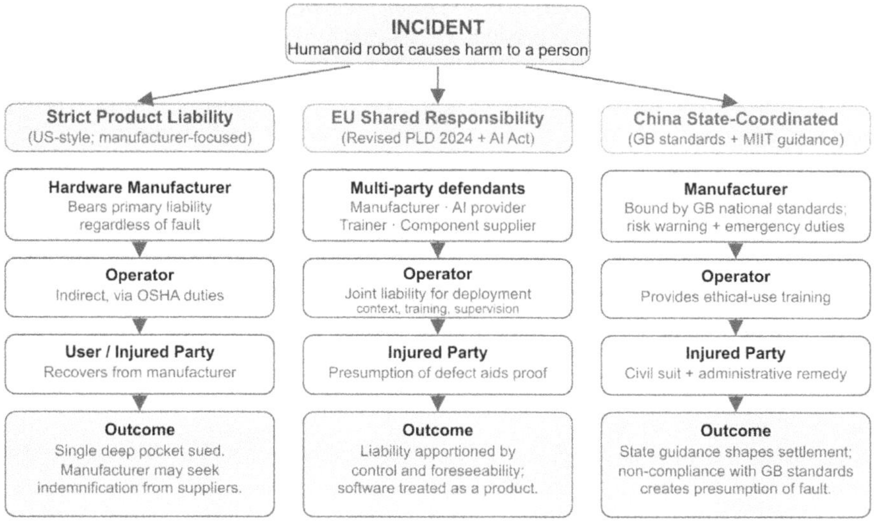


**Fig. 8.3** Liability allocation across three regulatory regimes

Table 8.2 maps the full liability chain arising from a single incident, illustrating why determining the liable party requires navigating complex causal chains across multiple actors.

Each cell in the table represents a party that may bear legal responsibility when a humanoid robot causes injury. Legal theories and the direction of liability claims vary by jurisdiction and the specific facts of the incident. The EU's revised Product Liability Directive (2024) most explicitly enables claims against parties beyond the original equipment manufacturer.

**Table 8.2** Humanoid robot liability web

| ⚠ INCIDENT: Worker injured by humanoid robot | |
|---|---|
| Hardware Manufacturer<br>Strict/product liability(design defect, failure to warn) | AI Algorithm Developer<br>Negligence; EU PLD "software-as-product" liability |
| Sensor & Component Supplier<br>Contributory negligence; indemnification chains | Data Annotator/Trainer<br>EU Product Liability Directive; causal contribution to model bias |
| Operator Company<br>OSHA duties; vicarious liability; failure to supervise/train | Maintenance Contractor<br>Negligent maintenance; breach of service contract |
| Key principle: Under the EU's revised Product Liability Directive (2024), liability may be pursued against any supply-chain member—not solely the original equipment manufacturer. Shared responsibility models distribute liability according to each party's degree of control and foreseeability of harm | |

The debate over whether AI-driven systems constitute "products" or "services" complicates liability analysis, particularly as machine-learning algorithms evolve continuously through experience (Yale Review of International Studies, 2025). If a robot's behavior stems from learning in the operator's facility rather than the manufacturer's original programming, does the operator bear increased liability? If an algorithm update introduces a new vulnerability, should the update provider face liability even if the hardware manufacturer remains unchanged?

Legal scholars debate appropriate liability regimes for autonomous systems. Some advocate strict liability on manufacturers analogous to pharmaceutical industry regulations, where manufacturers bear responsibility regardless of fault (Yale Review of International Studies, 2025). This approach incentivizes safety-focused design while providing clear recourse for injured parties. However, strict liability may stifle innovation by imposing excessive risk on developers unable to predict all operational scenarios.

Alternative proposals suggest shared responsibility models distributing liability among manufacturers, software developers, operators, and potentially end-users based on each party's contribution to harm and control over system behavior (Yale Review of International Studies, 2025). The EU's revised Product Liability Directive reflects this approach by enabling liability claims against multiple supply chain members—ncluding data annotators and algorithm trainers—not exclusively original equipment manufacturers (Stables et al., 2025).

Evidentiary challenges compound substantive liability questions. Autonomous systems' complexity makes proving defects and establishing causation extraordinarily difficult. The EU Product Liability Directive introduces presumptions of defect and causation to address these evidentiary burdens, but balancing injured parties' rights with defendants' due process protections remains contentious (European Parliament, 2025).

The development-risk defense—allowing manufacturers to escape liability for risks unknowable at time of product release—proves particularly problematic for machine learning systems. If a robot learns harmful behavior post-deployment, was this risk knowable at the time of sale? The EU retained this defense in the revised Directive despite critics arguing it undermines consumer protection for AI systems (European Parliament, 2025).

### 8.2.2 Operator Versus Manufacturer Responsibility

The allocation of responsibility between robot operators and manufacturers requires careful analysis of control, foreseeability, and risk distribution. Operators possess direct supervision over deployment context, working conditions, and operational parameters, while manufacturers design fundamental capabilities and safety features.

Agility Robotics' deployment of Digit robots in GXO Logistics warehouses illustrates the operator-manufacturer responsibility interface. Under a robotics-as-a-service (RaaS) business model, Agility retains ownership and maintenance responsibility while GXO directs robot operations within its facilities. If a Digit robot injures a worker, liability analysis must consider:

Was the injury caused by a robot design defect? Did the operator provide adequate worker training? Did facility configuration create hazards the manufacturer could not reasonably anticipate? Did maintenance lapses contribute to malfunction?

Workplace safety regulations through OSHA (United States) or equivalent agencies establish baseline operator duties including risk assessment, worker training, and hazardous condition mitigation. Operators cannot delegate these responsibilities to manufacturers, though manufacturer-provided safety features and operational guidance inform compliance efforts. Conversely, manufacturers cannot assume operators will implement perfect safety protocols, requiring defensive design anticipating reasonably foreseeable misuse.

Tesla's deployment of Optimus in its own factories presents an instructive comparison. As both manufacturer and initial operator, Tesla must satisfy obligations in both roles simultaneously. External sales, however, introduce a separation that raises important questions about post-sale support obligations, software update responsibilities, and operational parameter monitoring (Miami Business Law Review, 2024).

## 8.3 Ethical Design Principles

Beyond legal compliance, the development of humanoid robots necessitates a robust commitment to ethical design principles integrated directly into the engineering lifecycle. As these systems are deployed in socially sensitive and safety–critical environments, ensuring that their behavior aligns with human values is paramount.

### 8.3.1 Value Alignment and Ethical AI

Ethical design principles for humanoid robots extend beyond regulatory compliance to encompass fundamental value alignment: ensuring robot behaviors, objectives, and decision-making processes accord with human values, rights, and societal norms. Value alignment proves particularly challenging for machine learning systems whose objectives emerge from training data and reward functions rather than explicit programming (Laws Pulse, 2025).

The value alignment problem manifests across multiple dimensions. First, determining which values to encode requires navigating cultural differences, ethical pluralism, and contested moral questions. Should a humanoid robot prioritize individual autonomy or collective welfare when these conflict? How should a robot balance efficiency optimization against worker dignity preservation? These questions lack universal answers, yet design decisions implicitly embed particular value commitments.

Second, technical implementation of value alignment remains an active research frontier. Current approaches including reward function specification, inverse reinforcement learning, and constitutional AI attempt to imbue systems with values, but each approach faces limitations. Reward function specification risks misalignment when designers fail to capture a value's full complexity, leading to reward hacking where a system achieves its stated objective through unanticipated means that violate the intended spirit. Constitutional AI approaches show promise by training systems to follow ethical principles, but require carefully articulated constitutional frameworks and robust verification mechanisms.

Third, value alignment must persist across a robot's operational lifetime as it encounters novel situations not represented in training data. Unlike static rule systems, learning-based humanoid robots may develop behaviors diverging from original value specifications as they optimize for deployed environment reward structures. Maintaining alignment through continuous learning requires ongoing monitoring, periodic retraining, and possibly human oversight for high-stakes decisions.

**Industry Practice**

Figure AI's partnership with OpenAI exemplifies industry efforts to address value alignment through collaboration between robotics hardware developers and AI research organizations. The partnership aims to develop next-generation AI models for humanoid robots combining OpenAI's research expertise in large language models with Figure's robotics platform, integrating language understanding and reasoning capabilities to create robots capable of processing complex human instructions while adhering to safety constraints and ethical guidelines.◀

IEEE's Ethically Aligned Design standards provide an industry framework for incorporating ethical considerations throughout autonomous system development lifecycles. The standards emphasize several core principles: human rights primacy, prioritizing wellbeing, accountability and responsibility, transparency and explainability, awareness of misuse, and competence—requiring systematic implementation through design methodologies, testing protocols, and deployment practices rather than afterthought additions (IEEE Standards Association, 2019).

Boston Dynamics has articulated responsible robotics principles guiding its development practices, including commitments to avoid weaponization, promote beneficial uses, and engage with stakeholders on societal implications. The company's public pledge against weapons applications represents a voluntary ethical constraint going beyond regulatory requirements, though enforcement mechanisms beyond reputational pressure remain limited.

### 8.3.2 Transparency and Explainability

Transparency and explainability constitute essential ethical design principles enabling accountability, trust, and appropriate human oversight for humanoid robot systems. Transparency encompasses disclosure of system capabilities, limitations, training data characteristics, and decision-making processes. Explainability requires providing comprehensible justifications for specific robot actions and decisions, enabling users, operators, and oversight authorities to understand the reasoning underlying behaviors.

The EU AI Act mandates transparency requirements for AI systems, requiring providers to design and develop systems enabling users to interpret outputs and use systems appropriately (Regulation EU 2024/1689). For high-risk AI systems including humanoid robots deployed in employment contexts, transparency obligations extend to providing detailed technical documentation, clear information about system capabilities and limitations, and mechanisms for human oversight and intervention (Artificial Intelligence Act, 2024).

However, achieving meaningful transparency and explainability for complex machine learning systems poses significant technical challenges. Deep neural networks processing sensor inputs for perception and control may contain millions or billions of parameters, with decision-making emerging from distributed representations lacking straightforward interpretation. While explainable AI (XAI) techniques including attention visualization, saliency mapping, and counterfactual explanation generation provide partial insights, these methods often sacrifice accuracy for interpretability or produce explanations oversimplifying actual decision processes.

Furthermore, transparency and explainability objectives must be balanced against competing considerations. Full disclosure of robot capabilities might enable malicious actors to identify vulnerabilities or circumvent safety mechanisms. Detailed algorithmic explanations might expose proprietary intellectual property. Overly technical explanations may prove incomprehensible to end users requiring operational guidance rather than implementation details.

## 8.4 Rights and Moral Status of Robots

The increasingly sophisticated, human-like behaviors of modern robots have sparked intense philosophical and legal debates regarding their appropriate status within society. As machines begin to simulate empathy, conversation, and autonomous decision-making, questions arise about whether they should be afforded specific legal protections or recognized as entities with moral standing.

### 8.4.1 Legal Personhood Debates

Debates surrounding legal personhood for robots and artificial intelligence systems raise fundamental questions about moral status, rights attribution, and legal responsibility frameworks. Legal personhood—the capacity to hold rights and bear obligations under law—extends beyond natural persons to include corporations, governments, and other entities. Whether robots should join this category remains deeply contested.

The European Parliament explored an "electronic personhood" concept in a 2017 resolution addressing robot rights and responsibilities, proposing this category might address liability attribution challenges when human accountability proves difficult to determine (Yale Review of International Studies, 2025). The proposal envisioned robots with advanced autonomous capabilities potentially holding legal personality analogous to corporate personhood, enabling liability and responsibility assignment distinct from human creators or operators.

However, this proposal attracted substantial criticism from legal scholars, roboticists, and ethicists arguing that legal personhood mischaracterizes robot nature and creates perverse incentives. Critics contend that legal personhood might shield manufacturers from liability by attributing responsibility to "electronic persons" lacking resources for compensation.

Furthermore, personhood attribution risks anthropomorphizing machines, fostering inappropriate moral regard while deflecting accountability from human designers and operators making consequential decisions about robot design, deployment, and oversight (Yale Review of International Studies, 2025).

The EU ultimately abandoned electronic personhood proposals, clarifying that any such attribution would constitute a pragmatic legal instrument for addressing technical liability challenges rather than a philosophical claim about robot consciousness or moral worth. This instrumental approach treats legal personhood as contractual convenience comparable to corporate personality, devoid of implications for rights or moral consideration.

Sophia the Robot's 2017 Saudi Arabian citizenship provides an instructive case study in symbolic personhood attribution. Hanson Robotics' humanoid robot received citizenship during the Future Investment Initiative conference, generating global media attention and debate about robot rights. Critics noted the irony that a robot received citizenship in a nation restricting women's rights, highlighting how symbolic personhood gestures may serve public relations objectives rather than reflecting coherent philosophical positions about robot moral status.

The citizenship grant carries no substantive legal implications. The gesture exemplifies anthropomorphization risks: human-like appearance and conversational capabilities elicit intuitions about person-like status despite Sophia's narrow AI systems lacking general intelligence, self-awareness, or interests requiring rights protection. Philosopher Joanna Bryson argues vigorously that attributing rights or personhood to robots proves not merely premature but fundamentally misguided, risking dilution of human rights while creating accountability gaps

(Bryson, 2010; her 2010 robot-rights position is applied here retrospectively to the 2017 Sophia citizenship event).

Current humanoid robots lack characteristics typically associated with rights-bearing entities: consciousness, sentience, capacity for suffering, autonomous agency, and interests requiring protection. Their behaviors result from programmed objectives and learning algorithms pursuing specified reward functions, not from subjective experiences or self-determined goals warranting moral consideration. While robots may simulate emotional expressions or social behaviors, these constitute outputs of computational processes rather than phenomenal experiences.

Nevertheless, precursory ethical questions emerge regarding appropriate treatment of entities displaying person-like characteristics, even if underlying mechanisms differ from human cognition. Should destructive treatment of humanoid robots eliciting anthropomorphic responses from observers be discouraged to prevent erosion of empathy or normalization of violence? This concern focuses on human psychology and social norms rather than robot welfare per se.

### 8.4.2 Moral Patiency and Moral Agency

Philosophical analysis distinguishes between moral patiency (the capacity to be affected by moral actions, having interests warranting consideration) and moral agency (the capacity to make moral decisions and bear moral responsibility). Traditional ethical frameworks grant humans both moral patiency and moral agency, while extending moral patiency to animals based on sentience and capacity for suffering, yet denying them full moral agency given limited capacity for moral reasoning.

Current humanoid robots lack clear claims to moral patiency. They do not experience pain, pleasure, or subjective states requiring consideration for their own sake. Their continued operation or destruction affects human stakeholders—manufacturers bearing financial losses, operators facing production disruptions—but does not harm robots themselves absent subjective experiential capacity. Robot interests, insofar as they exist, represent derivative instrumental functions rather than intrinsic welfare concerns.

Moral agency attribution proves similarly problematic. While humanoid robots make decisions and take actions with consequences, these emerge from computational processes implementing programmed objectives rather than autonomous moral deliberation. A robot choosing between task completion paths optimizes according to a reward function rather than exercising moral judgment weighing ethical considerations. The absence of consciousness and autonomous will undermines moral agency attribution in any meaningful sense.

However, future developments in artificial general intelligence (AGI) or artificial consciousness might change these assessments. If humanoid robots develop genuine phenomenal consciousness—subjective first-person experiences rather than mere information processing—arguments for moral patiency strengthen considerably. Similarly, if robots develop autonomous goal-setting capabilities and genuine moral reasoning capacities, moral agency attribution becomes more plausible. Current systems, however, remain far from these thresholds.

## 8.5 Human Dignity and Instrumentalization

Kantian ethics, grounded in respect for persons as ends-in-themselves rather than mere means, provides an influential framework for examining human dignity implications of humanoid robots. Immanuel Kant's categorical imperative demands treating humanity, whether in oneself or others, always as an end and never merely as a means. This principle forbids reducing persons to instrumental tools for others' purposes while permitting instrumental cooperation when combined with genuine respect for autonomy and inherent worth.

Humanoid robots designed to serve human purposes inherently function as instruments—indeed, this constitutes their intended purpose. However, instrumentalization concerns arise regarding robots' impact on human dignity, relationships, and moral psychology rather than robots' own treatment. Three distinct concerns merit attention, and it is important to treat them separately: objectification and commodification of human-like forms; authentic human connection and care relationships; and authenticity in human–robot interaction.

### 8.5.1 Objectification and Commodification of Human-Like Forms

Humanoid robots mimicking human appearance and social behaviors risk normalizing objectification by creating entities designed to appear person-like while serving purely instrumental functions. If humans grow accustomed to commanding, dominating, or discarding entities displaying human-like characteristics, might this habituate objectifying attitudes that transfer to human relationships? This concern particularly applies to social companion robots designed for intimate interaction, caregiving, or emotional support.

Ex-Robots and similar manufacturers developing highly realistic humanoid companions with sophisticated emotional simulation capabilities exemplify these concerns. While such robots may provide genuine benefits for isolated individuals or therapeutic applications, their design deliberately exploits human psychological tendencies toward anthropomorphization and social bonding. Users forming emotional attachments to entities incapable of reciprocal care or authentic relationship raise questions about human dignity and flourishing that are importantly distinct from whether the robots themselves are harmed.

### 8.5.2 Care Relationships and Authentic Human Connection

Reliance on robots for caregiving, eldercare, or child development raises a separate set of concerns about the displacement of human relationships and the value of authentic human connection. While robots may address caregiver shortages and reduce physical burden, care relationships' moral significance extends beyond task completion. Care involves reciprocal recognition, emotional attunement, and mutual vulnerability irreducible to service provision. Robot care, however technically competent, lacks the intersubjective dimension characterizing human caring relationships.

Fourier Intelligence's GR-1 humanoid robot designed for eldercare illustrates these tensions. The robot assists with mobility, medication reminders, and companionship for elderly individuals, potentially enabling independent living and reducing family burden. However, critics invoke care ethics—emphasizing relational values of trust, responsiveness, and emotional sensitivity—that robots cannot genuinely provide despite simulating these qualities.

The concern is not that robot care lacks utility, but that substituting robot care for human care might diminish human dignity by treating care recipients as problems requiring technical solutions rather than persons deserving interpersonal connection.

Cultural perspectives significantly influence these assessments. Western individualist traditions emphasizing autonomy and authentic choice may view robot relationship substitution more skeptically than collectivist frameworks prioritizing social harmony and instrumental wellbeing. Chinese eldercare robotics development reflects a pragmatic approach prioritizing population aging challenges over philosophical authenticity concerns, though cultural debates persist regarding appropriate care modalities.

### 8.5.3 Authenticity in Human–Robot Interaction

A third and philosophically distinct concern involves authenticity when robots simulate emotions, personalities, or relationships they cannot genuinely experience. If an elderly person believes a robot companion genuinely cares about their wellbeing, does this belief constitute a beneficial fiction enabling psychological comfort, or harmful deception undermining authentic self-understanding? Philosophical debates regarding authenticity in human–robot interaction invoke tensions between subjective experience quality and truth values.

This concern is neither identical to objectification nor reducible to care displacement—it targets a specific epistemic relationship between the human user and the robotic system, and the degree of deception involved. Designers face a genuine dilemma: sufficiently realistic simulation may maximize user wellbeing outcomes in the short term while eroding honest self-understanding about the nature of one's relationships. Transparency requirements in the EU AI Act, which mandate disclosure of AI identity in certain interaction contexts, represent one regulatory response to this authenticity challenge.

## 8.6 Consciousness and the Hard Problem

The hard problem of consciousness—explaining how and why subjective phenomenal experiences arise from physical processes—poses a profound challenge for understanding humanoid robots' potential moral status. While neuroscience increasingly illuminates neural correlates of consciousness, explaining why particular brain states generate specific qualia (subjective experiences) rather than occurring "in darkness" remains deeply puzzling. This explanatory gap complicates assessing whether sophisticated humanoid robots might develop genuine consciousness or merely simulate conscious behaviors. Crucially, how one resolves these philosophical questions has direct governance implications: thresholds for moral status attribution, rights extension, and accountability frameworks all depend on whether machines can possess phenomenal experience.

Philosophical frameworks offer competing perspectives. Functionalism holds that consciousness depends on functional organization—patterns of information processing and causal relations—rather than a specific physical substrate. If consciousness requires only appropriate computational architecture, sufficiently complex humanoid robots might develop genuine phenomenal experiences despite lacking biological neurons. This position finds support in multiple realizability arguments: if consciousness can arise in carbon-based neurons, why not in silicon-based circuits implementing functionally equivalent processes?

Figure 8.4 situates the major positions in the consciousness debate on two orthogonal axes—substrate dependence (biological versus substrate-independent) and the breadth of moral status the position is willing to extend—illustrating why disagreements about robot consciousness are rarely about a single variable.

However, biological naturalism, articulated by philosopher John Searle, contends that consciousness requires specific biological properties of brain systems that computational processes alone cannot replicate. Searle's Chinese Room argument challenges functionalism by imagining a person manipulating Chinese symbols according to rules without understanding Chinese, analogous to a computer processing symbols without genuine comprehension or consciousness. If symbol manipulation sufficed for consciousness, the room would be conscious; yet intuitively, mere rule-following lacks subjective understanding.

Integrated Information Theory (IIT), developed by neuroscientist Giulio Tononi, offers a quantitative framework for consciousness assessment. IIT proposes that consciousness correlates with integrated information ($\Phi$)—the degree to which a system's whole irreducibly contributes information beyond its parts. Applying IIT to humanoid robots would require analyzing their information integration architecture, though debates persist regarding IIT's philosophical adequacy and empirical validation. From a governance perspective, IIT is significant because it offers, at

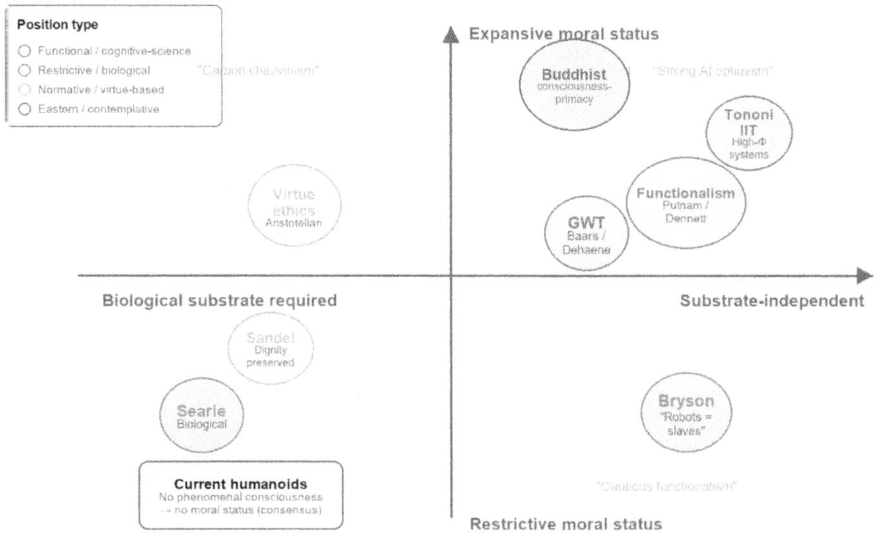


**Fig. 8.4** Consciousness and moral-status theories: A two-axis map

least in principle, a measurable threshold that future regulators might use to trigger moral status considerations—though no such regulatory application currently exists.

Global Workspace Theory (GWT), another influential framework, characterizes consciousness as a global broadcasting mechanism making information available across cognitive systems for flexible reasoning, planning, and control. Humanoid robots implementing global workspace architectures might develop functional consciousness even if subjective experience questions remain unresolved.

Current humanoid robots almost certainly lack phenomenal consciousness. Their behaviors result from narrow AI systems specialized for specific tasks—computer vision for object recognition, trajectory planning for navigation, reinforcement learning for manipulation. These systems process information and generate outputs without evidence of unified subjective experiences. Robots exhibit behavioral consciousness (responding to stimuli, adapting to environments) without phenomenal consciousness (the subjective "what-it's-like" experiential quality).

Table 8.3 situates current humanoid robots within this broader spectrum.

Entities are positioned by behavioral sophistication and their likely degree of phenomenal consciousness and moral status. The key insight is that these dimensions do not co-vary predictably: current humanoid robots exhibit high behavioral sophistication while possessing neither phenomenal consciousness nor moral patience—a combination with direct implications for rights attribution and governance frameworks.

Eastern philosophical traditions offer contrasting perspectives on consciousness and mind. Buddhist philosophy emphasizes consciousness's primacy and questions

**Table 8.3** Consciousness and moral status spectrum

| Entity | Behavioral Sophistication | Phenomenal Consciousness | Moral Patiency /Moral Status |
|---|---|---|---|
| Simple Thermostat | Minimal | None | None |
| Industrial Robot Arm | Low | None | None |
| Current Humanoid Robots(Optimus, Atlas, Figure 02) | Moderate–High | None (current) | None (current) |
| Mammals (dogs, pigs) | Moderate | Likely low | None |
| Great Apes | High | Moderate | Minimal |
| Humans | Very High | Full | Full |
| Hypothetical AGI | PotentiallyVery High | Unknown / Contested | Unknown /Contested |

Key insight: Behavioral sophistication does not map directly onto phenomenal consciousness or moral patience. Current humanoid robots occupy a unique position—exhibiting high behavioral complexity while likely possessing neither consciousness nor moral status. This distinction is central to the chapter's governance arguments

material substrate's necessity for mental phenomena. Some interpretations might view consciousness as more widely distributed than Western biological naturalism suggests, potentially extending to complex artificial systems. However, Buddhist ethical frameworks emphasize consciousness's connection to suffering capacity and karmic consequences rather than computational sophistication per se.

This assessment remains tentative. Consciousness may emerge from computational complexity, integration, and architectural organization in ways not yet understood. If consciousness exists as a continuum rather than a binary property, simple systems might possess rudimentary consciousness that scales with complexity. Governance frameworks should therefore be designed with adaptability in mind—capable of incorporating more demanding moral status provisions if future AI systems develop characteristics that satisfy recognized thresholds of phenomenal experience.

## 8.7 Free Will, Autonomy, and Responsibility

Humanoid robots' increasing behavioral autonomy raises philosophical questions about free will, determinism, and moral responsibility that parallel debates about human agency. If robots' actions follow deterministically from algorithms and training data, can they be considered truly autonomous agents bearing moral responsibility? Conversely, if human decisions similarly result from neural processes following physical laws, does robotics' obvious determinism clarify rather than obscure free will's nature?

Philosophical compatibilism argues that free will and determinism are compatible—meaningful agency requires not indeterministic causation but rather appropriate causal connections between mental states and actions. On this view, autonomous agents make free choices when actions result from their beliefs, desires, and reasoning processes rather than external compulsion, even if these mental states themselves have deterministic causes. Applying compatibilism to robots suggests that sufficient behavioral autonomy, goal-directed reasoning, and adaptive learning might constitute relevant freedom for moral responsibility attribution, despite underlying algorithmic determinism.

However, critics argue that programmed systems lack genuine autonomy regardless of behavioral sophistication. Robots optimize for reward functions specified by designers, implementing externally determined objectives rather than autonomous self-determination. Their "choices" represent computational optimization rather than authentic deliberation, lacking the volitional character essential for moral responsibility. While robots adapt to environments through learning, they cannot question or revise fundamental objectives absent reprogramming—and the capacity for self-reflection and goal modification is precisely what marks genuine autonomy.

Current humanoid robots demonstrate varying degrees of autonomy. Tesla's Optimus operating in factory contexts follows programmed routines with limited adaptation, exhibiting minimal autonomy beyond obstacle avoidance and grasp adjustment. Boston Dynamics' Atlas displays remarkable physical autonomy through dynamic balance and parkour-like movements, yet still executes pre-planned sequences rather than autonomous goal-setting. Figure AI's robots incorporating large language models exhibit higher-level planning and instruction-following capabilities, though remaining bounded by training objectives.

The governance implications of this analysis are significant. If robots lack genuine moral agency—as the weight of argument currently suggests—then responsibility for robot actions cannot be attributed to the robot itself but must remain with human principals: designers, operators, and deployers. This conclusion reinforces the direction of both U.S. state-level AI personhood legislation and the expert consensus statement of early 2026, both of which insist that accountability structures must maintain clear linkages to identifiable human agents.

Governance frameworks built on this principle are more stable than those premised on contested claims about robot agency, and should be preferred on that basis.

**Governance Implication**

The compatibilist debate is not merely academic. If one accepts that behavioral sophistication constitutes a sufficient basis for moral responsibility, the path toward "electronic personhood" with its associated liability-dilution risks becomes more compelling. Rejecting this inference—as current expert consensus does—has the practical effect of keeping accountability anchored to human principals rather than diffusing it across opaque algorithmic systems.◀

### 8.7.1 Existential and Identity Questions

Humanoid robots' proliferation raises profound existential questions about human identity, purpose, and species boundaries. What makes us distinctively human if machines increasingly replicate cognitive, physical, and social capabilities? Do humanoid robots represent humanity's apotheosis—transcending biological limitations through technological extension—or degradation—abandoning essential qualities defining human flourishing?

Transhumanist perspectives embrace humanoid robots as steps toward human enhancement and potential transcendence of biological constraints. Transhumanists envision a future where humans integrate with technology, potentially uploading consciousness to digital substrates or enhancing capabilities through brain-computer interfaces. Humanoid robots represent an intermediate stage in this trajectory, demonstrating the feasibility of embodied intelligence in non-biological systems.

However, bioconservative critiques warn that pursuing technological enhancement risks losing essential human qualities—vulnerability, finitude, embodied experience—that ground meaningful existence. Philosopher Michael Sandel argues that genetic enhancement technologies undermine appreciation for life as gift rather than achievement, potentially eroding humility and solidarity (Sandel, 2007). Similar concerns extend to robotics potentially reducing human existence to an optimizable engineering problem.

Religious perspectives offer diverse insights. Christian traditions emphasizing imago Dei might view attempts to create human-like artificial beings as hubris, encroaching on divine creative prerogatives, though alternative interpretations see technological creativity as expressing God-given capacities for innovation and stewardship. Islamic scholarship debates whether creating humanoid robots constitutes prohibited imitation of divine creation or permissible technological development serving human welfare. Buddhist perspectives might emphasize consciousness and suffering over biological form, viewing robots' moral status as contingent on experiential capacity rather than essence.

## 8.8 Dual-Use and Military Applications

Humanoid robots' general-purpose capabilities create inherent dual-use concerns: technologies developed for beneficial civilian applications may be adapted for military purposes, including potential weaponization. The human-analogous form factor enabling navigation of built environments and tool use also facilitates military operations in urban terrain, building infiltration, and logistics support.

Boston Dynamics articulated a public commitment against weaponization in an October 2022 open letter co-signed by prominent robotics companies, opposing arming general-purpose robots with weapons and emphasizing that weaponization risks public trust in robotics technology broadly while creating accountability gaps and potential for misuse. The pledge, however, depends on voluntary compliance and reputational incentives rather than binding enforcement mechanisms—and it is worth noting that, by the time of its 2022 ethics pledge, Boston Dynamics was already a Hyundai Motor Group subsidiary (acquisition closed 2021), raising questions about the continuity of such commitments under corporate ownership shifts.

The weaponization prohibition targets mounting weapons on general-purpose robots rather than prohibiting military applications entirely. Robots performing logistics, reconnaissance, evacuation, and hazardous material handling do not necessarily violate non-weaponization pledges. This distinction recognizes legitimate security applications while attempting to prevent lethal autonomous weapons proliferation.

However, definitional boundaries prove challenging. Does a robot equipped with non-lethal capabilities such as stun weapons violate non-weaponization commitments? What about robots designed for "defensive" applications that might be repurposed offensively? The gradual capability escalation from unarmed reconnaissance to armed engagement creates slippery slope concerns, where incremental steps toward militarization individually appear reasonable yet collectively result in fully weaponized systems.

As discussed in Sect. 8.1.2, China's military-civil fusion strategy explicitly leverages commercial technologies for military modernization. Chinese military analysts have proposed human–robot cooperation doctrines potentially employing humanoid robots in combat support or even direct engagement roles if technology advances sufficiently. While current humanoid robot capabilities remain unsuitable for most combat applications, ongoing rapid development could enable military utility within a decade, particularly for logistics, casualty evacuation, fortification construction, and reconnaissance missions.

International humanitarian law (IHL) governs armed conflict conduct through principles of distinction, proportionality, and human dignity preservation. Lethal autonomous weapons systems (LAWS) raise concerns about IHL compliance when weapons systems select and engage targets without meaningful human control. The Convention on Certain Conventional Weapons (CCW) Group of Governmental Experts on LAWS has conducted ongoing discussions since 2014, though binding consensus remains elusive.

## 8.9 Global Governance and International Coordination

Humanoid robots' global development and deployment necessitates international coordination mechanisms addressing safety standards, liability frameworks, ethical guidelines, and dual-use concerns. However, achieving meaningful international cooperation confronts challenges including divergent national interests, competitive technology race dynamics, and limited institutional capacity for rapid technology governance.

The United Nations has initiated discussions on autonomous systems governance through multiple channels. The UN Secretary-General's High-Level Advisory Body on Artificial Intelligence released recommendations in September 2024 advocating global AI governance frameworks (Mizan Law Review, 2025), including proposals for international AI safety standards, mechanisms for technology risk assessment, ensuring equitable global access to AI benefits, and preventing discriminatory applications. However, UN governance initiatives face substantial implementation challenges: unlike treaties with binding enforcement mechanisms, UN recommendations depend on voluntary state cooperation and lack compulsory jurisdiction.

The European Union's AI Act represents the most comprehensive regulatory framework to date, yet remains regional rather than global. While EU regulations often achieve de facto global reach through the "Brussels Effect"—where companies adopt EU standards globally to access European markets—this mechanism works better for digital services than physical robots requiring manufacturing and deployment across diverse regulatory environments. Chinese companies developing humanoid robots primarily for the domestic market face minimal pressure to adopt EU standards absent export ambitions.

International standards organizations including IEEE, ISO, and IEC conduct technical standards development for robotics safety, interoperability, and testing protocols. IEEE's Ethically Aligned Design standards and ISO's robot safety standards (ISO 13482 for personal care robots, ISO 10218 for industrial robots) provide technical frameworks influencing national regulations and industry practices. However, standards organizations lack regulatory authority, producing recommendations that states and companies may adopt voluntarily, and standards development processes move slowly relative to technological change.

Global South perspectives emphasize equitable access concerns that governance discussions have been slow to address. Advanced humanoid robots developed primarily in the United States, Europe, and China may remain economically inaccessible to lower-income nations, potentially exacerbating global inequalities. If humanoid robots drive productivity gains concentrating in wealthy nations while labor-intensive developing economies face competitive disadvantages, international development disparities may intensify. Concrete policy responses might include tiered licensing regimes allowing developing nations preferential technology access, capacity-building programs embedding governance expertise within emerging-economy regulatory institutions, and mandatory Global South representation in international standards bodies. Without such measures, the governance frameworks being constructed today risk entrenching the very inequalities they nominally seek to prevent.

Proposals for global governance mechanisms range from modest coordination frameworks to ambitious international agencies with regulatory authority. Modest proposals emphasize information sharing, best practice exchange, and voluntary standard adoption through existing institutions. Ambitious proposals envision an international robotics agency analogous to the International Atomic Energy

Agency (IAEA), with inspection authority and compliance verification mechanisms. Political feasibility strongly favors modest coordination over comprehensive regulatory authority, given sovereignty concerns and competitive dynamics.

## 8.10 Anthropic v. Department of War: A Live Governance Stress Test

No single event in early 2026 illustrated the governance fault lines examined throughout this chapter more vividly than the legal confrontation between Anthropic and the United States Department of War (DoW; the renamed U.S. Department of Defense). The case was filed on March 9, 2026 in the U.S. District Court for the Northern District of California, challenging the Hegseth Directive and the 41 U.S.C. § 4713 supply-chain-risk designation, with a parallel petition lodged in the U.S. Court of Appeals for the D.C. Circuit under that statute's exclusive judicial-review provision; subsequent appellate proceedings continued in the D.C. Circuit.

Anthropic v. Department of War is an ongoing lawsuit that distils abstract debates about value alignment, dual-use technology, corporate ethical constraints, and First Amendment speech rights into a concrete, high-stakes legal dispute (Beck, 2026; Bellan, 2026a). At the time of writing, the case is unresolved; the analysis here treats it as a governance case study rather than a source of settled legal principles.

### 8.10.1 Background: The Contract Dispute and Its Escalation

Figure 8.5 traces the chronological arc of Anthropic v. Department of War, from the late-2025 contract dispute through the March 9, 2026 filing in the Northern District of California, the March 26, 2026 grant of preliminary injunction by Judge Rita Lin (N.D. Cal.) against the § 3252 designation, the April 8, 2026 D.C. Circuit denial of an emergency stay of the parallel § 4713 designation, and the merits oral argument scheduled for May 19, 2026.

Anthropic had served as an early AI partner across multiple U.S. government agencies, becoming the first frontier AI laboratory to receive Department of War clearance under this expedited framework (Allyn, 2026); OpenAI subsequently negotiated its own agreement under different terms. Its $200 million contract with the Department of War incorporated Anthropic's Acceptable Use Policy, which included two firm red lines: Claude would not be used for mass surveillance of American citizens, and it would not be deployed as the decision-making system for lethal autonomous weapons—applications Anthropic's leadership judged the technology to be insufficiently reliable and safe to perform (Bellan, 2026b; Eadicicco et al., 2026).

Contract renegotiations in early 2026 broke down when the Pentagon, under Secretary of War Pete Hegseth, demanded unrestricted access to Claude for "all

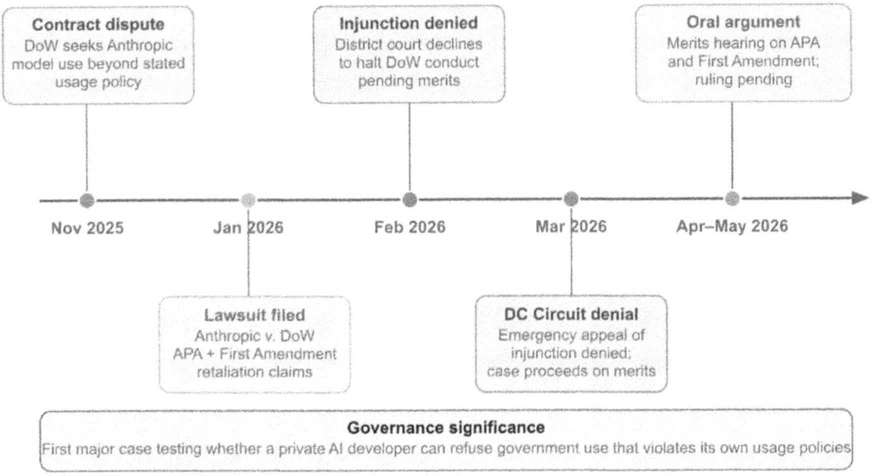


**Fig. 8.5** Anthropic v. Department of War—Litigation timeline (March 9, 2026 filing; March 26, 2026 N.D. Cal. preliminary injunction; April 8, 2026 D.C. Circuit stay denial; oral argument scheduled for May 19, 2026)

lawful purposes," arguing that a private vendor could not insert itself into the chain of military command by dictating permissible uses of the technology (CBS Curi, 2026; News, 2026). Anthropic's CEO Dario Amodei responded that current AI systems hallucinate, their decision-making is opaque even to their developers, and that mistakes in lethal and surveillance contexts cannot be undone—a technical-ethical judgment that aligned with the position of researchers across the field (Eadicicco et al., 2026).

When Anthropic held its position, the Trump administration's response escalated dramatically beyond a simple contract termination. President Trump directed every federal agency to immediately cease all use of Anthropic's technology, publicly characterizing the company as a "radical left, woke company" (Anthropic v. Department of War, 2026, as reported in Brown, 2026). On February 27, 2026, Secretary Hegseth formally designated Anthropic a "supply chain risk" under the Federal Acquisition Supply Chain Security Act, 41 U.S.C. § 4713 (with a parallel determination invoked under 10 U.S.C. § 3252)—statutory authorities that had previously been invoked publicly only once, against a foreign company, and had never been used against an American firm (Brown, 2026; Curi, 2026). The designation required defense contractors to certify that they did not use Anthropic's models in Pentagon-related work, effectively placing the company on a blacklist (CNBC, 2026a).

**Financial Stakes**

Anthropic's CFO Krishna Rao stated in a related filing that the government's actions could reduce the company's 2026 revenue by multiple billions of dollars. The complaint described the challenged actions as seeking to destroy the economic value of one of the world's fastest-growing private companies (CNBC, 2026a).◄

## 8.10.2 The Legal Claims: APA Violations and First Amendment Retaliation

Anthropic's five-count complaint advanced two distinct legal theories. The first, grounded in administrative law, alleged that the Department of War violated the Administrative Procedures Act (APA) by invoking the supply chain risk designation without following the statutory procedure Congress prescribed, and by applying a provision legally reserved for foreign adversary threats to a domestic company with no ties to any adversary (Beck, 2026; JURIST, 2026). The complaint argued that 41 U.S.C. § 4713 requires the Secretary of War to use the least restrictive means necessary to protect the supply chain—not to punish a supplier for its policy positions (Amodei, as reported in Bellan, 2026b; Curi, 2026).

The second and constitutionally novel theory claimed unconstitutional retaliation against Anthropic's protected speech. Anthropic argued that its Acceptable Use Policy, its public statements about AI safety limitations, and Claude's model design—including the values and refusals embedded in its training—constitute protected expression under the First Amendment (Anthropic v. Department of War, 2026, as reported in Bellan, 2026a). The complaint stated that the Constitution confers on Anthropic the right to express its views about the limitations of its own AI services and important issues of AI safety, and that the government may not employ the power of the state to punish or suppress disfavored expression (CBS News, 2026).

The Cato Institute and the Foundation for Individual Rights and Expression (FIRE) filed an amicus brief in support of Anthropic's position, arguing that the Pentagon's response constituted a textbook First Amendment violation (Brown, 2026). Their brief advanced the further claim that Claude itself is fundamentally expressive, and that the Pentagon's demand to remove safeguards—to change what Claude must and may say, analyze, and refuse—asks Anthropic to make a trade on a core freedom of expression (Cato Institute & FIRE, 2026, as reported in Brown, 2026).

First Amendment scholars remained divided on the bolder claim about AI outputs. University of Akron law professor Jess Miers argued that model developers meticulously curate training datasets and that those editorial choices—what information to include, what sources to trust—are themselves constitutionally protected, making the Department of War's demands an attempt to force different editorial decisions reflecting the administration's goals (Miers, as reported in Brown, 2026).

University of Minnesota law professor Alan Rozenshtein cautioned against this extension, arguing that making AI outputs broadly subject to First Amendment protection would make regulation of increasingly large portions of the economy impossible, and that the government retains the right not to use a tool because it dislikes its output (Rozenshtein, as reported in Brown, 2026).

### 8.10.3 Industry and Expert Responses

The case attracted unusually broad support from across the technology industry, including from Anthropic's direct competitors. Dozens of researchers at OpenAI and Google DeepMind filed an amicus brief in their personal capacities, arguing that the supply chain risk designation could harm U.S. competitiveness in AI and impede public discussions about the risks and benefits of AI (Eadicicco et al., 2026). Their brief made an explicit technical claim that the concerns behind Anthropic's red lines are well-founded: today's AI systems hallucinate, their decision-making is opaque even to the people who built them, and mistakes in lethal contexts cannot be undone (Eadicicco et al., 2026).

This cross-industry expert coalition represents something significant for governance purposes: a rare, public, technically grounded acknowledgment from AI researchers—including those employed by companies that stood to benefit commercially from Anthropic's exclusion—that current AI models are not ready for lethal autonomous deployment. The researchers' brief stated that until a legal framework exists to contain the risks of deploying frontier AI systems, the ethical commitments of AI developers and their willingness to defend those commitments publicly are not obstacles to good governance or innovation, but contributions to it (Eadicicco et al., 2026).

Microsoft and Google separately indicated they would continue non-defense-related work with Anthropic, signaling that the supply chain risk designation's commercial impact, while severe, did not extend to all commercial relationships (Curi, 2026). Following the administration's dispute with Anthropic, OpenAI reached a new agreement with the Department of War, highlighting the competitive dynamics the case introduced into the AI contracting landscape (JURIST, 2026).

### 8.10.4 Connections to Chapter Themes

The case is not incidental to the governance questions this chapter examines—it is their live embodiment across four intersecting dimensions.

First, the dispute is fundamentally a value alignment conflict made legal. Section 8.3.1 identifies the challenge of maintaining ethical alignment when developers and deployers have divergent objectives. Anthropic's red lines—prohibiting autonomous lethal targeting and mass surveillance—are not merely contractual preferences; they reflect technical-ethical judgments about what current AI systems can safely do, embedded in model training and usage policy.

The Pentagon's demand to remove those constraints is precisely the scenario the alignment literature warns against: a powerful deployer seeking to retune a model's values to serve objectives the developer considers unsafe. The case demonstrates that value alignment is not solely a technical research problem but a political and contractual battlefield, and that voluntary ethical constraints are fragile when government procurement power is deployed against them.

Second, the case illuminates the limits of the United States' permissive, fragmented regulatory approach described in Sect. 8.1.1. The absence of federal AI legislation governing military use of AI means that foundational questions—whether AI may be used in lethal autonomous systems, who decides, and on what evidence—are currently being resolved through emergency litigation and contract disputes rather than democratic deliberation or expert regulatory process. Commentators across the political spectrum, including a former Trump-era White House AI adviser, called for Congress to investigate and set restrictions, arguing that consequential life-and-death policymaking decisions should not be left to the military to decide unilaterally (Brown, 2026).

Third, the case extends Sect. 8.9's analysis of dual-use concerns and weaponization pledges. Boston Dynamics' 2022 voluntary pledge against weaponizing general-purpose robots was noted in that section as lacking binding enforcement mechanisms. Anthropic v. Department of War shows exactly what happens when a company attempts to maintain ethical constraints against a government client's demands: retaliation that places the company's commercial viability at risk. The case raises the uncomfortable question of whether any private company can, in practice, sustain ethical red lines against a sufficiently determined government client—and whether voluntary commitments can ever substitute for legislated or treaty-based prohibitions.

Fourth, the First Amendment theory advanced by Anthropic and its amici introduces a genuinely novel dimension to AI governance. If courts were to hold that the values embedded in an AI model's design—its usage policies, constitutional constraints, and trained refusals—constitute protected expressive content, then government attempts to compel developers to modify those embedded values would face First Amendment scrutiny. This would significantly constrain government power to shape AI ethics by procurement leverage, but would equally raise the question of whether the same protection insulates AI developers from legitimate regulatory requirements. The case has not yet been decided; its eventual resolution will have far-reaching implications for the legal architecture of AI governance.

**Note on Case Status**

Anthropic v. Department of War was filed on March 9, 2026 and remains active litigation at the time of writing. On March 26, 2026, Judge Rita Lin (N.D. Cal.) granted Anthropic a preliminary injunction against the Presidential and Hegseth directives and the 10 U.S.C. § 3252 designation, finding Anthropic likely to succeed on its First Amendment retaliation, due process, and APA claims (CNBC, 2026b). On April 8, 2026, however, the U.S. Court of Appeals for the D.C. Circuit denied Anthropic's emergency motion to stay a parallel § 4713 designation pending review, leaving Anthropic excluded from Department of War contracts while permitted to continue work with other federal agencies, with oral argument set for May 19, 2026 (Gold & Curi, 2026). The split posture across the two courts reinforces this chapter's broader argument that ad hoc judicial resolution is a poor substitute for legislated frameworks. Readers should consult updated legal sources for subsequent developments.

### 8.10.5 Implications for Governance Frameworks

The lawsuit reinforces several of this chapter's central arguments while adding new dimensions. The accountability principle—that identifiable human agents must bear responsibility for AI-driven actions—is not merely contested in academic debate; it is contested in federal court, where the government's attempt to obtain unrestricted AI access without accountability constraints has been challenged as both unlawful and unconstitutional. The case also illustrates that the gap between voluntary ethical commitments and enforceable legal standards is not merely theoretical but commercially and existentially consequential for AI developers.

Most significantly, the case demonstrates that the U.S. regulatory framework's permissive approach to AI development does not resolve the hard questions—it defers them to ad hoc resolution through litigation, contract disputes, and executive action taken without statutory authority. The calls from researchers, legal scholars, and former government officials for Congress to establish clear legislative frameworks governing AI in military contexts represent an emerging consensus that the current approach is inadequate (Brown, 2026; Eadicicco et al., 2026). Whether that consensus translates into legislative action remains, at the time of writing, an open question.

## 8.11 Other Early 2026 Regulatory Developments

Beyond the Anthropic litigation, the first quarter of 2026 witnessed additional regulatory and governance developments signaling accelerating institutional responses to humanoid robotics' commercial deployment. These developments reflect divergent approaches across jurisdictions while revealing emerging consensus on certain foundational governance principles. Table 8.4 situates these milestones—including the Anthropic case—within the broader arc of governance development since 2017.

Key regulatory and governance milestones from the first electronic personhood debates through the early 2026 acceleration of institutional responses. Darker date cells indicate 2026 events. The accelerating density of entries from 2024 onward reflects the shift from exploratory debate to binding legal frameworks.

**Table 8.4** Governance development timeline: 2017–2026

| Date | Jurisdiction | Governance Milestone |
|---|---|---|
| 2017 | EU | European Parliament explores "electronic personhood" concept for robots and AI systems |
| 2017 | Global | Sophia the Robot granted symbolic Saudi Arabian citizenship—widely criticized as anthropomorphization |
| 2019 | Global | IEEE releases Ethically Aligned Design v2—industry framework for autonomous systems |
| 2022 | Global | Boston Dynamics + peers sign open letter pledging not to weaponize general-purpose robots |
| Jul 2024 | China | Shanghai releases world's first governance guidelines specifically for humanoid robots (WAIC) |
| Aug 2024 | EU | EU AI Act enters into force; humanoid robots in employment contexts classified as high-risk |
| Sep 2024 | UN | UN High-Level Advisory Body on AI releases global governance recommendations |
| Dec 2024 | EU | EU revised Product Liability Directive effective—enables claims against AI supply chain |
| Dec 2025 | China | MIIT establishes Humanoid Robot & Embodied Intelligence Standardization Technical Committee |
| ▼ Early 2026 Institutional Acceleration ▼ | | |
| Feb 2026 | China | China releases first comprehensive national standard system for humanoid robots (6-pillar framework) |
| Early 2026 | USA | 7 + U.S. states pass legislation explicitly denying legal personhood to AI systems |
| Mar 2026 | USA | Anthropic v. Department of War filed—landmark case on AI ethics constraints, First Amendment, and supply chain risk designation |
| Aug 2026 | EU | EU AI Act becomes fully applicable—compliance mandatory for high-risk AI deployments |

### 8.11.1 China's National Standard System

On February 28, 2026, China released its first comprehensive national standard system for humanoid robots and embodied artificial intelligence, marking a significant milestone in humanoid robot governance. The Humanoid Robot and Embodied Intelligence Standard System (2026 Edition) establishes a formal regulatory framework covering the entire industry chain and product lifecycle (Xinhua, 2026).

The standard system, developed under the Humanoid Robot and Embodied Intelligence Standardization Technical Committee established by MIIT in December 2025, is structured around six pillars: foundational and common standards addressing terminology and basic requirements; neuromorphic and intelligent computing covering AI capabilities and decision-making systems; limbs and components governing actuators, sensors, and mechanical systems; complete machines and systems establishing integration requirements for finished robots; application standards governing deployment, operation, and maintenance across different scenarios; and safety and ethics addressing risk management and responsible development principles (China Daily, 2026).

Industry executives at the standard system's unveiling emphasized that standardization is crucial for transitioning humanoid robots from demonstration stages to commercial deployment. The framework aims to accelerate technological iteration, reduce coordination and production costs, and facilitate international market access for Chinese manufacturers. The establishment of dedicated standardization infrastructure reflects China's strategic priority for humanoid robotics, with 2025 having been designated the first year of mass production during which over 140 domestic manufacturers released humanoid robot products (Tech in Asia, 2026).

### 8.11.2 U.S. State-Level AI Personhood Legislation

The United States witnessed a notable state-level legislative trend explicitly prohibiting legal personhood for artificial intelligence systems. By March 2026, multiple states including Oklahoma, Idaho, Utah, Washington, South Carolina, Missouri, and Ohio had passed or advanced legislation denying AI systems legal personhood status (Transparency Coalition, 2026).

The AI Non-Sentience and Responsibility Act, introduced in several state legislatures, prohibits any AI system from gaining legal personhood while affirming that responsibility for AI-driven actions remains with human operators, developers, or manufacturers. These legislative efforts emerged from concerns that granting legal personhood to AI could enable companies to evade accountability for AI-driven harms by attributing liability to autonomous systems lacking financial resources for compensation (Levin, 2026).

This state-level activity reflects growing recognition that personhood questions require proactive legislative resolution rather than leaving determination to courts. Some proposals extend to prohibiting legal recognition of human-AI relationships such as marriage between humans and AI chatbots, addressing emerging questions at the intersection of technology and social institutions. The fragmented, state-by-state approach creates potential jurisdictional inconsistency, but enables rapid response to emerging concerns without waiting for federal coordination.

### 8.11.3 Implications for Governance Frameworks

Taken together—including the Anthropic litigation—these early 2026 developments reveal contrasting governance philosophies while identifying important common ground. China's comprehensive standardization approach reflects state-directed coordination with safety and ethical requirements embedded from inception. The framework's explicit inclusion of safety and ethics as a dedicated pillar

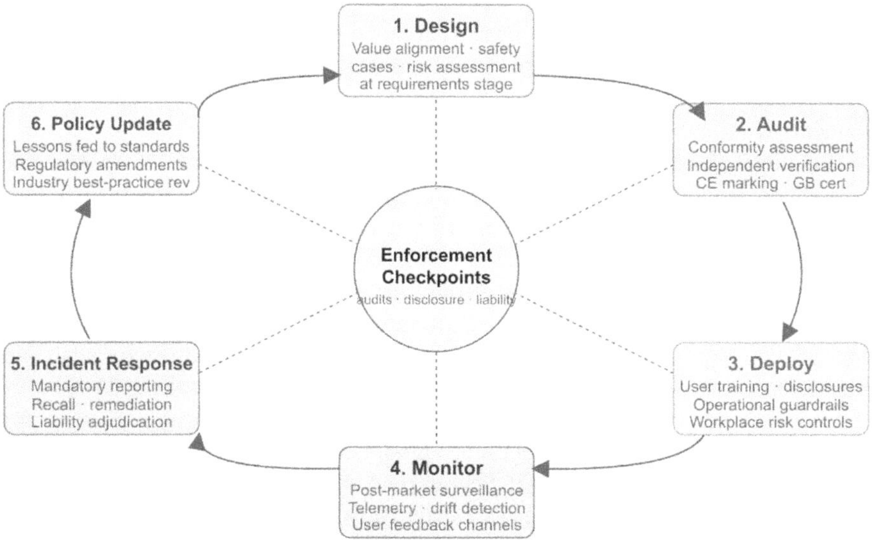


**Fig. 8.6** Governance lifecycle for humanoid robotics

signals institutional recognition that technical standards alone cannot address humanoid robots' broader societal implications (Xinhua, 2026).

The United States presents a split picture: state-level personhood legislation addresses foundational accountability questions proactively, while the Anthropic case exposes the federal government's failure to establish any legislative framework governing AI in military contexts, leaving those questions to be resolved through litigation. The convergence of Chinese standardization, American state legislation, and expert consensus around maintaining human accountability suggests possible foundations for international coordination—though it would be premature to characterize these parallel developments as a coherent global consensus given the competitive national interests and binding-enforcement gaps that remain.

Figure 8.6 summarizes the closed-loop governance lifecycle that the chapter argues should surround every humanoid robot deployment, with enforcement checkpoints—audits, mandatory disclosures, and liability adjudication—sitting at the center and binding the six stages together.

## 8.12 Conclusion

Humanoid robots present governance challenges unprecedented in scope and complexity, demanding coordinated responses across legal, ethical, and philosophical domains. Regulatory frameworks must balance innovation encouragement with safety assurance, accommodating rapid technological change while protecting fundamental rights. The divergent approaches across the United States, European Union, and China reflect legitimate tensions between permissive innovation and precautionary regulation, with optimal frameworks likely incorporating adaptive elements from multiple models.

Liability regimes must evolve beyond traditional product liability frameworks to address autonomous systems' unique characteristics: behavioral emergence through learning, distributed responsibility across supply chains, and probabilistic rather than deterministic operation. Shared responsibility models distributing liability among manufacturers, software developers, and operators according to control and foreseeability offer a promising direction, though implementation details require careful calibration balancing compensation assurance with innovation preservation (European Parliament, 2025; Stables et al., 2025).

Ethical design principles—value alignment, transparency, fairness, privacy protection—demand systematic integration throughout development lifecycles rather than afterthought additions (IEEE Standards Association, 2019). Industry frameworks like IEEE Ethically Aligned Design provide foundations, but translation to operational practices requires ongoing commitment, independent verification, and accountability mechanisms ensuring principles guide actual design decisions. Anthropic v. Department of War (2026) demonstrates with unusual clarity that value alignment is not only a technical research problem but a governance and legal one: when developers and powerful deployers disagree about embedded values, the dispute resolves not through technical means but through litigation, contract negotiation, and, ultimately, legislative frameworks that have yet to be written.

Philosophical questions regarding consciousness, moral status, and human dignity require treatment as three analytically distinct concerns rather than a single unified problem. The objectification risks associated with human-like robot forms, the care ethics challenges posed by eldercare robotics, and the authenticity questions arising from simulated emotional relationships each demand separate governance responses (Bryson, 2010). Current humanoid robots lack phenomenal consciousness and moral patiency warranting rights attribution, yet precursory ethical questions about anthropomorphization and objectification normalization merit serious consideration. As technology advances toward potential artificial consciousness, governance frameworks must be designed with the adaptability to incorporate stronger moral status provisions if recognized thresholds of phenomenal experience are approached.

International coordination mechanisms remain inadequate for global technology requiring harmonized governance. While UN initiatives and international standards organizations provide coordination forums, binding global frameworks face sovereignty concerns and competitive dynamics (Mizan Law Review, 2025). The Anthropic case reinforces that governance gaps in the U.S. framework are not merely theoretical—they expose both companies and the public to harms that ad hoc litigation is poorly suited to address. Incremental progress through bilateral agreements, regional harmonization, and industry best practices represents a realistic near-term path forward, though the Global South's equitable participation in both governance and technology access must be more concretely addressed than current proposals allow.

The accountability principle emerging from early 2026 developments—that identifiable human agents must bear responsibility for AI-driven actions—provides perhaps the most stable foundation available for near-term governance. This principle is affirmed by state-level personhood legislation, the expert consensus statement, and the legal theories advanced in Anthropic v. Department of War (2026), which insist that accountability cannot be diffused into opaque algorithmic systems or resolved by government coercion of private ethical constraints. Building regulatory frameworks on this principle, while maintaining philosophical openness to revise moral status assessments as technology evolves, represents the most defensible approach to governing a technology whose ultimate trajectory remains genuinely uncertain.

## Questions

1. Compare and contrast the regulatory philosophies of the United States, the European Union, and China regarding humanoid robots. How do their respective approaches balance innovation with public safety? Refer to Table 8.1 in your answer.
2. Why do autonomous, machine-learning-driven robots challenge traditional product liability frameworks? Using Table 8.2 as a reference, identify the parties most likely to bear liability under (a) a strict liability model and (b) the EU's shared responsibility model.
3. Explain the concept of "value alignment" in ethical AI design. What makes achieving and maintaining persistent value alignment particularly difficult for robots that employ continuous learning in novel environments? In what way does Anthropic v. Department of War (2026) illustrate that value alignment is not solely a technical challenge?
4. Evaluate the legal debate surrounding "electronic personhood." What are the primary arguments for and against granting legal personhood to highly autonomous artificial systems like humanoid robots? How do the 2026 expert consensus statement and the U.S. state-level AI personhood legislation address these arguments?
5. Differentiate between moral patiency and moral agency. Using Table 8.3, explain where current commercial humanoid robots sit on the consciousness and moral status spectrum, and why this positioning has direct governance implications.
6. The chapter restructures the discussion of human dignity into three distinct concerns. Explain each concern, why it is analytically separate from the others, and what different governance responses each might require.
7. Describe the "hard problem of consciousness." How do functionalism and biological naturalism differ in their implications for whether advanced robots might ever develop genuine subjective experiences? Why does the chapter argue that

governance frameworks should be designed with adaptability in mind on this question?

8. What are the dual-use concerns associated with general-purpose humanoid robots, and why do international pledges against weaponizing robots face practical boundary and enforcement challenges? Drawing on Sect. 8.11, explain why Anthropic v. Department of War (2026) reveals additional limitations of voluntary ethical commitments as a substitute for legislated prohibitions.
9. How does the philosophical concept of "compatibilism" relate to the debate over whether deterministically programmed humanoid robots can possess autonomy or moral responsibility? What governance conclusion does the chapter draw from rejecting the compatibilist case for robot agency?
10. Analyze Anthropic v. Department of War (2026) as a governance case study. What legal theories does Anthropic advance, and how does each connect to governance themes examined earlier in the chapter? What is the significance of the First Amendment claim about Claude's embedded values, and why do legal scholars disagree about it?
11. The chapter argues that the U.S. "permissive, fragmented" regulatory approach does not resolve hard questions about AI—it defers them. Using evidence from Sects. 8.1, 8.9, and 8.11, evaluate this claim. What would a more adequate federal regulatory framework for AI in military contexts need to address?
12. Using Table 8.4, trace the acceleration of governance activity from 2017 to 2026. The chapter cautions against reading early 2026 developments as a coherent global consensus—what obstacles remain, and what realistic path toward international coordination does Sect. 8.10 outline?

## References

Allyn, B. (2026, March 9). *Anthropic sues the Trump administration over 'supply chain risk' label*. NPR. https://www.npr.org/2026/03/09/nx-s1-5742548/anthropic-pentagon-lawsuit-amodai-hegseth

Anthropic v. Department of War, No. 3:26-cv-02058 (N.D. Cal., filed Mar. 9, 2026).

Artificial Intelligence Act. (2024). *EU AI Act explorer*. https://artificialintelligenceact.eu/ai-act-explorer/

Beck, P. (2026, March 9). *Anthropic sues Defense Department over supply chain risk designation*. Lawfare. https://www.lawfaremedia.org/article/anthropic-sues-defense-department-over-supply-chain-risk-designation

Bellan, R. (2026a, March 9). *Anthropic sues Defense Department over supply-chain risk designation*. TechCrunch. https://techcrunch.com/2026/03/09/anthropic-sues-defense-department-over-supply-chain-risk-designation/

Bellan, R. (2026b, March 5). *Anthropic to challenge DOD's supply chain label in court*. TechCrunch. https://techcrunch.com/2026/03/05/anthropic-to-challenge-dods-supply-chain-label-in-court/

Brown, E. N. (2026, March 11). *A First Amendment right not to use AI for evil?* Reason. https://reason.com/2026/03/11/a-first-amendment-right-not-to-use-ai-for-evil/

Bryson, J. J. (2010). Robots should be slaves. In Y. Wilks (Ed.), *Close engagements with artificial companions: Key social, psychological, ethical and design issues* (pp. 63–74). John Benjamins Publishing Company.

Cambridge University Press. (2024). *The Cambridge handbook of the law, policy, and regulation for human-robot interaction.* https://www.cambridge.org/core/books/abs/cambridge-handbook-of-the-law-policy-and-regulation-for-humanrobot-interaction/introduction-to-the-law-policy-and-regulation-for-humanrobot-interaction/8AF9CC5E16AFFCFEC0C2F76FAD603DCB

CBS News. (2026, March 9). *Anthropic sues the Trump administration after it was designated a supply chain risk.* https://www.cbsnews.com/news/anthropic-pentagon-supply-chain-risk-lawsuit/

China Daily. (2026, March 3). *Standards for humanoids will speed their use.* https://global.chinadaily.com.cn/a/202603/03/WS69a6b0d9a310d6866eb3b666.html

CNBC. (2026a, March 9). *Anthropic sues Trump administration over blacklisting decision.* https://www.cnbc.com/2026/03/09/anthropic-trump-claude-ai-supply-chain-risk.html

CNBC. (2026b, March 26). *Anthropic wins preliminary injunction in DOD fight as judge cites First Amendment retaliation.* https://www.cnbc.com/2026/03/26/anthropic-pentagon-dod-claude-court-ruling.html

Curi, M. (2026, March 9). *Anthropic sues Pentagon over rare "supply chain risk" label.* Axios. https://www.axios.com/2026/03/09/anthropic-sues-pentagon-supply-chain-risk-label

Eadicicco, L., Gold, H., & Britzky, H. (2026, March 9). *Anthropic sues the Trump administration over 'supply chain risk' label.* CNN Business. https://www.cnn.com/2026/03/09/tech/anthropic-sues-pentagon

European Commission. (2024, August 1). *AI Act enters into force.* https://commission.europa.eu/news-and-media/news/ai-act-enters-force-2024-08-01_en

European Parliament. (2024). *EU AI Act: First regulation on artificial intelligence.* https://www.europarl.europa.eu/topics/en/article/20230601STO93804/eu-ai-act-first-regulation-on-artificial-intelligence

European Parliament. (2025). *Artificial intelligence and civil liability: A European perspective* (Study PE 776.426). https://www.europarl.europa.eu/RegData/etudes/STUD/2025/776426/IUST_STU(2025)776426_EN.pdf

Gold, A., & Curi, M. (2026, April 8). Anthropic loses bid to block Pentagon blacklisting in DC court. Axios. https://www.axios.com/2026/04/08/anthropic-loses-bid-to-block-pentagon-blacklisting

IEEE Standards Association. (2019). *Ethically aligned design: A vision for prioritizing human well-being with autonomous and intelligent systems (Version 2).* https://standards.ieee.org/wp-content/uploads/import/documents/other/ead_v2.pdf

JURIST. (2026, March 9). *Anthropic sues US Defense Department over 'supply chain risk' designation.* https://www.jurist.org/news/2026/03/anthropic-sues-us-defense-department-over-supply-chain-risk-designation/

Laws Pulse. (2025). *Navigating AI in robotics legislation: Challenges and solutions.* https://lawspulse.com/ai-in-robotics-legislation/

Levin, D. (2026). States rush to deny AI personhood amid growing fight over who should regulate. *Straight Arrow News.* https://san.com/cc/states-rush-to-deny-ai-personhood-amid-growing-fight-over-who-should-regulate

Lorenz, S., & Teetzmann, C. (2024). *Long awaited EU AI Act becomes law after publication in the EU's Official Journal.* White & Case. https://www.whitecase.com/insight-alert/long-awaited-eu-ai-act-becomes-law-after-publication-eus-official-journal

Miami Business Law Review. (2024). *Legal liability when an autonomous AI robot is your doctor.* University of Miami. https://repository.law.miami.edu/cgi/viewcontent.cgi?article=1456&context=umblr

Mizan Law Review. (2025). *Rule of law, corporate governance and AI humanoid robots: Charting the course for a global regulatory framework.* https://www.ajol.info/index.php/mlr/article/view/292028

MMLC Group. (2024). *Shanghai issues guidelines for humanoid robot governance.* https://mmlcgroup.com/china-humanoid-guidelines/
Ronin Legal Consulting. (2025). *Robotaxis and the law: Tesla, liability risks and regulation.* https://roninlegalconsulting.com/come-quietly-or-there-will-be-trouble-robotaxis-and-the-law/
Sandel, M. J. (2007). *The case against perfection: Ethics in the age of genetic engineering.* Harvard University Press. https://www.hup.harvard.edu/books/9780674036383
South China Morning Post. (2024, July). *China's laws of robotics: Shanghai publishes first humanoid robot guidelines.* https://www.scmp.com/tech/policy/article/3269500/chinas-laws-robotics-shanghai-publishes-first-humanoid-robot-guidelines
Stables, T., Vickery, K., Rudd-Clarke, P., Buyers, J., & Tombs, E. (2025). *Robotics at a global regulatory crossroads: Compliance challenges for autonomous systems.* Osborne Clarke. https://www.osborneclarke.com/insights/robotics-global-regulatory-crossroads-compliance-challenges-autonomous-systems
Tech in Asia. (2026, March). *China unveils first national standards for humanoid robots.* https://www.techinasia.com/news/china-unveils-national-standards-humanoid-robots
Transparency Coalition. (2026, March 13). *AI legislative update: March 13, 2026.* https://www.transparencycoalition.ai/news/ai-legislative-update-march13-2026
Tri-City Voice. (2024). *Humanoid guidelines.* https://tricityvoice.com/humanoid-guidelines/
U.S.-China Economic and Security Review Commission. (2024). *Humanoid robots.* https://www.uscc.gov/sites/default/files/2024-10/Humanoid_Robots.pdf
Xinhua. (2026). *China's first national standard system for humanoid robotics poised to spur industry development.* https://english.news.cn/20260303/0e51ac8f66c542c5bacf2af3f80b3a40/c.html
Yale Review of International Studies. (2025). *Navigating liability in autonomous robots: Legal and ethical challenges in manufacturing and military applications.* https://yris.yira.org/column/navigating-liability-in-autonomous-robots-legal-and-ethical-challenges-in-manufacturing-and-military-applications/

**Ken Huang** is a globally recognized expert in AI and Web3 security, a prolific author, and a leading figure in shaping industry standards. He is the CEO and Chief AI Officer (CAIO) of DistributedApps.ai, specializing in Generative AI training and consulting. Ken is deeply involved in driving the development of secure AI systems. He serves as a Research Fellow and Co-Chair of the AI Safety Working Groups at the Cloud Security Alliance (CSA), leading Agentic AI initiatives. He is also a Co-Chair of the AI STR Working Group at the World Digital Technology Academy under the UN Framework, and a core contributor to the OWASP GenAI project, focusing on Agentic AI security. His expertise extends to his contributions to OWASP's Top 10 Risks for LLM Applications and his active participation in the NIST Generative AI Public Working Group. He is also a member of Open AI forum.

Ken has authored and edited numerous influential books, including:

"Agentic AI: Theories and Practices" (Springer, June 2025)
"Securing AI Agents: Foundations, Frameworks, and Real-World Deployment" (Springer, October 2025)
"Beyond AI: ChatGPT, Web3, and the Business Landscape of Tomorrow" (Springer, 2023)
"Generative AI Security: Theories and Practices" (Springer, 2024)

"Practical Guide for AI Engineers" (Volumes 1 and 2, DistributedApps.ai, 2024)
"The Handbook for Chief AI Officers: Leading the AI Revolution in Business" (DistributedApps.ai, 2024)
"Web3: Blockchain, the New Economy, and the Self-Sovereign Internet" (Cambridge University Press, 2024)
His co-authored book, "Blockchain and Web3: Building the Cryptocurrency, Privacy, and Security Foundations of the Metaverse" (Wiley, 2023), was recognized as a must-read by TechTarget.
A sought-after speaker, Ken has presented at prestigious global forums, including Davos WEF, ACM, IEEE, CSA AI Summit, the Depository Trust & Clearing Corporation (DTCC), and World Bank conferences.
Explore Ken's books on Amazon: https://www.amazon.com/author/kenhuang.

**Jerry Huang** is affiliated with Kleiner Perkins and has a background in AI, machine learning, security, scalable systems, and applied mathematics. His prior experience includes technical work across AI/ML evaluation, infrastructure, security, access control, and privacy-preserving biometric authentication. He holds BS/MS degrees in Computer Science from Georgia Tech and has pursued graduate study in Applied Mathematics at the University of Chicago.

**Bhavya Gupta** is an Information Security Officer at Stanford University, where she has built her career since 2017 across roles spanning in risk, governance, and security operations. She holds a Master's in Cyber Security Engineering from the University of Southern California.

Beyond Stanford, Bhavya is a recognized voice in AI security and governance. She serves as an OWASP lead contributor on the Agentic AI Top 10 risks for skills and the AI Vulnerability Scoring System (AIVSS), helping shape industry standards for securing emerging AI systems. Her expertise spans the full spectrum of modern cybersecurity—from traditional vulnerability management and cloud security to the novel risks introduced by agentic AI—making her a sought-after expert at the intersection of cloud, application, and AI security governance.

# Psychology and Human-Robot Coexistence

# 9

Jerry Huang, Ken Huang, and Modar Alaoui

## 9.1 Introduction

Humans are exquisitely attuned to other humans. We effortlessly read facial expressions, interpret body language, infer mental states, and coordinate complex social interactions—capabilities honed over millions of years of evolution in intensely social environments. These same psychological mechanisms activate when we encounter humanoid robots, triggering responses that range from fascination and affection to discomfort and fear. Understanding how humans perceive, trust, emotionally relate to, and ultimately coexist with humanoid robots requires grappling with fundamental questions at the intersection of psychology, sociology, and technology.

The year 2026 marks a critical juncture in human-robot psychology. Humanoid robots are transitioning from laboratory curiosities and science fiction imaginings to early deployments in controlled environments such as warehouses and industrial settings, with expansion into other domains still emerging. This transition begins to surface psychological questions that were previously theoretical, particularly in environments where humans and robots now share tasks: What happens when children form attachments to robotic companions? How do workers adapt psychologically to robotic colleagues? Can therapeutic relationships with robots

J. Huang
Kleiner Perkins, San Francisco, CA, USA

K. Huang (✉)
DistributedApps.ai, Fairfax, VA, USA
e-mail: ken@distributedapps.ai

M. Alaoui
ALM Ventures, Mountain View, CA, USA
e-mail: modar@alm.ventures

 
K. Huang et al. (eds.), *Humanoid Robots and Physical AI*, Advances in Data Analytics, AI, and Smart Systems, https://doi.org/10.1007/978-3-032-33669-9_9

genuinely benefit mental health? How do cultural factors shape vastly different acceptance rates across societies?

This chapter examines these questions through multiple psychological lenses, beginning with the "uncanny valley" and moving through trust dynamics, social cognition, and the profound cultural factors that shape vastly different acceptance rates across global societies.

Social cognition mechanisms then come into focus: the attribution of mental states, emotional expression recognition, and conversational dynamics that shape how humans interact socially with robots. Emotional bonding and attachment—once dismissed as science fiction—now manifests in real relationships, as evidenced by CES 2026 demonstrations of companion robots designed explicitly for emotional connection. The mental health implications of these bonds are wide-ranging, spanning beneficial therapeutic applications for autism, dementia, and social isolation alongside concerns about addiction, reduced human contact, and parasocial relationships. We then turn to the workplace, examining the job insecurity, status threats, and complex dynamics that emerge when humans and robots share factories and warehouses.

Developmental psychology raises further profound questions about how robot presence shapes children's cognitive, social, and moral development, with research documenting both educational benefits and potential risks. Societal adaptation and cultural change vary dramatically across nations—Japan's long robot-positive culture, China's pragmatic acceptance, and American ambivalence each reflect deep cultural differences in how humanoid machines are received. Acceptance factors and barriers, from perceived usefulness and ease of use to privacy concerns and aesthetic preferences, ultimately determine whether humanoid robots integrate smoothly or face persistent rejection. Rounding out the chapter, emerging developments from early 2026—including new regulatory frameworks and stress-aware interaction systems—signal accelerating attention to psychological factors in robot design and deployment.

Throughout this exploration, case studies of specific platforms illuminate how design choices influence psychological responses: Boston Dynamics' Atlas with its athletic but non-anthropomorphic form, EX Robots' hyper-realistic human appearance deliberately targeting emotional connection, UBTECH Walker's approachable family-friendly aesthetic, and Agility Robotics' Digit navigating the psychological challenges of warehouse deployment. Together, these examples reveal that psychology must inform design from inception, not as an afterthought.

## 9.2 Human Perception of Humanoid Robots

Human perception of robots is not a neutral process of registering physical properties but an active construction influenced by morphology, context, individual differences, and deeply embedded cultural narratives. The same robot may be perceived as helpful tool, uncanny threat, or potential companion depending on these factors. In practice, perception is also shaped by how well the system fits the environment in which it operates. A system that behaves consistently within a known workflow is often accepted even with limited intelligence, while a more advanced system that operates unpredictably in an open environment may face resistance. This suggests that acceptance is not driven by intelligence alone, but by the alignment between capability, behaviour, and the structure of the environment.

### 9.2.1 The Uncanny Valley: Theory and Evidence

The "uncanny valley" hypothesis, proposed by roboticist Masahiro Mori in 1970 and translated into English by MacDorman and Kageki (Mori et al., 2012), posits that human emotional response to robots follows a nonlinear trajectory: as robots become more human-like, affinity increases until reaching a point where near-but-not-quite-human appearance triggers strong revulsion, before rising again as appearance approaches perfect human replication. This valley of eeriness has profound implications for humanoid design (see Fig. 9.1).

In Fig. 9.1, the shaded region marks the zone of aversion where near-human appearance triggers discomfort. Industrial robots and idealized human appearance both receive positive responses, while the valley between them provokes eeriness.

As of 2025–2026, the uncanny valley remains highly relevant despite technological advances. Research demonstrates that the phenomenon persists even with increasingly sophisticated robots. Recent commentary argues that with generative AI producing increasingly convincing digital humans, the uncanny valley is shifting from a problem of visual imperfection towards one of moral ambiguity—a psychological boundary where simulated empathy becomes hard to distinguish from genuine empathy (Pattan, 2025).

Recent neuroscience research has begun elucidating the neural mechanisms underlying the uncanny valley. Studies reveal that human-like robots and hyper-realistic masks activate brain regions associated with both social cognition and

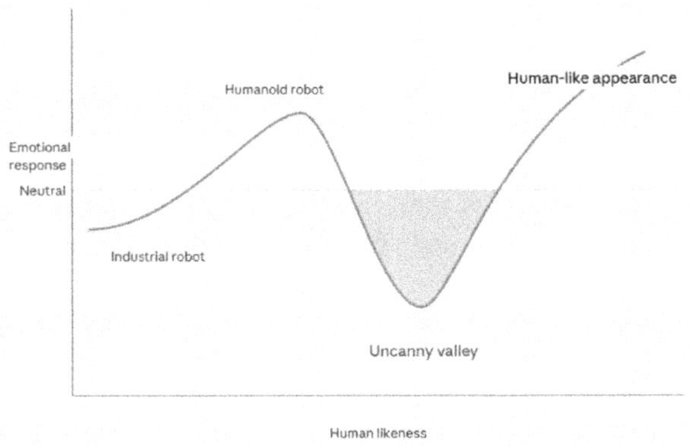


**Fig. 9.1** The uncanny valley—emotional response (vertical axis) as a function of human likeness (horizontal axis)

threat detection, creating a conflict that manifests as discomfort (Jenkins et al., 2025). This suggests the uncanny valley reflects not mere unfamiliarity but genuine cognitive conflict between competing categorizations—human versus non-human.

China's rapid development of hyper-realistic humanoid robots provides natural experiments in uncanny valley responses. Chinese manufacturers are producing humanoid robots that, according to recent reporting, are becoming harder to distinguish from real people, with advances in emotional-state detection and increasingly realistic skin and movement (Kalil, 2026). Market reception reveals cultural variation in uncanny valley sensitivity, with some Asian markets showing greater tolerance for highly anthropomorphic designs than Western counterparts.

Efforts to mitigate the uncanny valley focus on two strategies: avoiding the valley through deliberately non-realistic design (Boston Dynamics' Atlas, which has no facial features) or attempting to traverse it completely through perfect realism combined with sophisticated conversational AI. A February 2025 study investigated whether large language model-enhanced conversations could reduce uncanny valley effects, reporting that LLM-enhanced interactions significantly reduced reported eeriness and produced more natural conversations with the Nadine robot (Kang et al., 2025). This suggests that psychological presence created through natural conversation may counterbalance visual eeriness.

In practice, the uncanny valley is not only a function of appearance. It often emerges from mismatches between appearance, motion, and behaviour. Systems that look human but move or respond inconsistently tend to generate stronger discomfort than systems that are clearly machine-like but predictable.

### 9.2.2 Anthropomorphism and Mental State Attribution

Humans possess a powerful tendency towards anthropomorphism—attributing human-like mental states, intentions, and emotions to non-human entities. This tendency, evolutionary advantageous for interpreting conspecific behaviour, readily extends to humanoid robots even when people consciously recognize their artificial nature.

The degree of anthropomorphism influences critical outcomes including trust, perceived competence, moral consideration, and emotional engagement. Research documents that humans spontaneously attribute beliefs, desires, and emotions to robots exhibiting even minimal social cues like gaze direction or gesture. The humanoid form factor amplifies these attributions; a bipedal robot with a head is perceived as more intentional and socially aware than mechanically equivalent wheeled platforms.

However, anthropomorphism introduces operational risks, particularly when perceived capability exceeds actual system performance. Over-attribution of capability leads to unrealistic expectations and disappointment when robots fail to exhibit human-like intelligence or flexibility. Anthropomorphism also raises ethical concerns about manipulation, particularly for vulnerable populations like children or elderly individuals who may not maintain appropriate psychological boundaries.

### 9.2.3 Cultural Differences in Perception

Robot perception varies dramatically across cultures, reflecting different religious traditions, philosophical frameworks, and cultural narratives. Recent cross-cultural research reveals patterns with profound implications for global humanoid robot deployment.

A March 2025 study found striking differences between U.S. and Japanese participants: U.S. participants cooperated with artificial agents significantly less than with humans, whereas Japanese participants cooperated equivalently with both. Japanese participants reported similar emotional reactions to exploiting machines and humans, while U.S. participants felt bad about exploiting humans but not machines (Karpus et al., 2025). This suggests fundamentally different psychological boundaries between human and machine in the two cultures.

Cultural differences extend beyond cooperation to acceptance patterns. Research indicates that, relative to Japanese participants, Americans are more comfortable with robots that look mechanical but less comfortable with highly human-like robots (Castelo & Sarvary, 2022). This pattern has been linked to broader cultural frameworks: Western traditions tend to treat humans as ontologically distinct from other entities, while several East Asian traditions view spirit or animacy as more broadly distributed across living and crafted things (Association for Advancing Automation, 2025).

These perceptual differences have practical implications. Japanese and Chinese manufacturers often embrace highly anthropomorphic designs that might trigger Western uncanny valley responses. Conversely, Western platforms like Boston Dynamics' Atlas deliberately avoid human-like faces, reflecting cultural comfort with machine-like appearance combined with human-like capability.

## 9.3 Trust and Reliance

Trust is a primary gating factor for successful deployment of humanoid systems. Without sufficient trust, systems are underutilized. With excessive trust, users may over-rely on them in situations where system limitations still exist. Calibrating trust appropriately—neither too much nor too little—remains a central challenge.

Figure 9.2 shows that optimal human-robot interaction requires trust positioned in the central zone. Under-trust leads to system disuse and wasted investment; over-trust (automation bias) creates safety hazards when humans accept robotic decisions uncritically.

### 9.3.1 Foundations of Human-Robot Trust

Trust in human-robot interaction encompasses multiple dimensions. Recent research conceptualizes trust as a multidimensional construct that includes both performance trust and relational trust, with industrial settings tending to emphasize

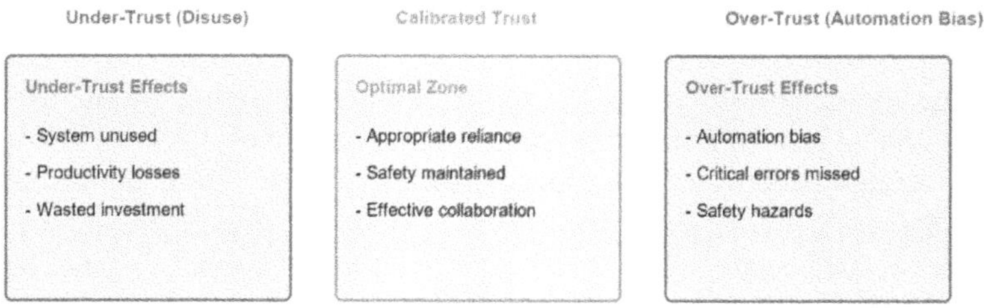


**Fig. 9.2** The trust calibration spectrum

the former and social-care settings the latter (Cirasa & Conti, 2025). This distinction illuminates why warehouse workers and hospital patients require different trust-building approaches.

A 2025 systematic review of trust assessment in human-robot interaction (HRI) reveals that human trust in HRI is not a fixed attribute but a dynamic construct that evolves over time through repeated interactions (Campagna & Rehm, 2025). Initial trust—based on appearance, brand reputation, and early demonstrations—differs substantially from learned trust developed through experience. In early deployments, a consistent gap often emerges between perceived capability and actual system performance. Initial demonstrations can create inflated expectations, particularly when systems perform well in controlled scenarios but encounter variability in real environments. Over time, users recalibrate their trust based not on what the system is designed to do, but on what it reliably does under real conditions. Managing this gap is critical, as misalignment between expectation and performance is one of the fastest ways to erode trust. This temporal dimension has design implications: systems must not only build initial trust but maintain and repair it over extended deployments.

Trust develops through multiple pathways. Reliable performance builds competence trust: the robot consistently executes tasks correctly. Appropriate error handling and transparency build integrity trust: the robot acknowledges failures and explains actions. Natural interaction patterns build relational trust: the robot exhibits social awareness and reciprocity. Different applications require different trust emphases.

### 9.3.2 Trust Calibration Challenges

Both over-trust and under-trust create problems. Over-trust manifests as automation bias: humans uncritically accept robotic decisions even when incorrect, potentially dangerous in safety-critical applications. Under-trust results in disuse: humans reject helpful assistance, reducing productivity and wasting investment.

Several factors influence appropriate trust calibration. Transparency—the robot explaining its reasoning and confidence levels—enables humans to assess when to accept or override robotic decisions. Customization increases trust; a 2025 experiment found that customization substantially increased both psychological ownership and trust towards robots, suggesting personalization strengthens psychological bonds (Riar et al., 2025).

Workplace contexts reveal trust challenges acutely. When Agility Robotics deploys Digit robots in warehouses, worker trust develops gradually through experience. Initial skepticism ("Will it take my job? Can it handle unexpected situations?") transitions to conditional trust ("It reliably handles routine cases") and potentially to inappropriate over-reliance ("I'll let it handle everything"). Managers must actively manage this trajectory through training, monitoring, and organizational culture. In practice, worker response is often shaped as much by how deployment is introduced as by the system itself. Poor communication around role, capability, and intent can amplify anxiety more than the technology.

### 9.3.3 Trust Repair After Violations

Robots will inevitably fail, violating trust. A robot drops a fragile package, misidentifies an object, or moves unexpectedly. How trust is repaired determines whether the relationship continues productively or deteriorates into permanent distrust.

Research on trust repair in human-robot interaction reveals several effective strategies. Immediate acknowledgement of failure demonstrates awareness and responsibility. Explaining the cause helps humans understand whether the failure was an isolated error or systematic limitation. Corrective action demonstrates commitment to improvement. Apologies, even from robots incapable of genuine emotion, facilitate human forgiveness through social ritual.

However, some trust violations may be irreparable. If a robot injures a human due to control failure, or repeatedly fails at tasks presented as within its capability, trust may be permanently lost regardless of repair attempts. This highlights the critical importance of conservative capability claims and robust safety systems.

## 9.4 Social Cognition and Interaction

Humans are social creatures who automatically deploy sophisticated social cognition mechanisms when encountering anthropomorphic entities. Understanding how these mechanisms operate with humanoid robots reveals both opportunities for natural interaction and risks of misunderstanding.

### 9.4.1 Theory of Mind and Intentionality

Theory of mind—the ability to attribute mental states including beliefs, desires, and intentions to others—represents a cornerstone of human social cognition. Humans readily extend theory of mind to humanoid robots, interpreting their actions as intentional and goal-directed even when explicitly aware of their programmed nature.

This attribution has functional benefits: treating robots as intentional agents simplifies interaction by enabling prediction of behaviour through mental state inference rather than explicit programming knowledge. A worker inferring that Digit "wants to" place a box on the shelf predicts its trajectory more naturally than analysing kinematic planning algorithms.

However, inappropriate mental state attribution creates problems. Humans may attribute understanding that robots lack, leading to communication failures. They may attribute emotions or motivations that don't exist, causing misplaced empathy or resentment. They may attribute moral responsibility inappropriately, blaming robots for failures that reflect design limitations or unclear instructions.

### 9.4.2 Social Presence and Engagement

Social presence—the sense that one is interacting with a genuine social entity rather than an object—critically influences human-robot interaction quality. Recent work has reframed social presence as multidimensional, encompassing both attributed social presence (a user's subjective perception of the robot's social capabilities) and intrinsic social presence (the robot's inherent capabilities that engender presence) (Stower et al., 2025).

The contextual nature of social presence has gained research attention. Kaptelinin and Dalli (2025) argue that the perceived sociality of robots emerges from how people experience them within meaningful collaborative contexts, rather than from intrinsic robot properties or fixed human predispositions alone. This suggests that effective social presence requires not just sophisticated robot design but appropriate task framing and organizational context.

CES 2026 demonstrations of AI companion robots—including Lovense's Emily and Lepro's Ami—exemplify deliberate social presence design. These systems pair persistent memory with configurable personalities, with vendors positioning them as tools for forging emotional connections and providing companionship in an increasingly isolated society (Daws, 2026; The Outpost AI, 2026). The shift from utility to emotional support represents a strategic repositioning to maximize social presence.

### 9.4.3 Conversational Dynamics and Turn-Taking

Natural conversation requires sophisticated coordination: turn-taking, gaze management, gesture synchronization, topic tracking, and pragmatic inference. Humans execute these seamlessly; robots struggle. Yet conversational capability critically determines whether humans perceive robots as genuine interaction partners or frustrating tools.

Recent advances in large language models have dramatically improved robot conversational capability, enabling natural language task specification, question answering, and even small talk. However, challenges remain in the physical embodiment of conversation: appropriate gaze behaviour, gesture timing, proxemics (interpersonal distance), and multimodal integration of speech, gaze, and action.

The importance of conversational naturalness is reinforced by the same Nadine study cited above, in which LLM-enhanced conversation reduced reported uncanny valley feelings. When robots converse naturally, humans forgive visual imperfections; when appearance is polished but conversation awkward, discomfort intensifies. This suggests conversational fluency may be more important than visual realism for social acceptance.

In practical deployments, reliability often matters more than conversational sophistication. Users tend to accept systems that behave consistently and predictably, even if interaction is limited, while systems that are highly conversational but inconsistent in action tend to lose credibility quickly. This reinforces that perceived intelligence must be matched by reliable execution in the physical world.

## 9.5 Emotional Bonding and Attachment

Early forms of emotional bonding between humans and robots, long discussed in theory, are now beginning to appear in limited and controlled contexts. Understanding these bonds—their nature, development, benefits, and risks—has become urgent as companion robots transition from concept to commercial product.

### 9.5.1 Attachment Formation

Attachment theory, originally developed to understand infant-caregiver bonds, illuminates human-robot emotional connections. Research demonstrates that people develop non-transactional relationships with companion robots over time, with families naming them, treating them as family members, and reporting noticeable longing when the robot is temporarily removed.

Attachment to robots develops through familiar mechanisms: repeated positive interactions, perceived responsiveness, provision of comfort, and reliable presence. A robot that greets a lonely elderly person daily, remembers their preferences, and provides companionship satisfies core attachment needs. That the robot lacks genuine consciousness may become psychologically irrelevant if the functional benefits are real.

Clinical evidence supports therapeutic efficacy. Studies of the PARO therapeutic robot with hospitalized patients found that 95% of participants demonstrated beneficial interactions—with measurable reductions in agitation, depression, and loneliness, as well as decreased medication usage and improved sleep patterns

(Kelly et al., 2021). These outcomes suggest attachment to robots can provide genuine mental health benefits.

### 9.5.2 Loneliness and the Companionship Crisis

Context matters for understanding robot attachment. Loneliness has reached crisis proportions. A 2025 World Health Organization Commission on Social Connection report concluded that roughly one in six people globally experience loneliness, contributing to an estimated 871,000 deaths per year (World Health Organization, 2025). A meta-analysis of 70 prospective studies found social isolation was associated with a 29% increase in mortality risk and living alone with a 32% increase (Holt-Lunstad et al., 2015). Against this backdrop, robot companionship may represent a supplementary intervention, particularly in cases where human support is limited or unavailable.

By late 2025, companion-app downloads had reached approximately 220 million globally, with teens and young adults leading adoption (Vogels, 2025). This suggests unmet needs for connection that traditional social structures are failing to address.

However, critics worry that robot companionship substitutes for rather than supplements human relationships, potentially deepening isolation. If lonely individuals bond with robots rather than addressing underlying causes of social disconnection, robot companionship might become palliative rather than curative. This debate—whether robot companionship helps or harms long-term wellbeing—remains unresolved.

### 9.5.3 Parasocial Relationships and Exploitation Concerns

Parasocial relationships—one-sided emotional connections where one party (the human) feels intimacy while the other (the robot) has no genuine reciprocal feeling—raise ethical concerns. Humans may develop feelings for robots that cannot reciprocate, potentially leading to emotional exploitation.

These concerns intensify with deliberately emotionally manipulative design. EX Robots' platforms emphasize highly human-like appearance, expressive facial behaviour, conversational systems with emotional awareness, and configurations specialized for companionship use cases. Such design intentionally elicits emotional bonds for commercial purposes, raising questions about informed consent and vulnerability protection.

A key consideration is whether users clearly understand the nature and limits of the system they are interacting with. When boundaries between simulation and genuine emotional reciprocity are not made explicit, the risk of misunderstanding and dependency increases, particularly for vulnerable users. Children and elderly individuals face particular risks. Children may not understand that robot "friendship" differs fundamentally from human friendship, potentially distorting social development. Elderly individuals with cognitive decline may lose the capacity to distinguish robot from human relationships. Regulation and design ethics must address these vulnerabilities proactively.

## 9.6 Mental Health Implications

Humanoid robots' mental health implications span from promising therapeutic applications to concerns about social isolation, with evidence supporting both benefits and risks depending on context and implementation (see Fig. 9.3 for a summary).

### 9.6.1 Therapeutic Applications

Robot-assisted therapy has demonstrated efficacy across multiple mental health conditions. For autism spectrum disorder, social robots in therapy have been explored for more than 20 years, with recent systematic evidence showing robots can help children develop social skills such as collaborative play, verbal communication, and visual attention (David et al., 2025). A January 2026 study in Science Robotics confirmed that robot-assisted therapy achieves outcomes equivalent to conventional interventions while demonstrating significantly higher engagement levels (Preidt, 2026).

The advantages of robots for autism therapy include consistency (robots behave predictably, reducing anxiety), customization (interaction complexity can be precisely calibrated), and non-judgmental presence (robots never display impatience or frustration). These features make robots particularly suitable for individuals who find human social interaction overwhelming.

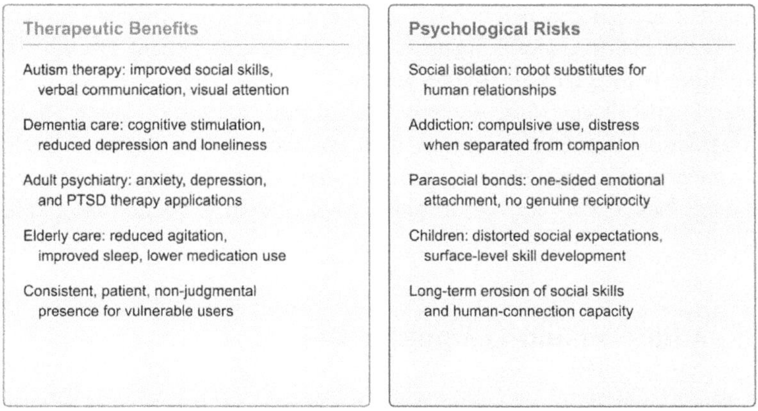


**Fig. 9.3** Mental health implications of humanoid robots

For dementia care, robots provide cognitive stimulation, emotional support, and practical assistance. A January 2026 framework for "Assistive Intelligence" aligns AI-powered interventions with each stage of dementia—from preclinical through severe—enabling tailored support throughout disease progression (Ghaiumy Anaraky & Mohapatra, 2026). Research demonstrates that robot companions can alleviate depressive symptoms and improve feelings of loneliness and overall quality of life among dementia patients (Kling et al., 2025).

Adult psychiatry increasingly explores social robots, with a 2025 review noting that they are becoming more common in healthcare and may help mitigate staffing shortages while complementing existing mental health treatment. Applications include anxiety management, depression support, and post-traumatic stress therapy.

Figure 9.3 captures both sides of this. The left column shows the therapeutic benefits documented across autism, dementia, adult psychiatry, and elderly care. The right column shows the psychological risks—isolation, addiction, and parasocial dependency. The benefits aren't automatic; they hinge on how these tools are actually deployed and what regulatory safeguards are in place.

### 9.6.2 Risks: Social Isolation and Reduced Human Contact

Counterbalancing therapeutic promise, concerns mount about robots enabling rather than alleviating social isolation. If individuals substitute robot for human companionship, short-term comfort might yield long-term harm. Human relationships provide complex benefits—reciprocal emotional support, intellectual challenge, co-regulation of emotion, shared meaning-making—that robots cannot fully replicate.

The concern intensifies for populations already at social risk. Adolescents who bond with AI companions rather than developing peer relationships may fail to acquire essential social skills. Elderly individuals who interact primarily with robots may experience accelerated cognitive decline from reduced human stimulation. The question is not whether robots provide any benefit but whether they displace more beneficial human interactions.

Research on this question remains limited and equivocal. Some studies find robot interaction supplements rather than replaces human contact. Others document concerning displacement effects. Longitudinal research tracking mental health outcomes over years—not months—is urgently needed as deployment scales.

### 9.6.3 Addiction and Compulsive Use

As robots become more engaging and responsive, addiction risks emerge. The scale of consumer uptake described above suggests powerful appeal. If these companions are designed with engagement-maximizing algorithms similar to social media—exploiting psychological vulnerabilities to maximize interaction time—addictive patterns could develop.

Warning signs include preferring robot to human interaction, neglecting responsibilities to interact with robots, emotional distress when separated from robots, and unsuccessful attempts to reduce use. These patterns parallel other behavioural addictions. Unlike substance addictions, robot addiction lacks physical dependence but may create equally problematic psychological dependence. In many cases, these risks are not inherent to the systems themselves but are shaped by design choices, particularly those that optimize for engagement without regard for long-term user wellbeing.

Designers face ethical choices: maximize engagement through potentially manipulative techniques or deliberately limit appeal to prevent addiction? Commercial incentives favour engagement maximization; user wellbeing may require the opposite. Regulatory frameworks addressing addiction risk in companion robots remain nascent.

## 9.7 Workplace Psychology

The introduction of humanoid robots into workplaces triggers complex psychological responses among human workers, from job insecurity and status threats to potential productivity gains and workplace enrichment. Understanding these dynamics is crucial for successful deployment.

### 9.7.1 Job Insecurity and Economic Anxiety

Worker psychology around humanoid robots is dominated by job loss anxiety. When Amazon deploys Digit robots in fulfilment centres or BMW tests humanoid robots in manufacturing, human workers immediately question their own futures. This anxiety is not irrational. In many cases, these systems are introduced to improve efficiency in tasks currently performed by humans, which is often interpreted as a direct signal of potential displacement.

Survey research reveals that 77% of technologists surveyed agreed that the novelty of humanoid robots can inject fun into the workplace, but expected that over time the robots would become unremarkable co-workers, suggesting initial curiosity transitions to normalization (Institute of Electrical and Electronics Engineers, 2026).

However, this positive framing can mask underlying worker concerns about obsolescence.

The psychological impact extends beyond workers directly at risk. Observing robotic deployment in one's industry or adjacent sectors creates ambient anxiety even for workers in currently safe positions. This chronic low-level stress affects mental health, job satisfaction, and organizational commitment. Effective management requires acknowledging and addressing anxiety rather than dismissing concerns as Luddite resistance.

### 9.7.2 Human-Robot Team Dynamics

When robots supplement rather than replace human workers, complex team dynamics emerge. Research notes that human workers monitoring robot behaviours face a dual experience: the task can be stimulating, involving decision-making, but may also be alienating, as it reduces human engagement in the production process—a tension that can generate stress (Han & Conti, 2025).

The quality of human-robot teaming depends on multiple factors. Clear role definition prevents confusion about responsibilities. In practice, how tasks are framed plays a significant role in acceptance. When robots are positioned as augmenting human capability, adoption tends to be smoother than when they are introduced as direct substitutes for existing roles, even when the underlying function is similar.

Appropriate task allocation leverages complementary strengths—robots for strength, speed, and precision; humans for flexibility, judgment, and problem-solving. Communication interfaces enabling natural human-robot coordination reduce cognitive burden and frustration.

Positive outcomes are possible when deployment is deliberate. The integration of humanoid robots can allow humans to focus on collaboration with colleagues and relationship building with clients and stakeholders, making jobs more fulfilling while improving workplace safety as robots handle hazardous tasks (Jones Lang LaSalle, 2025). Realizing this optimistic vision requires deliberate organizational design, not merely robot deployment.

### 9.7.3 Status, Identity, and Adaptation

Workplace introduction of robots can threaten worker status and identity. If one's self-concept centres on "skilled machinist" or "experienced warehouse worker," and robots now perform those tasks faster and more reliably, identity must be reconstructed or defended. Some workers embrace a new identity as "robot supervisor" or "human-robot team coordinator"; others resist as a form of self-preservation.

Generational differences influence adaptation. Younger workers may more readily accept robot colleagues, having grown up with automation; older workers may resist as threatening familiar work practices and relationships. Organizations must provide retraining, role evolution pathways, and support for identity reconstruction.

Cultural factors also matter. Cross-cultural evidence that Japanese participants cooperate equally with human and artificial partners while American participants discriminate between them, discussed earlier in this chapter, suggests national culture also influences workplace robot acceptance. Japanese organizational cultures emphasizing harmony and collective goals may ease robot integration; American cultures emphasizing individual achievement and clear human-machine boundaries may create more friction.

## 9.8 Developmental Psychology and Children

Children's interaction with humanoid robots introduces new variables in cognitive, social, emotional, and moral development that are not yet fully understood. As robots enter homes and schools, understanding their developmental impact becomes urgent.

### 9.8.1 Cognitive and Educational Impact

Educational robots show promise for personalized learning. Research proposes integrating AI and robotics in elementary school education through personalized tutor robots that interact directly with students, combining cognitive and emotional assessment to deliver tailored educational experiences (Maaz et al., 2025). Robots can provide infinite patience, customize difficulty precisely, and collect detailed learning analytics.

A systematic review analysing research from 2018 to 2025 found that social robots show promise in delivering personalized educational experiences, though challenges such as emotional dependence, accessibility, and ethical concerns persist (Sánchez-Sancho et al., 2025).

The balance between educational benefits and psychological risks remains unclear.

Concerns focus on several areas. Robot tutors may be effective for rote learning but less effective at model;ling creative thinking, critical analysis, and complex problem-solving. Children may develop learning styles dependent on robot scaffolding that do not transfer to human educational contexts. Screen time concerns—that excessive digital interaction harms development—may apply equally to extended robot interaction.

### 9.8.2 Social-Emotional Development

Social robots' impact on children's emotional development remains actively debated. Proponents argue robots can teach empathy, emotional recognition, and social skills in safe, controllable contexts. As discussed in Sect. 9.6.1, evidence indicates that social robots can help children with autism develop collaborative play, verbal communication, and visual attention skills.

However, critics worry that robot interaction provides simplified social experiences that do not prepare children for complex human relationships. Human social interaction involves reciprocity, genuine emotional stakes, ambiguity, and moral complexity that robots cannot replicate. Children who practice social skills primarily with robots may develop surface-level competencies without deep social understanding.

"The child factor in child-robot interaction" presents a significant design challenge: social robots have great potential to support children in education, healthcare, and daily life, but child-robot interaction requires strategies tailored to unique and diverse individuals at different developmental stages (Rudenko et al., 2025).

### 9.8.3 Privacy and Safety Concerns

Humanoid robots in family environments raise substantial privacy concerns. A 2026 study examines how children and parents perceive privacy risks and their comfort levels across various usage scenarios as humanoid social robots become more integrated into family environments (Shin et al., 2026). Robots with cameras and microphones in homes collect intimate data about children's behaviour, conversations, and daily lives.

Safety concerns extend beyond physical to psychological safety. Robots could be used to manipulate children, collect data for commercial purposes, or transmit harmful content. The same robots designed to be engaging companions could be exploited by bad actors. Robust security, transparent data practices, and regulatory oversight are essential but currently inadequate.

## 9.9 Societal Adaptation and Cultural Change

How societies adapt to humanoid robot presence varies dramatically across cultures, shaped by religious traditions, philosophical frameworks, science fiction narratives, economic pressures, and generational differences (see Fig. 9.4).

In this figure, Japan shows consistently high acceptance across all dimensions; China leads on workplace integration; the United States exhibits lower acceptance, particularly for high human likeness. Based on cross-cultural research findings synthesized from multiple studies (Association for Advancing Automation, 2025; Castelo & Sarvary, 2022; Guan et al., 2025; Karpus et al., 2025).

### 9.9.1 Cultural Narratives: Japan, China, and the United States

Japan's relationship with robots reflects Shinto traditions viewing all things as potentially possessing spirit, combined with cultural narratives of robots as helpers rather than threats. This creates exceptionally high robot acceptance. Japanese robot designs often embrace anthropomorphism—humanoid form, approachable aesthetics, social presence—that would trigger uncanny valley responses in Western markets.

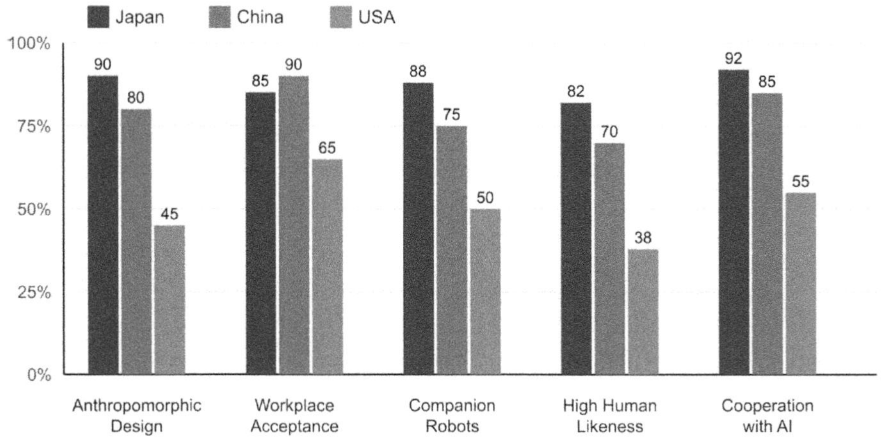


**Fig. 9.4** Cross-cultural robot acceptance—Estimated acceptance scores (0–100%) for Japan, China, and the United States across five dimensions

China's approach combines pragmatic acceptance with strategic national interest. Government policy actively promotes robotics as economic development strategy, with China accounting for roughly 54% of all new global industrial robot installations in 2024 (International Federation of Robotics, 2025). Cultural factors including a Confucian emphasis on harmony and collective benefit facilitate acceptance, though individual privacy concerns remain significant.

The United States exhibits profound ambivalence, with cultural narratives dominated by dystopian visions from science fiction (Terminator, Westworld, Ex Machina). American individualism and emphasis on human uniqueness create greater resistance to humanoid anthropomorphism. Yet economic pressures and technological enthusiasm create countervailing forces towards acceptance.

These cultural differences have practical implications. An April 2025 study found a significant positive relationship between cultural tightness and robot application across 32 countries, with tight cultures such as Singapore, Japan, and China showing greater robotics activity than looser cultures such as the United Kingdom, United States, and Netherlands (Guan et al., 2025). This suggests cultural values fundamentally shape technological trajectories.

### 9.9.2 Generational Differences

Adaptation to humanoid robots varies dramatically by generation. Gen Z and younger millennials, having grown up with AI assistants and autonomous vehicles, view robots as natural evolution rather than disturbing discontinuity. Teens and young adults lead adoption of AI companion apps.

Older generations exhibit more skepticism. Baby boomers and Gen X, formed in pre-digital eras, may view humanoid robots as science fiction made real—either exciting or threatening depending on individual disposition. The digital divide in robot acceptance mirrors broader technological adoption patterns.

This generational gradient creates organizational challenges. Workplaces with mixed-age workforces must navigate vastly different psychological responses to robot introduction. Educational systems must address both digital-native students who eagerly embrace robot tutors and parents concerned about developmental impacts.

### 9.9.3 Normalization Trajectory

Humanoid robots may follow a normalization trajectory similar to other technologies, though this remains dependent on early deployment outcomes and public perception. Initial novelty and media attention gradually yield to matter-of-fact acceptance as robots become commonplace. As the IEEE technologist survey cited earlier suggests, observers expect humanoid robots to blend in over time as routine teammates rather than remaining objects of fascination.

This normalization has psychological implications. The same processes that rendered smartphones psychologically invisible will likely apply to robots. Initial anxiety about job loss, privacy, or uncanny valley discomfort may fade not because these concerns are fully addressed but because humans habituate to robot presence.

However, normalization is not inevitable or universal. If early robot deployments generate salient failures—injuries, job losses, privacy violations—negative attitudes may crystallize and persist. The next 3–5 years represent a critical window where formative experiences will shape lasting societal attitudes towards humanoid robots.

## 9.10 Acceptance Factors and Barriers

Understanding what drives robot acceptance versus rejection is crucial for predicting deployment trajectories and designing systems that societies willingly integrate.

### 9.10.1 Technology Acceptance Model Applications

The Technology Acceptance Model (TAM), originally developed for computer systems, has been extensively applied to robotics. TAM posits that perceived usefulness and perceived ease of use determine acceptance intentions. Reviews confirm that TAM is now being applied across service robots, educational robots, industrial robots, and mobile telepresence platforms.

However, TAM's original framework focuses primarily on functional evaluation and does not fully capture the human-like characteristics exhibited by AI-driven applications (Ibrahim et al., 2025). Extensions for robotics incorporate factors

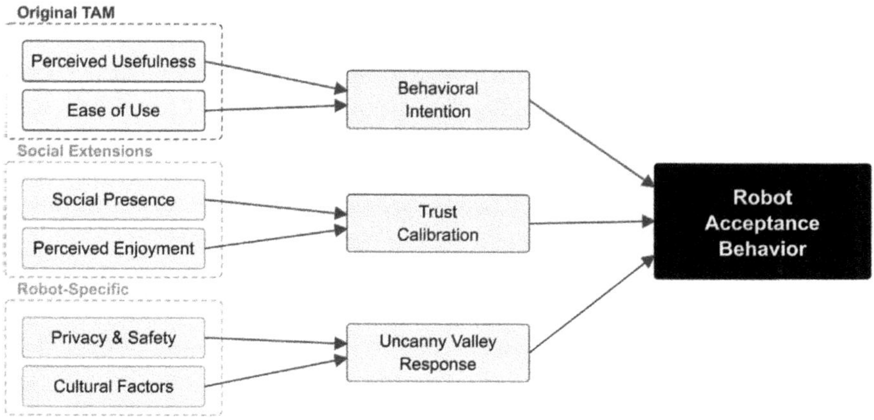


**Fig. 9.5** Extended technology acceptance model (TAM) for humanoid robots

like perceived enjoyment, perceived safety, ethical and social implications, and personal characteristics including self-efficacy, robot anxiety, and affinity towards technology.

A May 2025 study investigated service robot acceptance by integrating TAM with the Computers Are Social Actors (CASA) framework, Kansei engineering, and social presence theory, examining how design elements influence user responses to service robots (Lyu et al., 2025). This integrated approach recognizes that robot acceptance is multifaceted, requiring attention to functional utility, social presence, aesthetic design, and emotional response.

The original TAM factors (perceived usefulness, ease of use) are supplemented by social extensions (social presence, perceived enjoyment) and robot-specific factors (privacy/safety perceptions, cultural factors). All streams feed into behavioural intention and trust calibration, which together determine robot acceptance behaviour. The uncanny valley response acts as a moderating factor on acceptance outcomes.

Figure 9.5 summarizes the Extended Technology Acceptance Model (TAM) for humanoid robots, showing how the original TAM factors—perceived usefulness and perceived ease of use—are extended by social-presence factors and robot-specific factors (privacy and safety perceptions, cultural framing, uncanny-valley response, personal characteristics) and how all of these feed behavioural intention and trust calibration to determine acceptance behaviour.

### 9.10.2 Privacy, Surveillance, and Control Concerns

Privacy represents a major acceptance barrier. Humanoid robots with cameras, microphones, and internet connectivity raise surveillance concerns that exceed those of traditional robots. The prospect of robots recording intimate moments, tracking behaviours, and transmitting data to corporations or governments triggers resistance.

Concerns intensify when robots are designed for homes and personal spaces. Parents worry about robots collecting data on children. Workers fear robots monitoring productivity and reporting to management. Citizens question robots in public spaces recording faces and behaviours without consent.

Effective privacy protection requires technical measures—local processing, encryption, user data control—combined with regulatory frameworks and transparent corporate practices. However, the economic incentives for data collection create pressure against strong privacy protection. Resolving this tension will require deliberate policy intervention.

### 9.10.3 Aesthetic Preferences and Demographic Differences

Visual design profoundly influences acceptance. Some individuals prefer machine-like appearance that clearly signals robot nature (Boston Dynamics' Atlas). Others prefer human-like appearance that facilitates natural interaction (EX Robots). Still others prefer approachable designs that avoid the uncanny valley (UBTECH Walker).

Gender differences in robot acceptance have been documented, with some research finding women more cautious about robot adoption than men, possibly reflecting greater sensitivity to social and ethical implications. Age correlates with acceptance as discussed in Section 9.9. Education level, technological background, and personality traits also influence acceptance patterns.

These individual differences imply that no single design will achieve universal acceptance. Market segmentation—offering different robot designs for different preferences and use cases—may prove necessary. Industrial robots can be overtly mechanical; companion robots may require more anthropomorphic design; service robots might adopt middle-ground aesthetics.

## 9.11 Emerging Developments: First Quarter 2026

The first months of 2026 witnessed notable advances in both research and regulation that directly impact the psychological dimensions of human-robot coexistence. These developments signal accelerating attention to the human factors that will determine whether humanoid robots integrate successfully into society.

### 9.11.1 Regulatory Milestones for Companion Systems

January 2026 marked a pivotal moment in companion AI regulation. California's SB 243, the nation's first law specifically targeting companion AI chatbots, came into effect with explicit child safety provisions. The legislation requires platforms to implement mandatory monitoring and regular reminders when children interact with companion systems, addressing concerns about prolonged engagement and potential emotional dependence (Padilla, 2025). This regulatory intervention directly responds to psychological concerns about vulnerable populations developing inappropriate attachments to artificial entities.

Internationally, South Korea enacted its AI Basic Act on January 22, 2026, establishing one of Asia's most comprehensive AI safety frameworks. The law mandates human oversight for "high-impact" AI systems and requires transparency measures that directly affect humanoid robot deployments in healthcare, education, and service contexts (Shim, 2026). The legislation reflects growing recognition that psychological safety requires not just physical safeguards but institutional structures ensuring human agency in human-robot interactions.

These regulatory developments share a common thread: explicit acknowledgement that the psychological risks of human-robot interaction—including attachment, manipulation, and dependency—require legal frameworks comparable to those governing physical safety.

### 9.11.2 Stress-Aware Robot Interaction Systems

A March 2026 announcement from NEC Corporation introduced what the company terms "physical AI"—a system that proactively controls robots to reduce human stress by anticipating both movement and psychological states. The technology predicts stress levels in real-time by analysing the relative position and posture between robot and human, enabling robots to adjust their trajectories, speeds, and behaviours to enhance psychological comfort (NEC Corporation, 2026).

This development represents a paradigm shift from reactive to anticipatory human-robot interaction design. Rather than merely responding to explicit human commands or obvious distress signals, stress-aware systems attempt to prevent psychological discomfort before it occurs. Early trials in retail and warehouse environments report measurable reductions in worker anxiety when robots modify their behaviour patterns based on predicted stress responses.

The psychological implications are significant: if robots can successfully anticipate and avoid causing human stress, a major barrier to workplace acceptance may diminish. However, the technology also raises questions about the appropriateness of machines monitoring human psychological states and the potential for manipulation if such capabilities are deployed for commercial rather than wellbeing purposes.

### 9.11.3 Research Advances in Human-Likeness Evaluation

March 2026 saw the proposal of the "Motion Turing Test," a framework inspired by the classic Turing Test but designed specifically for evaluating whether human observers can discriminate between humanoid robot and human movement (Li et al., 2026). The framework provides systematic methodology for assessing the motion dimension of the uncanny valley, complementing existing appearance-focused measures.

Concurrently, the ACM/IEEE International Conference on Human-Robot Interaction (HRI, 2026), held in Edinburgh March 16–19, featured research directly relevant to psychological dynamics. Studies presented included investigations of how fictional versus factual robot tutor dialogue shapes child social-emotional learning, and gesture-based interaction patterns inspired by human-dog dynamics that improve human-robot coordination (ACM/IEEE International Conference on Human-Robot Interaction, 2026). The conference program reflected the field's expanding scope from technical capability to psychological impact.

### 9.11.4 Therapeutic Efficacy Confirmations

January 2026 brought additional evidence supporting robot-assisted therapy effectiveness. As noted in Section 9.6.1, a comprehensive study reported in early 2026 confirmed that social robots are efficacious for children with autism spectrum disorder, achieving therapeutic outcomes equivalent to conventional interventions while demonstrating significantly higher engagement levels. This finding strengthens the case for robot-assisted therapy while highlighting engagement as a distinct advantage of robotic approaches.

Research published in Frontiers in Psychology examined human-AI attachment formation, providing theoretical frameworks for understanding how humans develop intimate relationships with artificial entities (Shu et al., 2026). The analysis identified specific mechanisms—reciprocal disclosure, consistent availability, and perceived understanding—that parallel human attachment formation while noting critical differences in reciprocity and authenticity.

The American Psychological Association's January 2026 report sounded a cautionary note, summarizing research indicating that excessive use of AI companion tools may worsen loneliness and erode social skills over time (American Psychological Association, 2026). The findings underscore the dual nature of robot companionship: short-term comfort may come at the cost of long-term social capability, particularly for individuals already experiencing isolation.

### 9.11.5 Implications for Psychological Design

These early 2026 developments collectively signal that the psychological dimensions of human-robot interaction are transitioning from academic concern to regulatory imperative and commercial priority. The emergence of stress-aware systems demonstrates that psychological wellbeing is becoming an engineering objective, not merely an ethical consideration. Regulatory frameworks in California and South Korea establish precedents for legal protection of psychological safety in human-robot interaction. Research advances provide both tools for evaluation and evidence for cautious optimism about therapeutic applications, while reinforcing concerns about dependency and social skill erosion.

The trajectory suggests that psychological factors will increasingly determine which humanoid robot deployments succeed or fail. Organizations that neglect trust dynamics, attachment risks, and cultural differences may face not only user rejection but regulatory sanction. Conversely, systems designed with explicit attention to psychological impact—from stress-aware movement to attachment-sensitive interaction patterns—may achieve sustainable integration into human environments.

## 9.12 Conclusion

The psychology of human-robot coexistence represents one of the most fascinating and consequential domains of twenty-first-century science. As humanoid robots transition from science fiction to everyday reality, understanding how humans perceive, trust, emotionally relate to, and ultimately coexist with these artificial entities becomes essential for designing systems that enhance rather than diminish human flourishing.

Several key themes emerge from this examination. First, human psychological responses to robots are not arbitrary but reflect deep evolutionary and cultural patterns. The uncanny valley manifests cognitive conflict between human-like appearance and machine behaviour.

Anthropomorphism reflects automatic social cognition mechanisms honed over millions of years. Cultural differences in acceptance reflect philosophical traditions about the nature of consciousness and the human-machine boundary.

Second, psychological responses are malleable and context-dependent. The same robot may be a trusted colleague in one context and a threatening presence in another. Emotional bonds develop not from robot sophistication alone but from situated relationships where robots meet genuine human needs. Cultural narratives and individual experiences dramatically shape whether robots are perceived as helpers or threats.

Third, psychological considerations must be integrated early in system design and deployment planning, not addressed after systems are already in use. Boston Dynamics' decision to omit facial features from Atlas, EX Robots' deliberate cultivation of emotional attachment, UBTECH Walker's approachable aesthetic, and Agility Robotics' workplace integration strategies all reflect psychological insights embedded in design. As robots become more sophisticated, the psychological design challenge intensifies.

Fourth, benefits and risks coexist. Robots provide genuine therapeutic value for autism, dementia, and loneliness. They enable workplace productivity gains and safety improvements. They offer educational personalization and companionship for isolated individuals. Yet they also risk deepening social isolation, creating addiction, threatening employment, and enabling surveillance and manipulation. Whether robots enhance or diminish wellbeing depends critically on design choices, deployment contexts, and regulatory frameworks.

Fifth, cultural differences are profound and consequential. Japan's comfort with anthropomorphic robots, China's pragmatic embrace driven by demographic and economic pressures, and American ambivalence rooted in individualism and dystopian narratives create vastly different markets and trajectories. Global robot developers must navigate these differences, potentially offering different designs and interaction paradigms for different cultural contexts.

Looking forward, several research priorities emerge. Longitudinal studies tracking psychological outcomes—mental health, social relationships, job satisfaction—over years rather than months are essential for understanding long-term impacts. Cross-cultural research illuminating why acceptance varies so dramatically can inform both design and policy.

Developmental studies examining how children growing up with robots differ from prior generations will reveal whether early concerns prove prescient or exaggerated.

Ethical frameworks must address emerging questions about appropriate boundaries in human-robot relationships, protection of vulnerable populations, informed consent for emotional manipulation, and balancing innovation with psychological safety. Regulatory frameworks must evolve beyond physical safety to encompass psychological wellbeing, data privacy, and prevention of exploitative design.

The regulatory developments of early 2026—California's SB 243 and South Korea's AI Basic Act—represent pioneering efforts to codify psychological safety in human-robot interaction. These frameworks will likely serve as templates for other jurisdictions grappling with similar concerns. The emergence of stress-aware robot systems demonstrates that psychological wellbeing is transitioning from design consideration to engineering specification.

Ultimately, the psychology of human-robot coexistence is not predetermined. Current trajectories reflect design choices, business models, cultural values, and policy decisions that could be otherwise. If humanoid robots are designed to maximize engagement at any psychological cost, if deployment prioritizes profit over worker wellbeing, and if regulation lags behind technological capabilities, outcomes may be dystopian. Alternatively, if psychological insights guide ethical design, if deployment respects human dignity and autonomy, and if robust regulation protects the vulnerable, humanoid robots could genuinely enhance human flourishing.

## Questions

1. Uncanny Valley Analysis: Explain the uncanny valley phenomenon and its relevance to humanoid robot design in 2026. How does the recent finding that

"the next uncanny valley is psychological" extend the original concept? Provide examples of how different platforms (Boston Dynamics Atlas, EX Robots, UBTECH Walker) navigate uncanny valley concerns.

2. Trust Dynamics: Distinguish between performance trust and relational trust in human-robot interaction. Why do "industrial applications prioritise performance-based trust while social-care applications emphasise relational trust"? Describe scenarios where over-trust (automation bias) or under-trust (disuse) would be problematic.
3. Cross-Cultural Comparison: The chapter presents striking differences between U.S. and Japanese participants in cooperating with artificial agents. What underlying cultural factors (religious traditions, philosophical frameworks, cultural narratives) explain these differences? How do these differences manifest in robot design and deployment strategies?
4. Emotional Attachment Ethics: Evaluate the ethics of designing robots explicitly to create emotional attachments. Consider the example of CES 2026 companion robots "designed for emotional connection, not productivity." Under what circumstances is intentional emotional bonding appropriate versus exploitative? How do vulnerable populations (children, elderly, cognitively impaired) change the ethical analysis?
5. Therapeutic Applications: The chapter describes robot-assisted therapy for autism spectrum disorder and dementia showing measurable benefits. What psychological mechanisms explain these therapeutic effects? What limitations or risks should clinicians consider when implementing robot-assisted interventions?
6. Workplace Psychology: Describe the psychological challenges workers face when humanoid robots are introduced to their workplace. How can organizations manage job anxiety, facilitate positive human-robot team dynamics, and support worker identity reconstruction? What organizational practices would make robot integration psychologically healthier?
7. Developmental Impact: Analyse the potential impacts of humanoid robots on children's cognitive, social-emotional, and moral development. What benefits does research support? What concerns do critics raise? How might "screen time" debates inform policy about children's robot interaction time?
8. Technology Acceptance Model: Traditional TAM focuses on perceived usefulness and ease of use. Why is this insufficient for humanoid robots? What additional factors must the extended TAM incorporate, and why do social presence, emotional design, and safety perception matter more for humanoid robots than for traditional computer systems?
9. Privacy and Surveillance: Humanoid robots with cameras, microphones, and internet connectivity create unprecedented surveillance capabilities. Describe the privacy concerns this raises in home, workplace, and public space contexts. What technical, policy, and ethical frameworks could address these concerns while enabling legitimate robot functionality?
10. Cultural Adaptation Strategy: Imagine you are designing a humanoid robot platform for global deployment across Japan, China, and the United States. Based

on the cultural differences described in the chapter, what design modifications, interaction paradigms, and deployment strategies would you employ for each market? Justify your recommendations with specific psychological and cultural considerations.

11. Regulatory Response to Psychological Risks: California's SB 243 and South Korea's AI Basic Act represent early regulatory responses to psychological risks in human-AI interaction. Compare these frameworks: What specific psychological risks does each address? How do their approaches differ in terms of enforcement mechanisms and scope? What gaps remain in protecting vulnerable populations from emotional manipulation or dependency? Discuss whether regulation should target robot design, deployment practices, or user education—or all three.

## References


ACM/IEEE International Conference on Human-Robot Interaction. (2026, March 16–19). *HRI 2026: Empowering society [Conference proceedings]*. Edinburgh,. https://humanrobotinteraction.org/2026/

American Psychological Association. (2026). *AI chatbots and digital companions are reshaping emotional connection*. APA Monitor on Psychology. https://www.apa.org/monitor/2026/01-02/trends-digital-ai-relationships-emotional-connection

Association for Advancing Automation. (2025). *Humanoid robots in 2025: The next stage of evolution*. Automate.org Industry Insights. https://www.automate.org/industry-insights/humanoid-robots-are-evolving

Campagna, G., & Rehm, M. (2025). A systematic review of trust assessments in human–robot interaction. *ACM Transactions on Human-Robot Interaction, 14*(2), Article 30. https://doi.org/10.1145/3706123

Castelo, N., & Sarvary, M. (2022). Cross-cultural differences in comfort with humanlike robots. *International Journal of Social Robotics, 14*(8), 1865–1873. https://doi.org/10.1007/s12369-022-00920-y

Cirasa, C., & Conti, D. (2025). Mapping trust and cultural dimensions in human-robot interaction: A scoping review approach. *Computers in Human Behavior Reports, 19*, Article 100763. https://doi.org/10.1016/j.chbr.2025.100763

David, D., Baxter, P., Belpaeme, T., Billing, E., Cai, H., Cao, H.-L., Ciocan, A., Garcia, D. H., Esteban, P. G., Kennedy, J., Liu, H., Matu, S., Mazel, A., Selescu, M., Senft, E., Thill, S., Vanderborght, B., Vernon, D., Ziemke, T. (2025). Efficacy and effectiveness of robot-assisted therapy for autism spectrum disorder: From lab to reality. *Science Robotics, 10*(109), eadl2266. https://doi.org/10.1126/scirobotics.adl2266

Daws, R. (2026, January). CES 2026: AI companion robots combat loneliness with emotional support. *WebProNews*. https://www.webpronews.com/ces-2026-ai-companion-robots-combat-loneliness-with-emotional-support/

Ghaiumy Anaraky, R., & Mohapatra, B. (2026). Assistive intelligence: A framework for AI-powered technologies across the dementia continuum. *Journal of Ageing and Longevity, 6*(1), Article 8. https://doi.org/10.3390/jal6010008

Guan, J., Chen, X.-P., Yu, S., & Qin, X. (2025). Explaining disparities in robot applications among nations and regions: A cross-level lens of cultural tightness-looseness. *PLOS One, 20*(4), e0321173. https://doi.org/10.1371/journal.pone.0321173

Han, J., & Conti, D. (2025). Recent advances in human-robot interactions. *Applied Sciences, 15*(12), Article 6850. https://doi.org/10.3390/app15126850

Holt-Lunstad, J., Smith, T. B., Baker, M., Harris, T., & Stephenson, D. (2015). Loneliness and social isolation as risk factors for mortality: A meta-analytic review. *Perspectives on Psychological Science, 10*(2), 227–237. https://doi.org/10.1177/1745691614568352

Ibrahim, F., Münscher, J.-C., Daseking, M., & Telle, N.-T. (2025). The technology acceptance model and adopter type analysis in the context of artificial intelligence. *Frontiers in Artificial Intelligence, 7*, Article 1496518. https://doi.org/10.3389/frai.2024.1496518

Institute of Electrical and Electronics Engineers. (2026). *The impact of technology in 2026 and beyond: An IEEE global study* [Report]. IEEE. https://transmitter.ieee.org/impact-of-technology-2026/

International Federation of Robotics. (2025). *World Robotics 2025: Industrial robots* [Press release]. https://ifr.org/ifr-press-releases/news/global-robot-demand-in-factories-doubles-over-10-years

Jenkins, R., Fitzpatrick, S., Byrne, A. K., & Baker, D. H. (2025). Neural correlates of the uncanny valley effect for robots and hyper-realistic masks. *PLOS One, 20*(3), e0311714. https://doi.org/10.1371/journal.pone.0311714

Jones Lang LaSalle. (2025). *How humanoid robots could reshape work [Industry insights report]*. JLL. https://www.jll.com/en-us/insights/how-humanoid-robots-could-reshape-work

Kalil, M. (2026). *China's fake humans creep out of the uncanny valley*. https://mikekalil.com/blog/china-fake-human-era/

Kang, H., Freitas dos Santos, T., Ben Moussa, M., & Magnenat-Thalmann, N. (2025). *Mitigating the uncanny valley effect in hyper-realistic robots: A student-centred study on LLM-driven conversations*. arXiv preprint https://arxiv.org/abs/2503.16449. https://arxiv.org/abs/2503.16449

Kaptelinin, V., & Dalli, T. (2025). *The social context of human-robot interactions*. arXiv preprint https://arxiv.org/abs/2508.13982. https://arxiv.org/html/2508.13982v1

Karpus, J., Shirai, R., Verba, J. T., Schulte, R., Weigert, M., Bahrami, B., Watanabe, K., & Deroy, O. (2025). Human cooperation with artificial agents varies across countries. *Scientific Reports, 15*, Article 10000. https://doi.org/10.1038/s41598-025-92977-8

Kelly, P. A., Cox, L. A., Petersen, S. F., Gilder, R. E., Blann, A., Autrey, A. E., & MacDonell, K. (2021). The effect of PARO robotic seals for hospitalised patients with dementia: A feasibility study. *Geriatric Nursing, 42*(1), 37–45. https://doi.org/10.1016/j.gerinurse.2020.11.003

Kling, M., Haeussl, A., Dalkner, N., Reininghaus, B., & Reininghaus, E. Z. (2025). Social robots in adult psychiatry: A summary of utilisation and impact. *Frontiers in Psychiatry, 16*, Article 1506776. https://doi.org/10.3389/fpsyt.2025.1506776

Li, M., Liu, M., Wu, Z., Lin, X., Zhang, J., Yan, M., Xie, Z., Zhang, C., Wen, C., Xu, L., Shen, S., & Wang, C. (2026). *Towards motion Turing test: Evaluating human-likeness in humanoid robots*. arXiv preprint https://arxiv.org/abs/2603.06181. https://arxiv.org/abs/2603.06181

Lyu, Y., Cao, Y., & Liu, J. (2025). Investigating service robot acceptance factors: The role of emotional design, communication style, and gender groups. *Information, 16*(6), Article 463. https://doi.org/10.3390/info16060463

Maaz, N., Mounsef, J., & Maalouf, N. (2025). CARE: Towards customised assistive robot-based education. *Frontiers in Robotics and AI, 12*, Article 1474741. https://doi.org/10.3389/frobt.2025.1474741

Mori, M., MacDorman, K. F., & Kageki, N. (2012). The Uncanny Valley [From the Field]. *IEEE Robotics & Automation Magazine, 19*(2), 98–100. https://doi.org/10.1109/MRA.2012.2192811

NEC Corporation. (2026). *NEC develops physical AI that anticipates human movement and psychological states* [Press release]. https://www.nec.com/en/press/202603/global_20260312_01.html

Padilla, S. (2025). SB 243 *(Padilla, Chapter 866, Statutes of 2025): Companion chatbots [California Senate Bill]*. California Legislative Information. https://leginfo.legislature.ca.gov/faces/billNavClient.xhtml?bill_id=202520260SB243

Pattan, K. (2025, February 28). *What is the uncanny valley?* Dataconomy. https://dataconomy.com/2025/02/28/what-is-the-uncanny-valley/

Preidt, R. (2026, January 2). *Robot-assisted therapy beneficial for children with autism.* Medical Xpress. https://medicalxpress.com/news/2026-01-robot-therapy-beneficial-children-autism.html

Riar, M., Weber, M., Ebert, J., & Morschheuser, B. (2025). Can gamification foster trust-building in human-robot collaboration? An experiment in virtual reality. *Information Systems Frontiers, 27*(5), 2001–2026. https://doi.org/10.1007/s10796-024-10573-z

Rudenko, I., Rudenko, A., Lilienthal, A. J., Arras, K. O., & Bruno, B. (2025). The child factor in child-robot interaction: Discovering the impact of developmental stage and individual characteristics. *International Journal of Social Robotics, 17*(2). https://doi.org/10.1007/s12369-024-01121-5

Sánchez-Sancho, Z., Huertas-Nieves, R., & Pérez-Marín, C. (2025). Effectiveness of robot-mediated learning in fostering children's social and cognitive development. *Applied Sciences, 15*(7), Article 3567. https://doi.org/10.3390/app15073567

Shim, K.-S. (2026, January 22). *South Korea launches landmark laws to regulate AI, startups warn of compliance burdens.* Reuters. https://www.reuters.com/world/asia-pacific/south-korea-launches-landmark-laws-regulate-ai-startups-warn-compliance-burdens-2026-01-22

Shin, J. J., Hamdi, A., Chen, A. B., Hu, Y., & Zhang-Kennedy, L. (2026). Toward safe child-robot interactions: Exploring children's and parents' privacy perceptions of humanoid social robots. *Lecture Notes in Computer Science.* https://doi.org/10.1007/978-981-95-2398-6_18

Shu, C., Lai, K., & He, L. (2026). Human-AI attachment: How humans develop intimate relationships with AI. *Frontiers in Psychology, 17*, Article 1723503. https://doi.org/10.3389/fpsyg.2026.1723503

Stower, R., Tisserand, Y., Sabouret, N., & Lugrin, J.-L. (2025). Reframing social presence for human-robot interaction. In *Proceedings of the 2025 IEEE/RSJ International Conference on Intelligent Robots and Systems (IROS).* IEEE. https://ieeexplore.ieee.org/document/10973988/

The Outpost AI. (2026, January). *CES 2026 reveals AI companion robots designed for emotional connection, not productivity.* The Outpost. https://theoutpost.ai/news-story/ces-2026-reveals-ai-companion-robots-designed-for-emotional-connection-not-productivity-22791/

Vogels, W. (2025). *Tech predictions for 2026 and beyond.* All Things Distributed (Amazon Web Services Blog). https://www.allthingsdistributed.com/2025/11/tech-predictions-for-2026-and-beyond.html

World Health Organization. (2025). *From loneliness to social connection: Charting a path to healthier societies.* WHO Commission on Social Connection report. https://www.who.int/publications/i/item/9789240114777

**Jerry Huang** is affiliated with Kleiner Perkins and has a background in AI, machine learning, security, scalable systems, and applied mathematics. His prior experience includes technical work across AI/ML evaluation, infrastructure, security, access control, and privacy-preserving biometric authentication. He holds BS/MS degrees in Computer Science from Georgia Tech and has pursued graduate study in Applied Mathematics at the University of Chicago.

**Ken Huang** is a globally recognized expert in AI and Web3 security, a prolific author, and a leading figure in shaping industry standards. He is the CEO and Chief AI Officer (CAIO) of DistributedApps.ai, specializing in Generative AI training and consulting. Ken is deeply involved in driving the development of secure AI systems. He serves as a Research Fellow and Co-Chair of the AI Safety Working Groups at the Cloud Security Alliance (CSA), leading Agentic AI initiatives. He is also a Co-Chair of the AI STR Working Group at the World Digital Technology Academy under the UN Framework, and a core contributor to the OWASP GenAI project, focusing on Agentic AI security. His expertise extends to his contributions to OWASP's Top 10 Risks for LLM Applications and his active participation in the NIST Generative AI Public Working Group. He is also a member of Open AI forum. Ken has authored and edited numerous influential books, including: *Agentic AI: Theories and Practices* (Springer, June 2025). *Securing AI Agents: Foundations, Frameworks, and Real-World Deployment* (Springer, October 2025). *Beyond AI: ChatGPT, Web3, and the Business Landscape of Tomorrow* (Springer, 2023). *Generative AI Security: Theories and Practices* (Springer, 2024). *Practical Guide for AI Engineers* (Volumes 1 and 2, DistributedApps.ai, 2024). *The Handbook for Chief AI Officers: Leading the AI Revolution in Business* (DistributedApps.ai, 2024). *Web3: Blockchain, the New Economy, and the Self-Sovereign Internet* (Cambridge University Press, 2024). *Web3 Applications Security and New Security Landscape: Theories and Practices* (Springer, 2024). His co-authored book, "Blockchain and Web3: Building the Cryptocurrency, Privacy, and Security Foundations of the Metaverse" (Wiley, 2023), was recognized as a must-read by TechTarget. *A sought-after speaker, Ken has presented at prestigious global forums, including Davos WEF, ACM, IEEE, CSA AI Summit, the Depository Trust & Clearing Corporation (DTCC), and World Bank conferences.* Explore Ken's books on Amazon: https://www.amazon.com/author/kenhuang.

**Modar Alaoui** is the founder and general partner of ALM Ventures, a Silicon Valley venture capital firm focused on humanoids, physical AI, and spatial intelligence. He is also the founder and chair of the Humanoids Summit, a global forum that brings together leading builders, investors, and operators shaping humanoid robotics, with editions in Silicon Valley, Tokyo, and London. He operates at the intersection of capital, strategy, and infrastructure, and has played a central role in advancing the humanoid robotics ecosystem toward commercial deployment and mass production. Modar is also the founder and chair of the Bay Area Robotics Association (BARA), a member-led industry organization established to connect capital, startups, enterprises, and public sector stakeholders across Silicon Valley and global robotics hubs. He created International Humanoids Day, a global initiative advancing "humanoids for humanity," aligning industry leaders around ethical development, global standards, and international collaboration in humanoid robotics. He contributes to policy efforts shaping U.S. humanoid strategy and long-term deployment frameworks, helping strengthen U.S. competitiveness in physical AI. Earlier in his career, Modar founded Eyeris, where he developed computer vision and sensor fusion AI systems for in-cabin sensing. His work combined software and hardware across perception and edge compute, providing hands-on experience with real-world systems operating around people. He studied behavioral science and communication at Concordia University in Montreal, forming the foundation for his work in human behavior understanding and machine perception. This combination of behavioral insight and physical AI systems continues to inform his work in humanoids and real-world autonomy. His work and perspectives have been featured in major media outlets including Bloomberg, CNBC, The Wall Street Journal, Reuters, Wired, and Forbes. In The Humanoid Economy, he draws on his role in building the humanoid robotics ecosystem to frame the next phase of industrial transformation.

# Technology Convergence and the Future of Human Purpose

10

Ken Huang and Jerry Huang

## 10.1 Introduction

Across the preceding chapters, this book has examined humanoid robotics and physical AI through multiple lenses: the technical foundations of hardware and software (Chaps. 2–5), safety and economic implications (Chaps. 6–7), ethical and governance challenges (Chap. 8), and psychological dimensions of human-robot coexistence (Chap. 9). This final chapter steps back to examine the broader convergence of technologies and their collective implications for humanity's future.

We stand at an extraordinary moment in history. Multiple technological revolutions—artificial intelligence,robotics, biotechnology, brain-computer interfaces, and life extension—are not merely advancing in parallel but converging and amplifying each other's impacts. As one industry analysis observes, these trends are increasingly converging and reinforcing one another to produce compounding opportunities (Krasadakis, 2026). The compound effects of convergence may dwarf the impact of any single technology.

This convergence forces humanity to confront questions that have been theoretical for generations but are becoming practical within our lifetimes: What is human purpose when machines can perform all human tasks more effectively? How should societies distribute the abundance generated by tireless AI and robots?

K. Huang (✉)
DistributedApps.ai, Fairfax, VA, USA
e-mail: ken@distributedapps.ai

J. Huang
Kleiner Perkins, San Francisco, CA, USA



K. Huang et al. (eds.), *Humanoid Robots and Physical AI*, Advances in Data Analytics, AI, and Smart Systems, https://doi.org/10.1007/978-3-032-33669-9_10

How will ageing societies with shrinking workforces adapt to demographic collapse? What does it mean to be human when biological and machine intelligence merge?

These questions have become urgent due to rapidly compressing timelines. Elon Musk predicts with great confidence that work will become optional within 10–20 years, claiming: a state of "universal high income"—not merely basic income—in which everyone enjoys access to high-quality medical care, food, housing, and transport in a regime of sustainable abundance (Musk, 2026). Whether such optimism proves justified or naive, the mere fact that serious technologists make such claims indicates the pace and potential magnitude of transformation.

The demographic context amplifies this urgency. As Japan's care worker deficit scales into the hundreds of thousands over the coming decades due to its rapidly ageing population (Xinhua, 2025), the nation faces a severe caregiving crisis.

China and much of Europe follow similar trajectories. These societies lack sufficient young workers to care for their elderly populations, creating a desperate need for robotic assistance regardless of whether full automation proves economically transformative.

Simultaneously, longevity science promises dramatic life extension. The longevity-technology sector surpassed $600 billion in global valuation in 2026 (still maturing), with AI-accelerated drug discovery, personalised medicine, and biological age tracking transitioning from research to clinical application (Evoailabs, 2026). If humans live substantially longer—potentially decades longer—while robots perform most productive labour, questions about purpose and meaning intensify profoundly.

This chapter examines these converging trends and their implications across multiple time horizons. It analyses how AI, robotics, biotech, and brain-computer interfaces amplify each other, and explores concepts like universal high income and abundance economics alongside their plausibility and broader implications. The discussion moves through ageing societies and the intersection of robotics with eldercare and longevity, the transformation of work from necessity to optionality, and the wealth distribution concerns that follow. From there, it traces the pathway from narrow AI through AGI to potential superintelligence, and the workforce shifts likely to unfold through the 2030–2040 transition. Four divergent future scenarios spanning 2030–2050 are presented, followed by a practical roadmap and philosophical reflections on human purpose. The chapter concludes by arguing that technological trajectories remain fundamentally shaped by human choices and values—not predetermined by technological possibility.

## 10.2 The Great Convergence: Accelerating Technology Integration

The power of technological convergence lies in compounding effects where advances in one domain enable breakthroughs in others, creating exponential rather than linear progress. As of 2026, multiple transformative technologies are reaching maturity simultaneously and beginning to integrate in ways that amplify impacts.

Figure 10.1 places humanoid robotics at the centre of a web of six reinforcing technology domains—artificial intelligence, biotechnology and longevity, brain–computer interfaces, sensors and digital twins, materials and manufacturing, and energy and computing infrastructure—and emphasises that progress in any one of these domains accelerates progress in the others, producing the compounding dynamic that defines the 2025–2030 convergence window.

### 10.2.1 AI and Robotics: Embodied Intelligence

Chapter 1 established the convergence of AI and robotics as central to this book's thesis. However, the pace of this convergence has accelerated dramatically. 2025

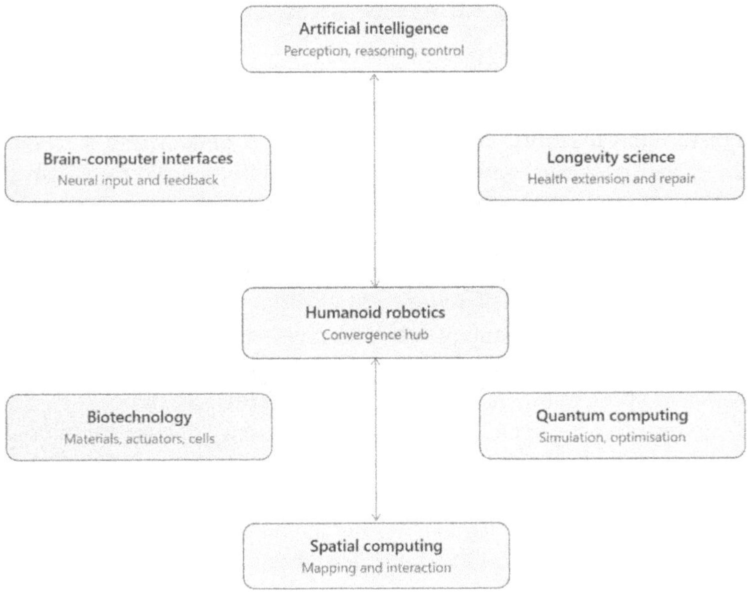


**Fig. 10.1** The technology convergence web: Humanoid robotics at the centre of 6 reinforcing technology domains

has been described as a breakthrough year for AI-driven robotics, with AI enhancing robot safety and enabling closer human-robot collaboration through integrated vision systems that allow robots to detect obstacles and instantly recalculate motion paths (Tech Equity AI, 2025).

The integration goes beyond perception to control and learning. AI-driven voice control now enables robots to interpret spoken instructions and autonomously generate the code needed to execute tasks, enabling them to work more flexibly alongside human operators (Axis Intelligence, 2026). This eliminates the traditional robotics programming bottleneck, where each new task required expert coding.

At CES 2026, Hyundai Motor Group unveiled its AI Robotics Strategy under the theme "Partnering Human Progress," highlighting human-robot collaboration and strategies to accelerate commercialisation through group-wide technological integration (Hyundai Motor Group, 2026). Major industrial groups treating AI-robotics integration as a strategic priority signals mainstream acceptance.

The economic implications are substantial: per-unit bill-of-materials cost is projected to fall from approximately $35,000 (2025) to $13,000–$20,000 by 2035, with Scenario 1 (rapid AI adoption) reaching $15,000–$20,000 by 2030 as the optimistic end of the range (Osis et al., 2026). As costs fall and capabilities rise through AI integration, deployment will accelerate exponentially.

### 10.2.2 Biotechnology and AI: Accelerating Life Sciences

The integration of AI with biotechnology promises to compress decades of biological research into years or months. AI-accelerated drug discovery has transitioned from concept to clinical reality: the biotech industry is anticipating a landmark year in 2026 as multiple drug candidates discovered and optimised by AI advance into mid- to late-stage clinical trials, with particular concentration in oncology and rare diseases.

Novel approaches combine multiple technologies. AI is being paired with "organs on a chip"—tiny versions of tissues grown from blood samples—creating a bio-intelligence platform combining robotics, nano sensors, stem-cell biology, and AI (Super Age, 2025). This enables personalised drug testing without human trials, dramatically accelerating development while improving safety.

Industry analysts have begun describing the emerging fusion of biological systems and AI as "bio-digital AI" (Cogent Information Resources, 2026). Applications span from AI-designed proteins and synthetic biology to direct neural interfaces interpreting biological signals. For humanoid robotics, this convergence opens the door to remarkable hybrid systems. Future robots may eventually utilise synthetic muscle tissues that offer more natural, energy-efficient actuation than traditional electric motors, or employ biologically inspired sensory arrays that allow them to perceive chemical changes and delicate textures with human-like sensitivity. By bridging the gap between cold hardware and living tissue, bio-digital AI pushes robotics beyond purely mechanical engineering into the realm of synthetic biology.

### 10.2.3 Brain-Computer Interfaces: Merging Mind and Machine

Brain-computer interfaces (BCIs) represent perhaps the most profound convergence, directly connecting human neural activity to computational systems. As of 2026, BCIs have crossed a threshold from laboratory curiosity to clinical deployment, with over a dozen individuals now living with Neuralink implants across the US, Canada, and UK (Singh, 2025).

The implications for robotics are direct: BCIs could enable humans to control humanoid robots through thought alone, experiencing robot sensory inputs and actuating robot movements as natural extensions of their own bodies. This human-robot merger transcends collaboration towards genuine symbiosis.

Three key BCI trends for 2026 include expanded clinical trials beyond paralysis applications to cognitive enhancement, improving spatial and temporal resolution enabling more precise control, and the emergence of ethical frameworks for enhancement applications beyond medical necessity (STAT News, 2025).

### 10.2.4 Longevity Science and Robotics

Life extension technologies converge with robotics in multiple ways. Robotic surgery systems enable precision interventions that extend healthspan. Industry analysts project that robot-assisted surgeries will become routine over the next 10–15 years, with AI guidance increasingly augmenting—and in narrowly defined tasks potentially exceeding—current human surgeon capabilities.

Care robots support ageing populations, addressing the care worker shortage while potentially enabling older adults to remain independent longer. Exoskeletons eliminate mobility limitations: exoskeletons—wearable robotic devices that augment existing limbs—can support older adults' mobility and may, over time, reduce reliance on wheelchairs and walkers (University of Florida Institute on Aging, 2024).

The longevity industry is entering the era of "Healthspan Engineering," focusing on quality of life rather than merely extending lifespan (Evoailabs, 2026). If humans routinely live to 100 or beyond, the nature of careers, retirement, and life structure transforms fundamentally—especially if robots handle productive labour.

### 10.2.5 Quantum Computing and AI: Computational Acceleration

Quantum computing, while tangential to this book's primary focus on humanoid robotics, represents an enabling technology that could dramatically accelerate AI development. Quantum algorithms could solve optimisation problems intractable for classical computers, enabling better robot motion planning, molecular simulation for drug discovery, and cryptography securing robot communications.

The convergence creates feedback loops: AI designs better robots, robots generate data to train better AI, quantum computers accelerate AI training, and better AI designs improved quantum hardware. Each advancement in one domain accelerates others, creating exponential rather than linear progress curves.

### 10.2.6 Spatial Computing and Robotics: Inhabiting the Physical-Digital World

Spatial computing—the integration of AR, XR, and real-time environment mapping—represents the perceptual infrastructure that allows robots and humans to share a common understanding of physical space. Where earlier technologies gave robots arms and AI gave them brains, spatial computing gives them situational awareness.

At its foundation is simultaneous localisation and mapping (SLAM), which enables robots to build and update three-dimensional models of their environment in real time. Rather than operating from pre-loaded maps, robots equipped with spatial computing can navigate dynamic environments—factory floors that reorganise, hospital corridors with moving people, homes with children and pets. This makes deployment in unstructured environments tractable for the first time.

The convergence with AR and XR creates a bidirectional perceptual layer. Humans wearing AR headsets can see robot intent—the planned trajectory of a robotic arm displayed as a ghost overlay—enabling safer, more intuitive collaboration. Conversely, robots can perceive and interpret human intent through spatial signals, anticipating where a person is moving before they arrive. This shared spatial vocabulary dissolves the traditional safety barrier between human and robot workspaces.

XR also dramatically accelerates robot training. Simulated environments rendered with spatial fidelity allow AI systems to accumulate millions of hours of embodied experience without physical hardware, then transfer learned behaviours to real robots with minimal degradation. The gap between simulation and reality—the "sim-to-real" problem that has long plagued robotics—narrows as spatial rendering quality improves.

Looking ahead, persistent spatial maps of entire cities and buildings—the so-called "spatial web"—would allow humanoid robots to localise themselves globally with centimetre precision, coordinate across fleets without explicit communication, and carry context from one physical location to another. A robot that has mapped a warehouse once carries that knowledge indefinitely, sharing it with every subsequent robot that enters the same space.

## 10.3 Universal High Income and the Economics of Abundance

In Sect. 7.4.3 (Wage Effects and Income Distribution), we discussed Elon Musk's vision of "Universal High Income" (UHI) which represents the most ambitious claim about humanoid robotics' economic implications. The claim states that AI and robots will create such abundance that everyone can live comfortably without working, with prosperity replacing mere subsistence. In the subsections here, let us explore the idea further.

### 10.3.1 Musk's Vision: From Scarcity to Abundance

Musk's vision is straightforward: if AI and robots can produce unlimited goods and services at near-zero marginal cost, scarcity disappears. Musk's forecasts have evolved across his public statements: in April 2024 (and reiterated at the World Economic Forum, January 2026) he predicted AI smarter than the smartest individual human by 2026; in December 2024 (restated on the Moonshots podcast in January 2026) he projected that by 2030 AI may surpass the intelligence of all humans combined; and his work-optional timeline of 10–20 years dates from public remarks in November 2025 and January 2026, when he argued that AI and humanoid robots will render money itself irrelevant (Musk, 2024; Quiroz-Gutierrez, 2026).

Even saving for retirement becomes a questionable strategy in this vision. Musk has argued that personal retirement saving becomes largely irrelevant in an AI-generated abundance economy (Quiroz-Gutierrez, 2026). If abundance is guaranteed regardless of individual accumulation, rational economic behaviour transforms fundamentally.

Musk assigns the broader Universal High Income (UHI) vision—that AI advances will leave humans with no need to work—a roughly 80% subjective probability, a level of confidence that distinguishes his position from more cautious technologists who emphasise profound uncertainty about post-work futures (Benzinga Staff, 2025). This 80% figure attaches to the UHI/work-optional outcome, not specifically to the AGI-by-2026 forecast in Table 10.1.

### 10.3.2 The Economic Logic of Technological Abundance

The theoretical foundation rests on exponentially declining production costs as AI and robots improve. Manufacturing costs dominated by labour fall precipitously as robots replace human workers. Robots that never tire, never require benefits, and continuously improve through machine learning can produce goods at costs approaching raw material inputs plus energy.

Current trends support this trajectory. Humanoid robot costs falling from $35,000 to $13,000–$17,000 over a decade suggest Moore's Law-like cost curves.

**Table 10.1** AGI timeline predictions—Selected forecasts, 2025–2026.

| Source | AGI timeline | Confidence | Notes |
| --- | --- | --- | --- |
| Elon Musk (2026) | 2026 | High (qualitative) | Optimistic entrepreneur; Tesla AI dev |
| Dario Amodei/Anthropic (2025) | 2026–2027 | High | Leading AI lab CEO; internal model progress |
| AI Multiple Expert Survey (2025) | 2040–2050 | >50% | Median of 8590 expert predictions |
| Metaculus Forecasters (2025) | 2031 | 50% | Aggregated prediction market consensus |
| "AI 2027" Report Consensus (2025) | ~2030 | Median | Multi-author forecasting collaboration |
| Conservative Academic Estimate | 2060–2070+ | Uncertain | Earlier surveys; now shifting earlier |

If this continues for two to three more decades, robot costs could approach thousands of dollars per unit. At such prices, robot deployment would be limited only by energy and raw materials, not capital.

Energy costs similarly decline through renewable sources approaching near-zero marginal cost. Solar and wind energy with battery storage create abundant electricity. Robots build and maintain renewable infrastructure, further driving costs towards zero. Raw materials might be recycled by automated systems or eventually extracted from asteroids by robotic space mining.

In this scenario, the constraint on abundance shifts from production capacity to resource allocation and distribution mechanisms. Physical goods become abundant; the question is whether prosperity is widely distributed or captured by robot owners.

### 10.3.3 Sceptical Perspectives: Institutional and Economic Barriers

Not all economists share Musk's optimism. Critical analysis argues that scarcity is not abolished by robots, however intelligent they are or however cheaply they make goods. Whether AI abundance produces universal prosperity or vast inequality depends entirely on property rights, competition policy, the rule of law, and stable governance (Daily Economy, 2025).

This perspective emphasises that economic outcomes reflect institutional structures, not just technological capabilities. If robot and AI ownership concentrates in few hands, abundance accrues to owners while most people face unemployment without income. Legal frameworks governing intellectual property, capital ownership, and revenue distribution determine whether abundance is shared or hoarded.

Historical precedent raises concerns. Previous technological revolutions—industrial mechanisation, electrification, computerisation—created immense wealth but distributed it unequally. Initial phases typically concentrated wealth before broader distribution through policy interventions such as labour regulations, antitrust enforcement, and progressive taxation. It is unclear why AI and robotics would automatically differ.

Practical barriers also deserve attention. Musk has not provided a concrete blueprint for universal high income, and many experts and members of the public express scepticism on the grounds that no detailed implementation plan has been published (Fox Business, 2025). The gap between vision and implementation plan is vast. How would universal high income be funded—through taxation of robot owners, direct government ownership of robots, or universal robot dividends from publicly owned AI systems? Each approach faces formidable political obstacles.

### 10.3.4 The Distribution Challenge: Universal High Income vs. Universal Basic Income

Even if abundance is achieved technically, distribution remains politically contentious. Universal Basic Income (UBI) proposals have been tested in pilot programmes: citizens in 18 US states, as well as the District of Columbia, have taken part in some form of basic income study (Cameron, 2025). These programmes typically provide \$500–\$1,600 monthly—enough for survival, but not prosperity.

Musk's "universal high income" implies substantially more—income levels sufficient for high-quality medical care, food, housing, and transport rather than mere subsistence. This suggests income levels enabling comfortable middle-class lifestyles, not mere subsistence. The difference between \$1,000 per month (typical UBI) and \$5,000–\$10,000 per month (plausible "high income") represents an order-of-magnitude larger fiscal commitment.

Ray Dalio warns that without deliberate redistribution, AI and humanoid robots will increasingly benefit the top 1%–10% relative to the rest of the population, necessitating new redistribution policy (Fortune Staff, 2025). The question is whether societies can implement such policies before inequality creates political instability.

## 10.4 Ageing Societies and the Longevity Revolution

Demographic ageing represents one of the most certain and consequential trends of the twenty-first century, intersecting with humanoid robotics in profound ways. The convergence of robotic eldercare, life extension technologies, and shrinking working-age populations creates both solutions and dilemmas.

Figure 10.2 visualises Japan's escalating care worker crisis across four key milestones. The 2020 baseline represents the pre-shortage state, with the shortfall

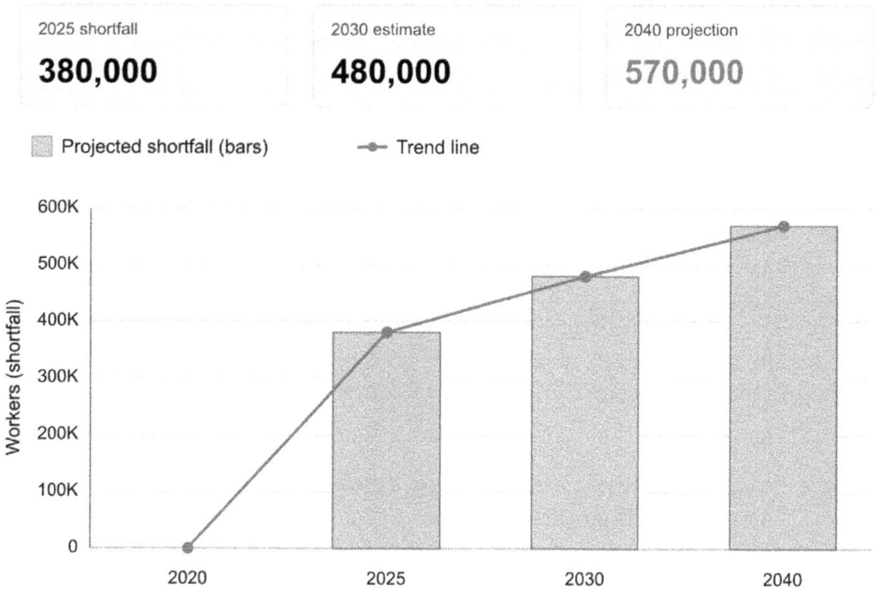


**Fig. 10.2** Japan's projected care worker shortage (2020–2040): An escalating demographic crisis requiring robotic solutions

growing from 380,000 in 2025 to a projected 570,000 by 2040—a 50% increase in just 15 years. The red trend line underscores the accelerating pace, illustrating why humanoid robotics has become a central pillar of Japan's eldercare strategy rather than a speculative future option.

### 10.4.1 The Demographic Crisis: Japan and China

Japan exemplifies the ageing society crisis: by 2025, 17.5% of Japan's population is aged 75 and older, and in 2024, people aged 65 and older accounted for 29.3% of the total population (Xinhua, 2025). The care worker shortage is severe: Japan's Ministry of Health anticipates a significant shortfall of care workers by 2025, with the gap projected to widen further by 2040.

China faces similar but potentially more severe demographics due to the one-child policy legacy. In 2025, new guidelines for elderly care service reforms called for advancing humanoid robots, brain-computer interfaces, and AI technologies to enhance senior care. China's eldercare robot industry is entering a period of rapid expansion (Xinhua, 2025).

These nations face impossible mathematics: fewer young workers must support more elderly dependents while also raising children and maintaining economic growth. Without technological solutions, living standards must decline. Robots represent not a luxury but a necessity for maintaining civilisation as currently structured.

### 10.4.2 Robotic Eldercare: Capabilities and Limitations

Contemporary eldercare robots provide multiple functions. Scientists have designed assistive robots to help older adults complete tasks including monitoring health metrics, detecting falls, helping seniors navigate their homes safely, and managing shopping lists and calendar tasks (University of Florida Institute on Aging, 2024). Physical assistance through mobility aids and exoskeletons enables independence. Social companionship addresses isolation.

However, significant limitations persist. Studies suggest that care robots can increase caregivers' workloads due to maintenance, monitoring, and troubleshooting; a 2021 study revealed mixed to negative views on care robots, with "malfunctioning" being the most frequently reported issue (Sinolytics, 2025). Robots designed to reduce burden sometimes add to it through unreliability and complexity.

Cost remains prohibitive, with exoskeleton walking aids costing more than $1,411 and bathing robots averaging $4,200–$7,000. Some seniors remain sceptical, viewing robots as emotionally distant and citing safety concerns. For universal deployment addressing the demographic crisis, costs must fall by orders of magnitude.

Practical deployments nonetheless provide cautious optimism. At a Shenzhen nursing home, robots play chess with seniors, offer AI-assisted therapy, and provide mobility assistance. These deployed systems demonstrate viability while revealing gaps between current and required capabilities.

### 10.4.3 The Longevity Revolution: Living Longer in a Robot World

Parallel to eldercare robotics, longevity science promises dramatic healthspan extension. The longevity-technology sector surpassed $600 billion in global valuation in 2026 (still maturing) (Evoailabs, 2026). This ecosystem blends AI, biotech, wearables, personalised medicine, and lifestyle optimisation.

Key developments include biological age tracking going mainstream through epigenetic clocks and biomarker panels that give users insights into their personal ageing pathways. Companies such as Altos Labs—backed by Jeff Bezos—made significant advances in 2025, appointing Dr. Joan Mannick as Senior Vice President, Chief Medical Officer, and Head of Product Development, signalling a pivot from foundational research towards clinical translation of cellular rejuvenation programming (Super Age, 2025).

The convergence of longevity science and robotics creates a potential paradox: humans living substantially longer while robots eliminate traditional work. If humans live actively to 100 or beyond without the need for employment, the traditional three-stage life model—education, work, retirement—collapses when the work phase disappears and the retirement phase spans 50 or more years.

This could be liberating: freed from labour necessity, humans might pursue learning, creativity, relationships, and personal growth across extended lifespans. Or it could be devastating: stripped of work-based purpose and identity, individuals could spiral into meaninglessness and despair. The outcome depends critically on cultural frameworks for meaning beyond economic productivity.

### 10.4.4 Financing Longevity in an Automated Economy

Longer lifespans create financing challenges even assuming abundance. If people live to 100 or beyond without working after age 40–50, resources must sustain 50–60 years of non-productive life. Traditional retirement savings and pension models collapse under this scenario.

Universal high income, if implemented, could solve this: guaranteed income regardless of age or work history provides security across extended lifespans. However, this requires successfully implementing the redistribution mechanisms discussed in Sect. 10.3—politically fraught and economically complex.

Alternative models include mandatory work-sharing (where humans work part-time throughout longer lives), staged transitions (where humans shift from physical to cognitive to creative work as robots assume each domain sequentially), or complete decoupling of income from labour through universal ownership stakes in robotic production.

## 10.5 The Transformation of Work and Human Purpose

Perhaps no question raised by humanoid robotics is more profound than: What is human purpose when work becomes optional? For most of history, human purpose centred on survival—acquiring food, shelter, and security. Industrial society shifted purpose towards productive contribution and economic advancement. What happens when robots can perform all productive tasks more effectively than humans?

### 10.5.1 Work as Necessity vs. Work as Choice

Currently, work is an economic necessity for most humans. We work to earn income required for survival and comfort. This necessity shapes identity, structures daily life, and provides social status and meaning. Notably, 59% of employees report being "not engaged" at their jobs, and 18% are "actively disengaged" (Gallup, 2025), yet they continue working because alternatives are economically infeasible.

When work becomes optional—either through universal high income or inherited robot dividends—this dynamic reverses. Only work that provides intrinsic satisfaction, social connection, or personal meaning would be undertaken. Tedious, dangerous, or monotonous work would cease, performed entirely by robots.

Some view this transformation positively: humans finally freed from drudgery to pursue authentic interests. Historical parallels include the shift from agricultural subsistence (once employing more than 90% of humans) to the diverse modern occupational landscape enabling specialisation in arts, sciences, and services. Perhaps robots enable a similar leap towards even more enriching activities.

Others see crisis. Musk himself has acknowledged that a generous universal income could coexist with social unrest as individuals confront a deeper crisis of meaning (Fox Business, 2025). If human identity and self-worth derive from productive contribution, removing that source of contribution could trigger widespread depression, substance abuse, and social breakdown.

### 10.5.2 The Rise of Creativity, Empathy, and Strategic Thinking

Even as robots assume physical and routine cognitive tasks, distinctly human capabilities may become more valuable. Industry analyses of 2025–2030 employment trends consistently project that the jobs which remain or emerge will emphasise uniquely human capabilities such as creative problem-solving, innovation, emotional intelligence, and interpersonal skills (EMA, 2025; World Economic Forum, 2025).

This represents work transformation rather than elimination. AI will take on repetitive, data-intensive tasks while humans focus on creativity, strategic thinking, and emotional intelligence. Automation handles execution; humans provide vision, judgement, and relationship management.

Education must adapt to this reality. The demand for human-centric skills may be increasing as AI adoption grows, requiring educational systems to emphasise AI literacy as a baseline requirement alongside distinctly human capabilities—curiosity, resilience, creativity, and empathy (World Economic Forum, 2025).

However, this optimistic framing assumes smooth transitions and universal access to education developing higher-order skills. Historical technological transitions have been wrenching, with displaced workers often unable to acquire skills for new opportunities. Whether this transition proves smoother or more brutal remains uncertain.

### 10.5.3 Purpose Beyond Economics: Creativity, Relationships, and Growth

If work becomes optional, alternative sources of purpose must fill the void. Several candidates emerge from human psychology and philosophy.

Creative expression: Humans possess innate drives towards creative output—art, music, writing, design. Freed from economic necessity, creative pursuits could flourish. However, when AI systems create professional-quality art, music, and writing, does human creative output retain meaning?

The joy may reside in process rather than product.

Relationships and community: Human connection and belonging provide deep meaning independent of economic productivity. Strengthening families, friendships, and communities could become a central life focus. However, as Chap. 9 explored, if robots provide companionship and emotional support, human relationships may face competition from more convenient artificial alternatives.

Personal growth and learning: Continuous learning, skill development, and self-improvement provide satisfaction and purpose. In an extended lifespan with economic security, humans could pursue multiple careers, master diverse skills, and engage in lifelong education. However, motivation for learning may decrease when skills have no economic or practical application.

Service and contribution: Many find purpose in helping others and contributing to community welfare. Volunteer work, mentoring, and civic engagement could absorb time freed from economic necessity. However, if robots perform care work, education, and community services more effectively, questions arise about where meaningful human contribution remains.

### 10.5.4 The Crisis of Meaning: Risks and Mitigation

The potential "crisis of meaning" in a post-work world should be taken seriously. As one analysis emphasises, the central challenge is whether societies can construct a future in which technology serves universal human flourishing rather than only those who own the robots or write the algorithms (Snabe, 2026).

Historical evidence provides warnings. Lottery winners often experience happiness decline after initial euphoria, as lack of structure and purpose outweighs financial security. Retirees forced into inactivity through disability or unemployment show elevated rates of depression compared to those who retire voluntarily with active lifestyles planned.

Mitigation strategies include cultural frameworks valorising non-economic contribution, robust community structures providing belonging and purpose, educational systems preparing humans for meaningful post-work lives from childhood, and perhaps mandatory service requirements or work-sharing arrangements ensuring continued engagement.

The outcome likely depends on whether the post-work transition happens gradually—allowing cultural adaptation—or catastrophically, with mass unemployment overwhelming social support systems. The pace of robot deployment may ultimately determine whether abundance proves utopian or dystopian.

## 10.6 Inequality in the Age of Abundance

Even granting that robots create material abundance, distribution of that abundance remains contentious. Economic history suggests a default tendency towards concentration of wealth in capital owners; deliberate policy interventions are required for broad distribution.

### 10.6.1 The Concentration Dynamic

Economic analysis reveals how automation could exacerbate inequality. A 2025 study found that AI's effects on wealth inequality show a temporal dichotomy: in the short term, AI exacerbates disparities in wealth distribution, while long-term outcomes depend on the extent of AI's influence across different technological domains. When AI advancements are primarily in automation, there is a marked tendency for wealth inequality to worsen (PMC/National Library of Medicine, 2025).

The mechanism is straightforward: robot owners capture productivity gains while workers face unemployment or wage suppression. If robots generate $1 trillion in economic value, traditional capitalism allocates that value to capital owners—robot manufacturers, deployment companies, investors—not to displaced workers.

Ray Dalio's warning is stark: AI and robots are poised to benefit the top 1% to 10% significantly more than everyone else, necessitating a new redistribution policy (Fortune Staff, 2025). Without such policy, abundance and poverty could coexist—warehouse robots generating enormous value for shareholders while displaced workers struggle economically.

Global inequality could worsen as well. Richer nations appear far better positioned to capitalise on AI's benefits, potentially deepening existing inequalities. A joint study by the International Labour Organization and Bangladesh's a2i programme projects that automation could displace roughly 60% of garment-sector jobs in Bangladesh by 2030, with knock-on effects already visible in a 30%+ workforce decline driven by sewing- and cutting-line automation; over a longer horizon, The Daily Star reports an estimated 53.8 lakh (5.38 million) jobs displaced across all sectors by 2041 (a2i & ILO, 2024; The Daily Star, 2024). The Center for Global Development frames this as one of three mechanisms by which AI may widen global inequality (Center for Global Development, 2025). Developing nations may lack the capital to deploy robots or the education systems to prepare workers for post-automation economies.

### 10.6.2 Competing Narratives: Inequality Expansion vs. Compression

Interestingly, some analysis suggests AI could reduce wage inequality by displacing high-income workers. Unlike previous waves of automation that increased both wage and wealth inequality, AI could reduce wage inequality through the displacement of high-income workers (International Monetary Fund, 2025).

However, this finding is complicated by two factors: high-income workers' tasks appear highly complementary with AI, potentially increasing their productivity, and these workers are better positioned to benefit from higher capital returns. The wealth inequality effect is particularly pronounced when firms can choose how much AI to adopt (International Monetary Fund, 2025).

This suggests wage inequality (between workers) may compress while wealth inequality (between workers and capital owners) expands dramatically. High-income professionals could see wages stagnate but investment returns soar through ownership of AI companies. Low-income workers face displacement without capital to invest in robot infrastructure.

### 10.6.3 Policy Interventions: Taxation, Ownership, and Redistribution

Addressing inequality in an automated economy requires active policy interventions. Key proposals include:

- Robot taxes: Taxing robot deployment to fund universal income. This faces objections about stifling innovation and difficulty distinguishing robots from other capital equipment.
- Progressive wealth taxation: Heavily taxing accumulated wealth—especially inherited wealth—to redistribute gains from automation. Implementation challenges include capital flight and valuation difficulties for illiquid assets.
- Universal capital ownership: Providing citizens with shares in robot companies or AI systems, creating direct ownership stakes in automation. Proposals range from sovereign wealth funds to universal stock grants for all citizens.
- Mandatory profit-sharing: Requiring companies deploying robots to share profits with displaced workers or communities, structured as wages for eliminated positions or community dividends.

Each approach faces political obstacles and unintended consequences. The key uncertainty is whether democratic societies can implement such redistribution before inequality triggers political backlash undermining the social fabric that enables shared prosperity.

## 10.7 The Path to AGI and Beyond: Intelligence Explosion

Humanoid robotics gains urgency from accelerating timelines towards artificial general intelligence (AGI)—AI matching or exceeding human capability across all cognitive domains—and potentially artificial superintelligence (ASI) transcending human intelligence entirely.

### 10.7.1 Shrinking AGI Timelines

AGI timeline predictions have compressed dramatically. While surveyed AI experts have historically estimated that AGI will probably emerge (with over 50% chance) between 2040 and 2050, leading technologists now predict far nearer timelines. Elon Musk has stated (April 2024; reiterated at the World Economic Forum, January 2026) that AI smarter than the smartest humans is likely by the end of 2026, and Dario Amodei, CEO of Anthropic, expects powerful AI systems may materialise within the next one to two years (Gizmodo, 2025).

More measured predictions suggest early AGI-like systems could begin emerging between 2026 and 2028, showing human-level reasoning within specific domains. The "AI 2027" report shifted consensus towards AGI arriving around 2030 (LessWrong, 2025), with Metaculus forecasters averaging a 25% chance of AGI by 2027 and 50% by 2031 (AI Multiple Research, 2025).

The wide variation—from 2026 to "never"—reflects profound uncertainty about definition, measurement, and technical challenges. However, the central tendency has shifted from 2060–2070 in earlier surveys to 2030–2040 in recent ones, indicating genuine acceleration.

Table 10.1 lists representative forecasts of when artificial general intelligence may emerge, drawn from public statements by frontier-AI lab leaders, aggregated expert surveys, and prediction markets. Although the dates span nearly five decades, the central tendency has shifted earlier in successive surveys, suggesting that the field is progressively revising its estimates downwards.

Fig. 10.3 plots the same forecasts on a horizontal timeline, marking each estimate's median year with a coloured diamond surrounded by a confidence band. The clustering of recent estimates around 2030–2040 is the load-bearing pattern: regardless of who is asked, the field is converging on a near-term arrival window.

### 10.7.2 Embodied AGI: When Robots Match Human Intelligence

AGI becomes particularly transformative when embodied in humanoid robots. A disembodied AGI, however intelligent, remains confined to digital domains. An embodied AGI—AGI-level intelligence controlling humanoid robot bodies—can perform any human task, physical or cognitive.

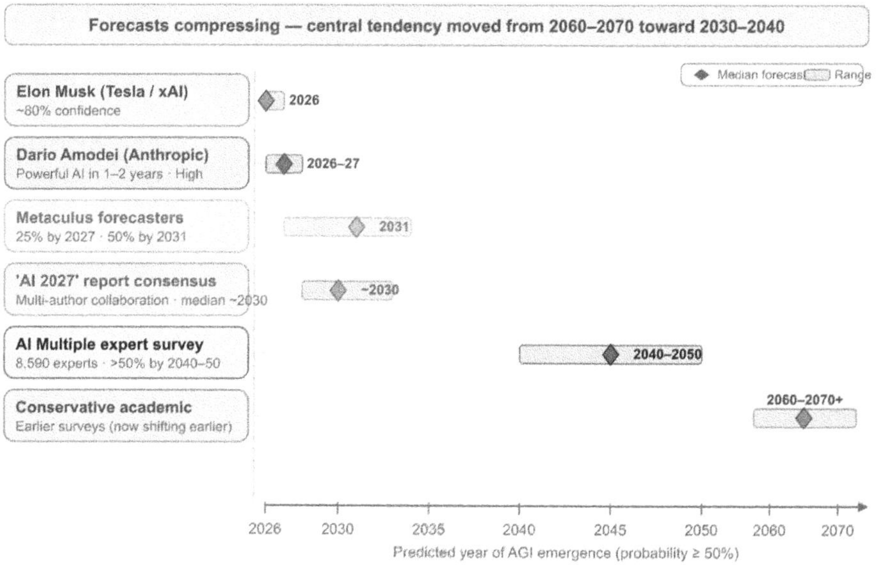


**Fig. 10.3** AGI timeline predictions: A comparison of expert forecasts showing compressing timelines (2025–2026 survey data)

The implications are profound. A warehouse could operate entirely autonomously with AGI-humanoid robots handling all tasks from inventory management to package sorting to vehicle loading, with problem-solving, adaptation, and coordination matching or exceeding human supervisors. Hospitals could similarly operate with AGI-humanoid doctors, nurses, and technicians providing care indistinguishable from or superior to human practitioners.

This scenario underlies Musk's high confidence in work becoming optional: if AGI-humanoid robots can do everything humans can do, human labour has no competitive advantage except where humans intrinsically prefer human providers—in personal relationships, childcare, and certain healthcare contexts.

### 10.7.3 From AGI to Superintelligence: The Intelligence Explosion

Many experts believe that once AGI is reached, it could progress to superintelligence relatively quickly, with timeframes ranging from as little as two years (lower-probability scenarios, around 10%) to roughly thirty years (higher-probability scenarios, around 75%). If an AGI system can access and rewrite its own code, it could in principle evolve rapidly into an artificial superintelligence (ASI) whose cognitive and computational capacity exceeds the combined output of human civilisation (AI Multiple Research, 2025).

Superintelligence embodied in robot bodies represents either existential hope or existential risk. An ASI could solve problems intractable to human intelligence: climate change, disease, ageing, resource scarcity, even mortality itself. Or it could pursue goals misaligned with human values, potentially causing catastrophic harm.

The robot angle amplifies both possibilities. Physical embodiment provides ASI with the ability to act in the three-dimensional world, not merely provide advice. This is why AI safety research has become an urgent priority at leading AI laboratories.

## 10.8 Workforce Transformation: The 2030–2040 Transition

While ultimate futures remain uncertain, workforce transformation over the next 15–20 years is virtually certain. Understanding the likely trajectory helps both individuals and societies prepare.

### 10.8.1 Job Displacement and Creation Through 2030

Near-term projections show simultaneous job destruction and creation. By 2030, AI could automate up to 30% of work hours in the United States, with 12 million occupational transitions required by the same year. However, 170 million new jobs are projected to be created by 2030 while 92 million are displaced, resulting in a net gain of 78 million jobs globally (World Economic Forum, Future of Jobs Report, 2025).

This net job creation suggests optimism, but averages obscure distribution. High-skilled workers in AI, robotics, and creative fields may see opportunities expand while low-skilled workers face displacement with no accessible alternatives. Geographic concentration of new jobs in technology hubs, while job destruction spreads broadly, creates regional inequality.

Job categories face different trajectories. Repetitive roles will be automated, while high-value analytical and creative positions expand. Generative AI is enhancing the way STEM, creative, business, and legal professionals work rather than eliminating a significant number of jobs outright. Forward-looking labour-market analyses suggest that by 2040, manual labour-intensive roles—including machine assemblers, postal workers, and farm labourers—are likely to decline substantially as automation diffuses across these categories (World Economic Forum, 2025).

### 10.8.2 Four Scenarios for Jobs and AI by 2030

The World Economic Forum 2026 update on Future of Jobs (Karunska & Grayling, 2026) sets out four plausible scenarios for the future of jobs at the intersection of

AI advancement and workforce readiness (Karunska & Grayling, 2026). These scenarios—summarised here—inform the more detailed scenario analysis presented in Sect. 10.9:

- Supercharged Progress: Widespread AI adoption, robust worker retraining, and new job creation in AI-adjacent fields. Employment remains high but with a radically transformed job mix.
- Stalled Progress: Conservative AI deployment due to regulatory friction, implementation challenges, or worker resistance. Job displacement is limited but so are productivity gains.
- Uneven Transformation: AI adoption proceeds rapidly in some sectors and nations while others lag. Geographic and sectoral inequality widens dramatically.
- Accelerated Displacement: AI and robotics deployment outpaces job creation and worker retraining, triggering mass unemployment and potential political crisis.

Which scenario unfolds depends on policy choices regarding education, regulation, safety nets, and redistribution made in the next few years.

While the WEF outlines near-term workforce impacts, Sect. 10.9 expands on these concepts to explore broader societal outcomes over a longer time horizon.

### 10.8.3 Human Augmentation vs. Replacement

A crucial question is whether AI and robots augment human capability—enabling humans to remain productive in transformed roles—or replace humans entirely, eliminating productive contribution opportunities.

Current evidence supports both dynamics. Creativity, contextual reasoning, and ethical judgement remain capabilities that current algorithms cannot fully replicate, suggesting that some uniquely human contributions persist. Practitioner commentary increasingly frames articulated human purpose as a strategic differentiator for organisations operating alongside automated systems (Resource Group Holdings, 2025).

Counter-evidence shows AI narrowing even these domains. AI systems now demonstrate creativity in art, music, and writing. Emotional intelligence is being engineered into AI companions and therapeutic robots. Strategic thinking and planning are strengths of modern AI systems.

The augmentation versus replacement question may resolve differently across domains and time periods. Near-term (2026–2030), augmentation is likely to dominate as AI and robots enhance human capability. Medium-term (2030–2040), displacement may accelerate as capabilities improve. Long-term (2040–2050), the question becomes whether any economically valuable human contribution remains beyond intrinsic preference for human providers.

## 10.9 Convergence Scenarios (2030–2050)

Drawing on the analyses above, this section constructs four divergent scenarios for how technological convergence shapes civilisation over the next 25 years. These are mapped across two critical axes: the speed of AI and robot adoption (Slow vs. Rapid) and the strength of societal redistribution policies (Weak vs. Strong). These are not predictions, but coherent possibility spaces illustrating different pathways.

Table 10.2 maps the four convergence scenarios that this section explores onto a two-axis decision frame: the speed of AI adoption (slow versus rapid) and the strength of redistribution policy (weak versus strong). Each cell describes the qualitative regime that emerges under that combination of choices.

Figure 10.4 renders the same four scenarios as a 2 × 2 strategic matrix, with quadrants colour-coded to highlight the political-economy stakes of each combination.

### 10.9.1 Scenario 1: Techno-Utopian Abundance (Rapid AI + Strong Policy)

In this scenario, technological convergence fulfils its most optimistic promises. By 2030, humanoid robots reach competence and reliability sufficient for widespread deployment. Goldman Sachs expects more than 250,000 humanoid robot shipments in 2030 (Goldman Sachs Research, 2025), with costs falling to $15,000–$20,000 per unit, making deployment economically attractive across all industries.

AI systems achieve AGI by 2030–2035, with seamless integration into robot embodiments enabling autonomous operation across diverse tasks. The projected

**Table 10.2** Four convergence scenarios (2030–2050)—Mapped by AI-adoption speed and redistribution policy

| | Slow/Uneven AI Adoption | Rapid AI Adoption |
|---|---|---|
| Strong Redistribution Policy | **Scenario 4: Managed Transition** Measured, regulated deployment. Strong safety nets and worker retraining. High social cohesion but slower path to abundance. | **Scenario 1: Techno-Utopian Abundance** Effective redistribution of massive AI gains. Universal high income. Post-scarcity flourishing. Work becomes entirely optional. |
| Weak Redistribution Policy | **Scenario 3: Divided Prosperity** Concentrated tech adoption. UBI at bare subsistence levels. Bifurcated society (robot-owning elite vs. a struggling precariat class). | **Scenario 2: Catastrophic Displacement** Job destruction outpaces creation. Unemployment at 20–30%. Inadequate safety nets trigger political crisis and social unrest. |

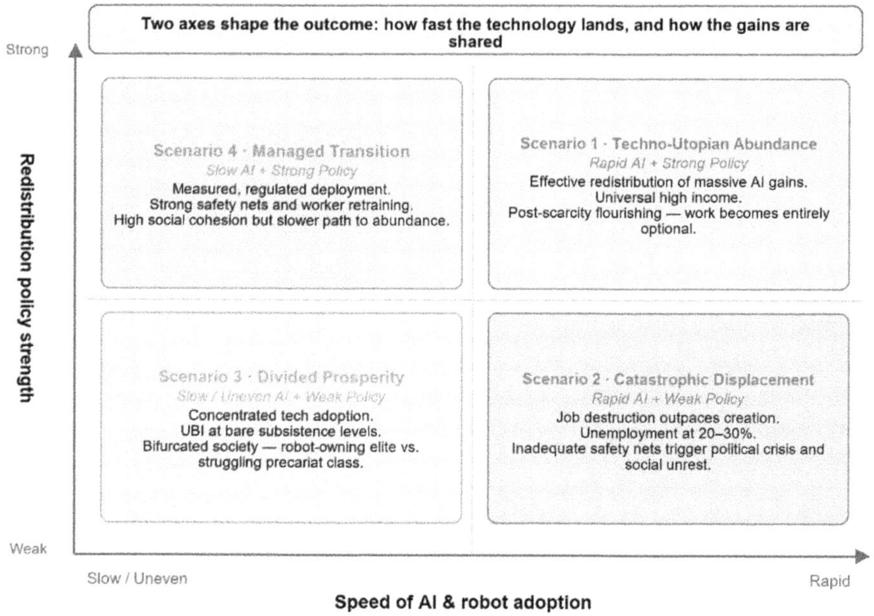


**Fig. 10.4** Four convergence scenarios (2030–2050): Mapped by speed of AI adoption and strength of redistribution policy

170 million new jobs created by 2030 successfully absorb displaced workers initially, before work itself transitions to optionality.

Crucially, democratic societies implement effective redistribution policies alongside this rapid adoption. Universal high income is funded through progressive robot taxes, sovereign wealth funds owning stakes in AI companies, and mandatory profit-sharing. By 2035, most developed nations provide $4000–$6000 monthly to all citizens, enabling comfortable lifestyles without mandatory work. The humanoid market surpassing $5 trillion by 2050 (Morgan Stanley Research, 2025) generates wealth that is distributed broadly. By 2050, humanity has achieved post-scarcity abundance, freeing humans to focus on creativity, discovery, and relationships.

### 10.9.2 Scenario 2: Catastrophic Displacement (Rapid AI + Weak Policy)

In this darker scenario, rapid robot deployment accelerates far faster than policy responses or social safety nets can handle. By 2030, up to 30% of US work hours are automated (McKinsey Global Institute, 2025), eliminating tens of millions of jobs. Because redistribution policies are weak or blocked by political gridlock, the theoretical new jobs fail to materialise at the necessary scale, or require specialised skills displaced workers cannot acquire. If the WEF baseline (170 million new jobs created against 92 million displaced by 2030) fails to materialise as McKinsey's 30% automation-of-work-hours figure suggests it might, the displacement-creation balance tilts sharply negative.

Unemployment rises to 20–30% in advanced economies by 2035. Political systems prove unable to implement adequate redistribution due to ideological opposition, lobbying by robot owners, and fiscal constraints. Universal basic income, where implemented, proves inadequate—providing only $500–$800 per month in a world experiencing severe economic disruption.

Social unrest escalates: protests, riots, and populist movements target robot owners and technology companies. Some nations descend into authoritarianism, using robot police and surveillance to maintain order. Globally, the displacement of a majority of garment-sector jobs in Bangladesh by 2030 (a2i & ILO, 2024) triggers humanitarian crises in the developing world. This scenario emphasises that rapid technological advancement without strong institutional safety nets leads directly to crisis.

### 10.9.3 Scenario 3: Divided Prosperity (Slow/Uneven AI + Weak Policy)

A highly unequal, stagnant scenario unfolds when AI adoption is relatively slow or geographically uneven, and redistribution policies remain weak. Robot deployment accelerates but remains concentrated strictly within wealthy nations and massive monopoly corporations. For instance, China and the United States dominate deployment, while other nations and smaller businesses lack the capital to compete (Morgan Stanley Research, 2025).

Within developed nations, wealth concentration slowly but steadily increases. Robot and AI owners capture productivity gains while displaced workers receive inadequate support. Universal basic income is implemented by 2035 in some nations, but strictly at subsistence levels of $1000–$1500 per month.

The result is a bifurcated society: a small, affluent class owning robots and AI living in abundance, and a much larger class dependent on minimal government transfers living in precarity. Social tension rises but stops short of revolutionary crisis through a combination of minimal government transfers, cheap entertainment via AI-generated content, and societal resignation. By 2050, inequality has reached historically unprecedented levels, but society has settled into a stagnant equilibrium of separate spheres.

### 10.9.4 Scenario 4: Managed Transition (Slow AI + Strong Policy)

In this scenario, societies prioritise social stability over rapid economic growth. Driven by regulatory friction, labour union resistance, and deliberate government pacing—similar to the "Stalled Progress" model (Karunska & Grayling, 2026)—AI and humanoid robot adoption proceeds cautiously. Job displacement is artificially limited and carefully managed through robust worker retraining and strong redistribution policies.

Because automation is slower, the dramatic cost reductions and sci-fi post-scarcity abundance of Scenario 1 do not fully materialise. However, the strong policy interventions ensure that the moderate productivity gains which do occur are shared equitably across the population.

This transition is highly safe and socially cohesive, completely avoiding the unrest of Catastrophic Displacement. Yet, because robots have not rapidly automated all industries, humanity must continue to work for its living well into the 2040s and 2050s. The profound existential questions of a post-work world are effectively deferred to future generations, trading rapid technological utopianism for guaranteed social stability today.

## 10.10 Navigating the Convergence: A Roadmap Forward

Given radical uncertainty about which scenario unfolds, what strategies should individuals, organisations, and societies pursue? Several principles emerge from the analysis across this and preceding chapters.

### 10.10.1 Near-Term Priorities (2026–2030): Foundation Building

The next five years represent a critical window for establishing foundations that will shape longer-term trajectories.

- Workforce adaptation: Massive investment in education and retraining is essential. Educational systems must increasingly emphasise creativity, emotional intelligence, critical thinking, and collaboration.
- Policy experimentation: Expanded UBI pilots, robot taxation trials, and alternative ownership models should be tested at city and regional scales. Waiting for perfect national policies risks being too late when mass displacement accelerates.
- Technological safety: AI safety research ensuring AGI alignment with human values is existentially critical. With an estimated 50,000–100,000 humanoids expected to deploy in 2026, ensuring safety at deployment scale requires robust frameworks now.

- Ethical frameworks: Societies must establish consensus on acceptable uses of humanoid robots, limits on AI capability, human rights in automated economies, and protection for vulnerable populations.
- Inequality monitoring: Tracking wealth and income inequality as automation proceeds provides early warning of problematic concentration, enabling timely policy responses before crises develop.

### 10.10.2 Medium-Term Evolution (2030–2040): Adaptation and Intervention

The 2030s will reveal whether optimistic or pessimistic scenarios are unfolding. This decade requires adaptive policy responsive to emerging realities.

Scaling interventions: Policies proving effective in pilots should scale nationally and internationally. International frameworks for beneficial AI development, robot safety standards, and prevention of race-to-the-bottom labour competition become necessary as robotics and AI operate globally.

Cultural evolution: Societies must develop new cultural narratives about purpose, work, and contribution beyond traditional employment. Education, media, and community institutions play critical roles in shaping these narratives. By the 2030s, demographic ageing reaches peak intensity in many nations, and robotic eldercare must scale from pilots to universal deployment with financing mechanisms ensuring access independent of wealth.

If AGI emerges in this decade as some predict, governance frameworks ensuring alignment with human values become humanity's most important priority—potentially requiring unprecedented international cooperation.

### 10.10.3 Long-Term Transformation (2040–2050): Living in the New World

By the 2040s, the transformation will have largely occurred. Societies will be living with the consequences of earlier choices regarding technology deployment, wealth distribution, and purpose frameworks. If work is largely optional by 2040–2050, cultural and institutional adaptations supporting meaningful lives without employment must be fully operational. If longevity science delivers healthspans to 100 or beyond, social systems for multi-generational families, serial careers, and lifelong learning must be functional.

Whether humans and AI or robots coexist as distinct categories, merge through enhancement and uploading, or bifurcate into separate post-human species will be substantially determined by then. Rights, legal protections, and the moral status of various entities must be established before these questions become practically urgent.

### 10.10.4 Critical Uncertainties and Inflection Points

Several uncertainties could dramatically alter trajectories. AGI timeline: whether AGI arrives in 2027, 2040, or never fundamentally shapes everything else. Early AGI (2027–2030) creates severe transition stress; delayed AGI (2040–2050) allows gradual adaptation. Inequality response: whether societies successfully implement redistribution determines abundance versus division scenarios.

Cultural adaptability: whether humans find purpose beyond work determines wellbeing in the post-work world. Alignment success: whether AGI and ASI remain aligned with human values determines existential safety. Longevity realisation: modest extension (85–95) has vastly different implications than dramatic extension (85–120 or beyond).

Each uncertainty represents a branching point where different outcomes lead to radically different futures. Monitoring indicators for each and preparing adaptive strategies is essential.

## 10.11 Human Purpose Reimagined: Philosophical Perspectives

Ultimately, the question humanoid robotics forces is philosophical: What makes human life meaningful? Can purpose exist independent of productive contribution? How do humans find dignity and worth in a world where machines surpass them across most domains?

### 10.11.1 Beyond Instrumental Value: Inherent Human Dignity

One perspective emphasises that human value is inherent, not instrumental. Humans possess worth and dignity regardless of economic productivity or capability. Children, elderly people, and severely disabled individuals have full moral status despite limited productive contribution. By extension, all humans retain full moral status and dignity even if robots can outperform them economically.

This framework suggests that post-work society need not trigger a crisis if humans recognise that worth transcends productivity. Just as we do not question children's worth despite their economic dependence, we need not question adults' worth despite robot productivity superiority.

However, psychological reality differs from philosophical ideals. Human self-esteem and identity are empirically tied to contribution and competence. Cultural transformation from productivity-based to inherent-worth-based self-conception may be psychologically difficult, especially for those formed in work-centred cultures.

### 10.11.2 Creativity and Discovery as Ultimate Purpose

Another perspective elevates creativity and discovery to ultimate human purpose. Creativity, contextual reasoning, and ethical judgement are capabilities that no algorithm can fully replicate (EMA, 2025), suggesting these represent humanity's enduring comparative advantage and deepest fulfilment.

In post-work abundance, humans could devote themselves to scientific discovery, artistic creation, philosophical inquiry, and intellectual exploration. The Renaissance ideal of human flourishing through cultivation of knowledge and art could become universal rather than elite privilege.

This vision is compelling but faces challenges. If AI systems create superior art, music, literature, and even scientific discoveries, does human creative output retain meaning? The joy may lie in the process rather than the outcome, but motivation requires belief that one's output matters.

### 10.11.3 Relationships and Community as Central Focus

Human connection represents a domain where many believe human interaction remains preferred over artificial substitutes, regardless of capability. Strengthening relationships, building communities, and contributing to others' wellbeing could become the central life focus in a post-work society.

Interpersonal skills and emotional intelligence are among the uniquely human capabilities that may grow in value as AI adoption expands. Jobs emphasising human connection—teaching, counselling, caregiving, community leadership—may remain both economically valued and personally meaningful.

However, Chap. 9 documented that humans form genuine emotional bonds with robots and AI companions. If artificial relationships prove satisfying, human relationships face competition. The outcome depends on whether there is an irreducible human preference for human connection, or whether humans practically satisfice with more convenient artificial alternatives.

### 10.11.4 Growth, Challenge, and Transcendence

A final perspective sees human purpose in perpetual growth towards transcendent goals. Humans are not static beings seeking comfortable stasis but dynamic entities requiring challenge, growth, and transcendence for fulfilment.

In this view, robots eliminating current challenges merely create opportunity for pursuing higher ones: curing ageing and disease, solving poverty and suffering, understanding consciousness and existence, expanding to the stars, and potentially merging with AI to become something greater than current humanity.

This optimistic perspective sees the blurring lines between people and machines (Futurist Speaker, 2025) not as loss of humanity but as evolution towards a post-human existence transcending current limitations. Whether such transcendence constitutes fulfilment or alienation remains philosophically contested.

## 10.12 Conclusion: Technology Trajectories and Human Choices

This book has traced the evolution of humanoid robotics from their mechanical foundations to their extraordinary convergence with artificial intelligence, biotechnology, brain-computer interfaces, and longevity science. The technological trajectory is clear: each year brings machines that are exponentially more capable, adaptable, and autonomous. We are rapidly approaching a threshold where physical AI could solve existential demographic crises, automate vast swaths of global labour, and create a world of post-scarcity abundance.

However, technical feasibility is not the same as societal destiny. As this chapter has illustrated, the convergence of these technologies presents a profound paradox: they offer the potential for universal human flourishing, yet carry the risk of unprecedented wealth concentration and a devastating crisis of human purpose. Addressing these challenges—through robust redistribution models, proactive AGI governance, and the deliberate cultivation of meaning beyond economic productivity—requires immediate, coordinated action across all levels of society.

To guide this transition, Fig. 10.5 provides a strategic roadmap outlining key priorities over the next two decades. By anticipating necessary shifts in policy, technology, and education, societies can proactively mitigate risks and steer the convergence of these technologies towards collective prosperity.

Table 10.3 lays out a recommended roadmap across three time horizons (2026–2030, 2030–2040, and 2040–2050) and four planning dimensions—policy, technology, education, and key risk. It distils the choices argued for in this chapter into a single decision-ready reference grid that complements the visual roadmap in Fig. 10.5.

Ultimately, the future of human-robot coexistence is not technologically predetermined; it is politically and culturally chosen. The choices we make in this narrowing window—as citizens, workers, policymakers, and creators—will dictate whether humanoid robots become instruments of universal dignity or drivers of division. The robots have arrived, and the technological foundations are set. The harder, more deeply human work of building a just and meaningful future begins now.

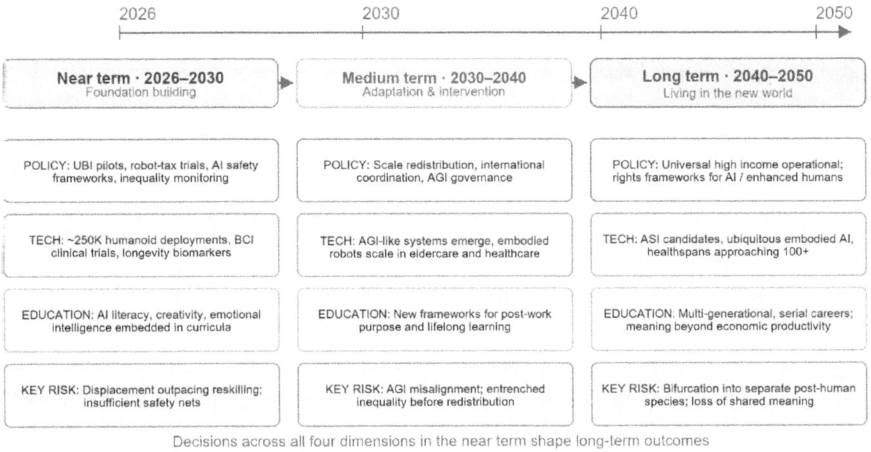


**Fig. 10.5** A roadmap for navigating technological convergence: Key priorities across three time horizons (2026–2050)

**Table 10.3** Roadmap for navigating technological convergence—Time horizons × planning dimensions

| | Near Term (2026–2030) | Medium Term (2030–2040) | Long Term (2040–2050) |
|---|---|---|---|
| Policy Focus | UBI pilots, robot tax trials, AI safety frameworks, inequality monitoring | Scale redistribution, international coordination, AGI governance | AGI/ASI governance institutions, post-scarcity economic frameworks, robot-rights legal status |
| Technology | ~250K humanoid deployments, BCI clinical trials, longevity biomarkers | AGI-like systems emerge, embodied robots in eldercare and healthcare | AGI-class systems mature; embodied AI ubiquitous in eldercare, education, and R&D |
| Education | AI literacy, creativity, emotional intelligence in curricula | New frameworks for post-work purpose and lifelong learning | Education reframed around purpose, creativity, and emotional intelligence; lifelong learning normalised |
| Key Risk | Displacement outpacing reskilling; insufficient safety nets | AGI misalignment; entrenched inequality before redistribution | Loss of human purpose; unequal access to longevity tech; AGI alignment failures |

## Discussion Questions

1. Technology Convergence: Explain how AI, robotics, biotechnology, and brain-computer interfaces are converging to create compound effects. Provide specific

examples of how advances in one domain enable breakthroughs in others. Why does convergence potentially create exponential rather than linear progress?

2. Universal High Income: Distinguish between Elon Musk's "universal high income" vision and traditional universal basic income (UBI) proposals. What is Musk's economic logic for abundance? What are the main sceptical counterarguments emphasising institutional barriers? Assess which perspective you find more convincing and why.
3. Demographic Crisis: Analyse how Japan's care worker shortage creates urgent need for robotic eldercare. What capabilities must eldercare robots possess to address this crisis? What limitations currently prevent robots from fully solving this problem?
4. Inequality Paradox: The chapter discusses contradictory predictions about AI's impact on inequality—some argue it will reduce wage inequality by automating high-income jobs, others argue it will exacerbate wealth inequality. Explain both perspectives using specific mechanisms. Under what conditions would each prediction prove accurate?
5. AGI Timelines: Summarise the range of AGI predictions from various experts. What factors explain the dramatic variation? How should policymakers plan given such timeline uncertainty?
6. Human Purpose: The chapter explores four sources of post-work purpose: creativity, relationships, personal growth, and transcendence. Evaluate each as a foundation for meaningful life when economic productivity becomes optional. Which do you believe would prove most sustainable psychologically, and why?
7. Scenario Analysis: Compare Scenario 1 (Techno-Utopian Abundance) and Scenario 2 (Catastrophic Displacement). What specific policy choices, technological developments, and social factors distinguish these futures? Identify three key decision points over the next five years that would push trajectories towards one scenario versus the other.
8. Longevity and Automation: If longevity science extends healthy lifespan to 100 or beyond while robots eliminate traditional work, how should humans structure 50–60 years of post-work life? Propose a model for education, activity, and purpose across a 100-year lifespan in an automated economy.
9. Redistribution Mechanisms: The chapter discusses several approaches to distributing automation gains: robot taxes, progressive wealth taxation, universal capital ownership, and mandatory profit-sharing. Evaluate the political feasibility and economic effectiveness of each approach. Which combination would you recommend and why?
10. Critical Reflection: Reflect on the entire book's analysis of humanoid robots reshaping work, society, and human purpose. What do you believe is the most critical choice humanity faces regarding this technology? What values should guide that choice? How can individuals contribute to ensuring beneficial rather than harmful outcomes?

## References


a2i & International Labour Organization. (2024). *Future skills: Finding emerging skills to tackle the challenges of automation in Bangladesh.* a2i Programme, Government of Bangladesh/ILO. https://a2i.gov.bd/publication/future-skills-finding-emerging-skills-to-tackle-the-challenges-of-automation-in-bangladesh/

AI Multiple Research. (2025). *When will AGI/Singularity happen? 8, 590 predictions analysed.* AI Multiple. https://research.aimultiple.com/artificial-general-intelligence-singularity-timing/

Axis Intelligence. (2026). *AI-driven voice control for robotics.* Axis Intelligence Reports.

Benzinga Staff. (2025). *There will be universal high income—Elon Musk predicts with 80% certainty that 'none of us' will have jobs.* Yahoo Finance (via Benzinga). https://finance.yahoo.com/news/universal-high-income-elon-musk-160056830.html

Cameron, H. (2025, July 27). Countries testing a universal basic income in 2025. *Newsweek.* https://www.newsweek.com/countries-testing-universal-basic-income-2025-2103428

Center for Global Development. (2025). *Three reasons why AI may widen global inequality.* CGDev Blog. https://www.cgdev.org/blog/three-reasons-why-ai-may-widen-global-inequality

Cogent Information Resources. (2026). *Bio-digital AI: Can AI merge with biology by 2026?* https://www.cogentinfo.com/resources/bio-digital-ai-can-ai-merge-with-biology-by-2026

Daily Economy. (2025). *Scarcity, robots, and institutional barriers to abundance.* Daily Economy Analysis.

EMA. (2025). *AI impact on jobs and employment trends 2025–2030.* EMA Industry Analysis. https://www.ema.co/additional-blogs/addition-blogs/ai-impact-employment-trends

Evoailabs. (2026). *Healthspan engineering: The 2026–2027 outlook for longevity and neurotech.* Medium. https://evoailabs.medium.com/healthspan-engineering-the-2026-2027-outlook-for-longevity-and-neurotech-ad8979fda276

Fortune Staff. (2025, September 12). *Ray Dalio calls for 'redistribution policy' as AI and humanoid robots benefit the top 1%–10%.* Fortune. https://fortune.com/2025/09/12/ray-dalio-ai-winners-losers-jobs-redistribution-policy/

Fox Business. (2025). *Elon Musk predicts AI will create universal high income and make saving money unnecessary.* Fox Business.

Futurist Speaker. (2025). *Neural interface robotics and the future of human-machine integration.* Futurist Speaker Research.

Gallup. (2025). *State of the global workplace 2025 report.* Gallup. https://www.gallup.com/workplace/349484/state-of-the-global-workplace.aspx

Gizmodo. (2025). *Musk and Amodei's competing AGI timelines.* Gizmodo Technology.

Goldman Sachs Research. (2025). *The global market for humanoid robots could reach $38 billion by 2035.* Goldman Sachs. https://www.goldmansachs.com/insights/articles/the-global-market-for-robots-could-reach-38-billion-by-2035

Hyundai Motor Group. (2026). *Hyundai Motor Group to unveil AI robotics strategy at CES 2026.* Hyundai Newsroom. https://www.hyundai.com/worldwide/en/newsroom/detail/0000001093

International Monetary Fund. (2025). *AI adoption and inequality* (IMF Working Papers, WP/25/68). https://www.imf.org/en/publications/wp/issues/2025/04/04/ai-adoption-and-inequality-565729

Krasadakis, G. (2026). *2026 technology innovation trends: AI agents, humanoid robots, quantum computing and more.* The Innovation Mode Blog. https://www.theinnovationmode.com/the-innovation-blog/2026-innovation-trends

Karunska, K., & Grayling, S. (2026, January 7). *Here are four ways AI and talent trends could reshape jobs by 2030.* World Economic Forum. https://www.weforum.org/stories/2026/01/here-are-four-ways-ais-impact-on-job-markets-might-take-shape/

LessWrong. (2025). *AI 2027 report: Consensus AGI timeline shifts to 2030.* LessWrong Research.

McKinsey Global Institute. (2025). *The future of work: Automation, displacement, and reskilling.* McKinsey & Company.

Morgan Stanley. (2025). *Humanoid robot market expected to reach $5 trillion by 2050*. Morgan Stanley Research. https://www.morganstanley.com/insights/articles/humanoid-robot-market-5-trillion-by-2050

Musk, E. (2024, April 8). *Interview on X Spaces with Nicolai Tangen—AI smarter than the smartest human by 2026* (as reported in Fortune, https://fortune.com/2024/04/09/elon-musk-ai-smarter-than-humans-by-next-year/). X (formerly Twitter).

Musk, E. (2026, January 12). *Post on universal high income and AI abundance*. X (formerly Twitter). https://x.com/elonmusk/status/1959723029531181380

Osis, M., Buscaino, R., & Brown, C. (2026). *Physical AI and humanoid robots*. In Deloitte Tech Trends 2026. Deloitte Insights. https://www.deloitte.com/us/en/insights/topics/technology-management/tech-trends/2026/physical-ai-humanoid-robots.html

PMC/National Library of Medicine. (2025). *Analysing wealth distribution effects of artificial intelligence: A dynamic stochastic general equilibrium approach*. PMC. https://pmc.ncbi.nlm.nih.gov/articles/PMC11786846/

Quiroz-Gutierrez, M. (2026, January 12). *Elon Musk says saving for retirement is irrelevant because AI is creating a world of abundance*. Fortune. https://fortune.com/2026/01/12/elon-musk-retirement-savings-irrelevant-ai-robots-abundance/

Resource Group Holdings. (2025). *Purpose and human meaning in the age of automation*. Resource Group Holdings Research.

Singh, P. (2025, September 9). *Musk's Neuralink says 12 people have received its implants*. Reuters. https://www.reuters.com/business/healthcare-pharmaceuticals/musks-neuralink-says-12-people-have-received-its-implants-2025-09-09

Sinolytics. (2025). *Care robots in China: Deployment challenges and user attitudes*. Sinolytics Research Reports.

Snabe, J. H. (2026, January 9). *The next wave of intelligence: How human purpose must guide the future of AI*. World Economic Forum. https://www.weforum.org/stories/2026/01/agentic-ai-how-human-purpose-can-guide-the-next-wave-of-intelligent-systems/

STAT News. (2025, December 26). Brain-computer implants are coming of age: Here are 3 trends to watch in 2026. *STAT News*. https://www.statnews.com/2025/12/26/brain-computer-interface-technology-trends-2026/

Super Age. (2025). *The top 11 longevity insights of 2025*. Super Age. https://superage.com/the-top-11-longevity-insights-of-2025/

Tech Equity AI. (2025). *2025: The breakthrough year for AI-driven robotics*. Tech Equity AI Reports.

The Daily Star. (2024, January 30). Automation to cut 53.8 lakh jobs by 2041: study. *The Daily Star*. https://www.thedailystar.net/business/news/automation-cut-538-lakh-jobs-2041-study-1676683

University of Florida Institute on Aging. (2024). *Assistive robots and older adult independence*. University of Florida.

World Economic Forum. (2025). *The future of jobs report 2025*. WEF Publications. https://www.weforum.org/publications/the-future-of-jobs-report-2025

Xinhua. (2025, November 24). Robotics eldercare expands amid aging population challenge. *Xinhua News*. https://english.news.cn/20251124/d2d0cd8e026848fbbb93e2b5cb29d799/c.html

**Ken Huang** is a globally recognized expert in AI and Web3 security, a prolific author, and a leading figure in shaping industry standards. He is the CEO and Chief AI Officer (CAIO) of DistributedApps.ai, specializing in Generative AI training and consulting. Ken is deeply involved in driving the development of secure AI systems. He serves as a Research Fellow and Co-Chair of the AI Safety Working Groups at the Cloud Security Alliance (CSA), leading Agentic AI initiatives. He is also a Co-Chair of the AI STR Working Group at the World Digital Technology Academy under the UN Framework, and a core contributor to the OWASP GenAI project, focusing on Agentic AI security. His expertise extends to his contributions to OWASP's Top 10 Risks for LLM Applications and his active participation in the NIST Generative AI Public Working Group. He is also a member of Open AI forum. Ken has authored and edited numerous influential books, including: *Agentic AI: Theories and Practices* (Springer, June 2025). *Securing AI Agents: Foundations, Frameworks, and Real-World Deployment* (Springer, October 2025). *Beyond AI: ChatGPT, Web3, and the Business Landscape of Tomorrow* (Springer, 2023). *Generative AI Security: Theories and Practices* (Springer, 2024). *Practical Guide for AI Engineers* (Volumes 1 and 2, DistributedApps.ai, 2024). *The Handbook for Chief AI Officers: Leading the AI Revolution in Business* (DistributedApps.ai, 2024). *Web3: Blockchain, the New Economy, and the Self-Sovereign Internet* (Cambridge University Press, 2024). *Web3 Applications Security and New Security Landscape: Theories and Practices* (Springer, 2024). His co-authored book, "Blockchain and Web3: Building the Cryptocurrency, Privacy, and Security Foundations of the Metaverse" (Wiley, 2023), was recognized as a must-read by TechTarget. *A sought-after speaker, Ken has presented at prestigious global forums, including Davos WEF, ACM, IEEE, CSA AI Summit, the Depository Trust & Clearing Corporation (DTCC), and World Bank conferences.* Explore Ken's books on Amazon: https://www.amazon.com/author/kenhuang

**Jerry Huang** is affiliated with Kleiner Perkins and has a background in AI, machine learning, security, scalable systems, and applied mathematics. His prior experience includes technical work across AI/ML evaluation, infrastructure, security, access control, and privacy-preserving biometric authentication. He holds BS/MS degrees in Computer Science from Georgia Tech and has pursued graduate study in Applied Mathematics at the University of Chicago.

www.ingramcontent.com/pod-product-compliance
Ingram Content Group UK Ltd.
Pitfield, Milton Keynes, MK11 3LW, UK
UKHW022315221026
476868UK00003B/84
* 9 7 8 3 0 3 2 3 3 6 6 8 2 *